ROUTLEDGE LIBRARY EDITIONS:
ECONOMIC GEOGRAPHY

Volume 9

THE CAPITALIST SPACE ECONOMY

THE CAPITALIST SPACE ECONOMY

Geographical Analysis after Ricardo, Marx and Sraffa

ERIC SHEPPARD & TREVOR J. BARNES

Routledge
Taylor & Francis Group

LONDON AND NEW YORK

First published in 1990

This edition first published in 2015
by Routledge
2 Park Square, Milton Park, Abingdon, Oxon, OX14 4RN

and by Routledge
711 Third Avenue, New York, NY 10017

Routledge is an imprint of the Taylor & Francis Group, an informa business

© 1990 Eric Sheppard and Trevor Barnes

British Library Cataloguing in Publication Data
A catalogue record for this book is available from the British Library

ISBN: 978-1-138-85764-3 (Set)
eISBN: 978-1-315-71580-3 (Set)
ISBN: 978-1-138-81406-6 (Volume 9)
eISBN: 978-1-315-74774-3 (Volume 9)
Pb ISBN: 978-1-138-81411-0 (Volume 9)

Publisher's Note
The publisher has gone to great lengths to ensure the quality of this reprint but points out that some imperfections in the original copies may be apparent.

Disclaimer
The publisher has made every effort to trace copyright holders and would welcome correspondence from those they have been unable to trace.

THE
CAPITALIST
SPACE ECONOMY

Geographical Analysis
after Ricardo, Marx and Sraffa

Eric Sheppard
&
Trevor J. Barnes

with a contribution from
Claire Pavlik

London
UNWIN HYMAN
Boston   Sydney   Wellington

Published by the Academic Division of
Unwin Hyman Ltd
15/17 Broadwick Street, London W1V 1FP, UK

Unwin Hyman Inc.
955 Massachusetts Avenue, Cambridge, MA 02139, USA

Allen & Unwin (Australia) Ltd
8 Napier Street, North Sydney, NSW 2060, Australia

Allen & Unwin (New Zealand) Ltd
in association with the Port Nicholson Press Ltd
Compusales Building, 75 Ghuznee Street, Wellington 1, New Zealand

First published in 1990

British Library Cataloguing in Publication Data

Sheppard, Eric
The capitalist space economy: analysis after Ricardo, Marx and Sraffa
1. Economics. Theories
I. Title   II. Barnes, Trevor J.   III. Pavlik, Claire
330.1

ISBN 0-04-330401-X
ISBN 0-04-330402-8 pbk

Library of Congress Cataloging in Publication Data

Sheppard, Eric S.
The capitalist space economy: Geographical Analysis after
Ricardo, Marx and Sraffa/Eric Sheppard and Trevor J. Barnes:
with a contribution from Claire Pavlik.
    p.   cm.
Includes bibliographical references and index.
ISBN 0-04-330401-X: $70.00. – ISBN 0-04-330402-8 (pbk): $24.95
1. Geography, Economic.   2. Capitalism.
I. Barnes, Trevor J.   II. Pavlik, Claire.   III. Title.
HF1025.S475 1990
330.9–dc20   90-40989 CIP

Typeset in 10 on 12 point Bembo by Computape (Pickering) Ltd,
Pickering, North Yorkshire
and printed in Great Britain by Cambridge University Press, Cambridge

Contents

List of tables   Page xiii
List of figures   xv
Preface and acknowledgements   xvii

1   Introduction   1
    Introduction   1
    1.1   From political economy to neoclassicism and back   2
    1.2   analytical political economy: three views   8
          1.2.1   Neo-Ricardianism   8
          1.2.2   Fundamental Marxists   10
          1.2.3   Analytical Marxism   11
    1.3   Questions of method   12
    1.4   Outline of the book   13
    Note   15

PART I   THE BASICS   17
2   The capital controversies   19
    Introduction   19
    2.1   The reswitching debate   19
          2.1.1   The aggregate neoclassical model of production   19
          2.1.2   The impossibility of measuring capital   23
          2.1.3   Denying the neoclassical parables: Reswitching   23
          2.1.4   Politics and the distribution of income   27
    2.2   Implications for economic geography   27
    Summary   29
    Note   30

3   The value controversies   31
    Introduction   31
    3.1   Commodities and their value   32
    3.2   Embodied labour values   34
    3.3   The labour theory of value   35
          3.3.1   Social necessity   36
          3.3.2   Empirical measurement   38
          3.3.3   Marxian exploitation   39
    3.4   Further concepts   41
    3.5   Labour values vis-à-vis exchange values   42
          3.5.1   Prices of production   44
          3.5.2   The transformation problem   45
          3.5.3   Controversies over the transformation problem   46
    3.6   Does joint production imply negative labour values?   51

3.7   Qualitative arguments for a labour theory of value   54
3.8   Geographical implications: Defining regions   56
Note   58

PART II   THE PROFIT-MAXIMIZING SPACE ECONOMY   59

4   *Production prices in a competitive space economy*   61
    Introduction   61
    4.1   The price circuit   63
          4.1.1   Spatial prices and trading patterns   65
          4.1.2   Wage–profit frontiers   70
          4.1.3   The location of industry   72
          4.1.4   Circulation and exchange   74
          4.1.5   The inter-regional trade balance   74
    4.2   Incorporating production and circulation time   76
          4.2.1   Capital advanced in production   77
          4.2.2   Circulation time   80
          4.2.3   Implications for the geography of commodity production   82
    Summary   83
    Notes   84
    Appendix: Existence of trading and pricing equilibrium, and potentials 84

5   *Reswitching in a space economy*   88
    Introduction   88
    5.1   An example of reswitching   88
          5.1.1   Background   88
          5.1.2   A two-region economy   90
          5.1.3   A single open region within a space economy   92
    5.2   Intra-regional location and reswitching   92
          5.2.1   Fixed transportation costs   94
          5.2.2   Endogenous transportation costs   97
    5.3   Inconsistencies in neoclassical economic geography   99
    Summary   102
    Notes   103

6   *Incorporating natural resources: rent theory*   104
    Introduction   104
    6.1   Nature and scarcity   105
    6.2   Rent and scarcity within political economic theory: two views   108
          6.2.1   Marxist rent theory   109
          6.2.2   Neo-Ricardian rent theory   114
          6.2.3   Towards a reconciliation   119
          6.2.4   Summary   122
    6.3   Transport costs, multi-commodity production and differential rent 123
          6.3.1   Extensive differential rent   123
          6.3.2   Intensive differential rent   126
    6.4   Differential rent and non-land resource sites   127

6.5   Absolute/monopoly rent   128
      6.5.1   Absolute rent   129
      6.5.2   Monopoly rent 1   132
      6.5.3   Monopoly rent 2   133
Summary   135
Notes   136

7   *The city: incorporating the built environment*   137
Introduction   137
7.1   The theoretical context   137
7.2   The intra-urban location of the basic sector and rent   138
      7.2.1   Optimal production sector location excluding land rents   139
      7.2.2   Optimal location of plants with land rents   141
7.3   Rent and the built environment as fixed capital   142
7.4   Residential location   148
      7.4.1   The Garin-Lowry model   148
      7.4.2   Population   148
      7.4.3   The retail sector   150
      7.4.4   The housing sector   152
      7.4.5   Wages, rationality and class conflict: Some reconsiderations   154
Summary   156
Note   157
Appendix: A matrix version of the Garin–Lowry model   157

8   *The labour value circuit*   160
Introduction   160
8.1   The geography of labour value and exploitation   160
      8.1.1   Labour values   161
      8.1.2   Exploitation   163
      8.1.3   Implications for class alliances   165
      8.1.4   Circulation of labour value   167
8.2   Unequal exchange and regional development   168
      8.2.1   A two-region economy   169
      8.2.2   A multi-regional economy   171
      8.2.3   Empirical studies   173
Summary   175

9   *The quantity circuit and capital accumulation*   177
Introduction   177
9.1   The geography of production under dynamic equilibrium   178
      9.1.1   The rate of capital accumulation   181
9.2   Incorporating workers' savings and capitalists' luxuries 184
      9.2.1   The consumption–growth frontier   187
9.3   The instability of dynamic equilibrium   188
      9.3.1   Responses to instability   190
Summary   193
Appendix: Characteristics of the dynamic equilibrium   194

PART III DISEQUILIBRIUM: CLASS CONTRADICTION AND
STRUGGLE   199
10 *Class and space*   201
    Introduction   201
    10.1 Class-in-itself and class-for-itself   201
    10.2 Class-in-itself   202
          10.2.1 Definitions of class   202
          10.2.2 Class structure   210
          10.2.3 Class interests   212
    10.3 Class-for-itself   213
          10.3.1 Class formation and consciousness   213
          10.3.2 Class struggle   218
    Summary   223
    Note   223

11 *Location and inter-class conflict*   224
    Introduction   224
    11.1 Workers vs capitalists   224
          11.1.1 Geographical conflicts over income and consumption levels 224
          11.1.2 Power relationships: domination at the workplace   228
          11.1.3 Property relationships: communally owned means of
                  production   230
    11.2 Workers vs landlords   231
          11.2.1 Conflict over rent levels   232
          11.2.2 Power and property relationships: community organizations 233
          11.2.3 Property ownership: the housing class debate   234
    11.3 Capitalists vs landlords   236
          11.3.1 Conflicts over rent   236
          11.3.2 Power relationships: conflict over land tenure   238
          11.3.3 Conflict over land ownership: capitalists, landlords and
                  fictitious capital   240
    Summary   242

12 *Location and intra-class conflict*   244
    Introduction   244
    12.1 Unintended consequences and intra-class conflict   244
    12.2 Intra-class conflict among capitalists   246
          12.2.1 The relocation decision   247
          12.2.2 The decision to introduce technical change   248
          12.2.3 The decision to specialize and trade   250
    12.3 Intra-class conflict among workers   253
          12.3.1 Direct conflict: working–class segmentation   254
          12.3.2 Indirect conflict: responses by capitalists   256
    12.4 Intra-class conflict among landowners   258
          12.4.1 Conflict over differential rent   258
          12.4.2 Conflict over monopoly rent 2   259
    Summary   260
    Note   261

PART IV   DISEQUILIBRIUM: TECHNICAL CHANGE AND
ORGANIZATION   263

13   *Strategies for reducing production costs*   265
     Introduction   265
     13.1   Cost reduction strategies   265
            13.1.1   Types of technical change   267
            13.1.2   The rate and direction of technical change   268
            13.1.3   Location and the direction of technical change 271
            13.1.4   Impact of technical change   273
            13.1.5   Relocation and wage contracts   275
            13.1.6   The case of resources   275
     13.2   Fixed costs   276
     13.3   Reducing circulation time   277
     Summary   280
     Notes   281

14   *Strategies for organizational restructuring*   282
     Introduction   282
     14.1   Oligopoly and the rate of profit   283
            14.1.1   The debate over profit rate differentials   283
            14.1.2   Incorporating profit rate differentials   284
            14.1.3   Inter-regional profit rate differentials   286
            14.1.4   The accumulation of investment funds   288
     14.2   The organization of production   288
     14.3   Investment strategies   291
     Summary   294
     Note   295

15   *Conclusions*   296
     15.1   Summary of the argument   296
     15.2   Extensions   298
     15.3   Reflections   300

Glossary   303

Bibliography   311

Name Index   324

Subject Index   326

List of tables

3.1   Economic magnitudes in exchange, labour and use value realms   43
3.2   Empirical correlations of labour values and exchange values   50
10.1   The relationship between class-in-itself and class-for-itself   211
10.2   A pay-off matrix showing the potential conflict between
individual and collective action   215

List of figures

1.1	Ricardo's corn model	3
1.2	Marx's model	3
1.3	The neoclassical model	6
2.1a	Neoclassical wage–profit frontiers	21
2.1b	A continuum of neoclassical wage–profit frontiers	22
2.2	Reswitching with heterogeneous capital goods	24
2.3	Capital reversing and reswitching	25
2.4	Capital reversing in the absence of reswitching	26
4.1	Inter-regional input–output matrix	67
4.2	A wage–profit frontier	71
4.3	Wage–profit frontiers for three location patterns (A, B & C)	73
5.1	Example of reswitching in a space economy	91
5.2	A single-centred region	93
5.3	Production around a central market	95
5.4	Transport costs and differential rents	96
5.5	A wage–profit–rent frontier	96
5.6	Reswitching: transport costs exogenous	98
5.7	Reswitching: transport costs endogenous	99
6.1	Straight-line wage–profit frontiers (price proportional to labour value); each representing a different level of land fertility	110
6.2	Three wage–profit frontiers (illustrating Ball's theory of DRII)	113
6.3	Wage–profit frontiers for three different lands	116
6.4	Three wage–profit frontiers and an intensive rent curve illustrating Sraffa's theory of intensive rent	118
6.5	Intensive rent with a positively sloping wage–profit frontier	120
6.6	Changing levels of intensive rent with changes in the technique of production	121
6.7	Potential conflicts among capitalists, workers and landlords shown on a wage–profit–rent frontier	134
7.1	Wage–profit frontiers for a building with a physical life of three years	146
9.1	A wage–growth frontier	182
9.2	A consumption–growth frontier	188
9.3	Profits and growth for two location patterns	189
11.1	Two wage–profit frontiers representing different labour processes	229
13.1	Production methods in sector n, region j	267
13.2	Innovation: searching about method x	269
13.3	Imitation of production method s	270
13.4	Direction of innovation and selection: two regions, two sectors	273
13.5	Direction of innovation and selection: two sectors, same region	273

Preface and acknowledgements

This book was conceived in the late 1970s when, by a happy coincidence, we met at the Department of Geography, University of Minnesota (Sheppard as Assistant Professor, Barnes as graduate student), and found that the other was beginning to explore the same issue of analytical political economy. Unfortunately, the gestation period of ideas does not display the same regularity as in the biological world. Our actual decision to combine the themes of Sheppard's articles and working papers with Barnes' PhD thesis to form a book was made in 1983. In the next six years the book emerged in dribs and drabs, despite frequent cajoling, jibing and exhorting by both authors. In the disproportionate time taken to complete this work, some of the reasons that first motivated our project no longer seem as important. Political economy as an approach is much more widely practised and understood in economic geography than it was ten years ago, and the criticisms of neoclassical economics, its rival, have been trenchant and convincing. The purpose of the book is more modest now. It is to introduce economic geographers, from the interested undergraduate to the professional, to a particular approach to political economy: the analytical one. In so doing, we are certainly not suggesting that this approach is the best, far less the only one. But it is one approach. And, as we try to show, it is a perspective that has a number of attributes to recommend it in the theoretical task of reconstructing a political economy that takes place and space seriously. We realize, however, that a number of readers will be unfamiliar with an analytical treatment of political economy, so we have compiled a glossary of the main terms used, which will be found at the end of the book.

Although in every sense this book is a collective product, we have agreed to disagree on some issues. Barnes maintains doubts about the labour theory of value, which he argues smacks of essentialism, while Sheppard is suspicious of the contextualist position, and attendant relativism, that Barnes proposes in its place. Over the last year, however, as the frequency of our discussions has increased, there has been a softening on both sides.

In pursuing the task of a geographical reconstruction of political economy we are treading the same path that a number of other geographers have taken. Although it is perhaps invidious to single out one person among the many on whose work we have drawn, we would like to acknowledge the special importance of David Harvey's writings. Without prejudice to him, David Harvey's work has been inspirational for our own. In constructing our arguments we invariably went back to Harvey, not necessarily agreeing with his answer, but always admiring how he posed and approached the question. We also have more direct acknowledgements to make: Michael Chisholm, Claire Pavlik and Paul Plummer each read portions of the book with a sharp critical eye, and their comments helped us to clarify our argument in many places. In Chapter 5 we drew heavily upon work carried out by Claire Pavlik in her MA thesis and we would like to thank her for allowing us to use it here. Also some of the arguments in Chapter 13 are taken from a joint research project in which Sheppard has been engaged with David Rigby and Michael Webber. We would like to thank both of them for permitting us to incorporate such work here. In addition, parts of Chapters 1, 4 and 7, along with some of the figures found there, were previously published in *Environment and Planning A* and *Antipode*. Furthermore, we would like to thank Cambridge University Press and the respective authors for allowing us to base Figure 2.3 on Figure 4.6 in Harcourt (1972) and reproduce Figures 6.4 and 6.5 from Mainwaring (1984). Finally, we would like to acknowledge the various institutions that provided financial and other assistance to us: Sheppard is grateful to the Institute of International Applied Systems Analysis at Laxenberg, Austria, and to Melbourne University for providing both financial assistance and conducive surroundings in which to work. In addition, he would like to thank the University of Minnesota for giving him a single quarter leave. Barnes would like to acknowledge the financial help afforded to him as a graduate student by the William F. Stout Fellowship, and the University of Minnesota Dissertation Fellowship. He would also like to recognize the rich intellectual environment of the Department of Geography, University of British Columbia, where much of the writing of his portion of the book was undertaken. Finally, we dedicate the book to our families: Claire, Helga, Joan, Kirstin, and Michael.

1 Introduction

Introduction

During the last fifteen years there have been some significant changes within economic geography. In the mid-1970s theoretical discussion within the discipline was dominated by ideas taken from mainstream, neoclassical economics. Certainly, there was an underground movement critical of that orthodoxy (Harvey, 1973; Massey, 1973; Sayer, 1976), but these attacks generated as much heat as light, and provided little in the way of a positive alternative theoretical framework. Since then, however, there has been a concerted attempt by a number of economic geographers to construct just such an alternative based on ideas of political economy. Furthermore, this approach has quickly achieved recognition, and now provides a coherent explanatory thesis that is taken at least as seriously as any other in economic geography.[1]

Writing within the spirit of this recent movement, our book is intended to explicate, clarify and extend the conceptual foundations of the political economic paradigm within economic geography. It is therefore a book about ideas and beliefs, an attempt to present the conceptual skeleton that underlies any political economic approach to economic geography, rather than a presentation of empirical research. None the less, it is a conceptual skeleton that can be fleshed out by existing case studies (which we point to throughout the book), and also can point to fruitful directions for further empirical work.

As a model for the kind of account we wish to present, we pay homage to David Harvey's (1982) *Limits to Capital*, from which we drew many insights and ideas. Yet our argument is different in style from his, and indeed from many others whose ideas we share. For at the core of this book is the belief that a clear account of the political economic approach is obtained by interrogating our subject matter with an analytical (mathematical) logic. We argue that this serves four closely interrelated purposes.

First, the adoption of an analytical language enables us to present a sustained critique of the dominant rival paradigm, neoclassical economic geography. In the rhetoric of debate, adherents of neoclassical ideas tend to dismiss their political economic critics on the basis that they are 'unscientific', 'woolly', or 'biased', irrespective of the (usually considerable) substantive merit of the opposing arguments. By harnessing an analytical logic in the service of political economy, we argue that a coherent and damaging critique of neoclassical theory, and the beliefs it propagates in economic geography, can be sustained. As such, the use of analytical methods provides the possibility of an 'internal critique' of neoclassical economic geography – internal in the sense that the criticism is framed in the same language and mode of thought as the position criticized (Chapters 2 and 5).

Second, by subjecting the claims of political economic theorists (both geographers and non-geographers) to this same analytical examination, we can logically separate sustainable ideas and concepts from those that are inconsistent and contradictory. From this analytical sieve we then provide a systematic and internally consistent account of the capitalist space economy from a political economic perspective (Chapters 3, 4 and 6).

Third, by introducing space and place into the very centre of this analytical account, we argue that a number of central theoretical claims made by previous writers in economic theory must be at least qualified if not rejected (Chapters 4, 6, 10, 11 and 13). In this sense our conclusion echoes that of Harvey's (1986, p. 142), who writes that 'the incorporation of space has a numbing effect upon the central propositions of *any* corpus of social theory'.

Finally, the analytical framework enables us to address a number of pedagogical tasks within the book: to provide an introduction to the paradigm of political economy for those students unfamiliar with the approach, but who have the patience to follow a deductive argument; to construct a wide-ranging model of the capitalist space economy incorporating commodity production, natural resources, the built environment, class formation, technical change and industrial organization; to show radical scholars how their work fits with others; to demonstrate to non-geographers how space is integral to political economy; and, finally, to cajole neoclassical adherents into thinking more critically about their own approach – and more favourably about our own!

As a preface to our substantive arguments and analysis, in this introductory chapter we present, first, a historical overview of the rise, fall and rise again of political economy in economics; second, a discussion of the leading variants of contemporary analytical political economy; third, a review of the book's method in light of this analytical turn; and, finally, a synopsis of the book's structure and argument.

1.1   From political economy to neoclassicism and back

In this section we discuss the emergence of political economy in the first part of the nineteenth century, the very different nature of neoclassical economics that followed it, and the subsequent revival of interest in a revamped analytical political economy during the last twenty-five years. Our characterizations are of necessity brief, and more systematic accounts of the issues and debates we allude to in this section are taken up in Chapters 2 and 3. In this chapter we simply wish to establish the intellectual and historical context of the political economic approach that we will pursue in the remainder of the book.

The political economic approach is primarily associated with the classical economists of the eighteenth and nineteenth centuries, and, in particular, with the

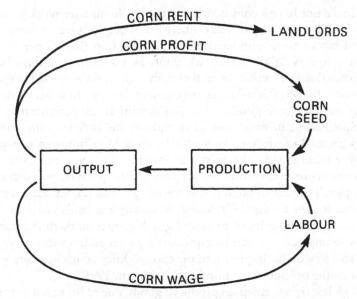

Figure 1.1 Ricardo's corn model

writings of David Ricardo and Karl Marx. Central to both Ricardo's and Marx's schemes were issues of first, value, that is, the fundamental determinants of relative prices; second, production and reproduction; and third, the distribution of income among social classes. The similarities and differences between Ricardo and Marx are seen in Figures 1.1 and 1.2, respectively.

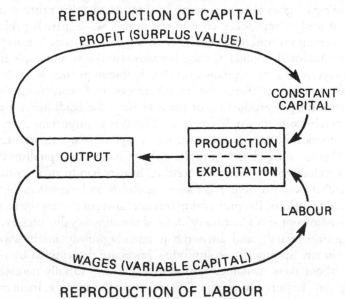

Figure 1.2 Marx's model

Ricardo did not have a consistent value theory. In his later work he employed some kind of labour theory of value where prices of goods are determined by the amount of labour time embodied in their production (the 93 per cent labour theory of value, as Stigler called it), while in his earlier writings he used a so-called corn theory of value. In each case, though, value is created in the process of production. In fact, Ricardo went to considerable lengths in his work to avoid discussing non-produced goods, which, by definition, are peripheral to his value theory. Specifically, in order not to complicate his analysis with scarce, non-produced goods Ricardo used an analytical trick. He eliminated non-produced goods by assuming that the prices of commodities produced on, say, the non-produced commodity land are set only by the price of production on the marginal plot. The crucial point is that the marginal land is *not scarce*, reflected in the fact that it pays no rent. Of course, intramarginal lands pay rent, but such rents do not affect the price of produced goods grown on them (Pasinetti, 1981, ch. 1). By setting scarcity and non-produced goods aside in this way, Ricardo was able to 'carry on his inquiry into the commodities which he really wanted to investigate – the produced commodities' (Pasinetti, 1981, p. 7).

Although Ricardo focused on produced goods, one of his major contributions to nineteenth-century political debate was in making an argument about the owners of non-produced goods, namely, the landlords. In particular, Ricardo argued that in the sphere of distribution landlords represent an obstacle to the reproduction of produced goods, and thereby the creation of value. This is best illustrated by Ricardo's corn model (Figure 1.1). In this one-commodity world, capital (which is made up only of corn seed), labour and land are combined to produce the output corn. From the total output of corn, part goes towards wages, while part goes towards profits. In turn, part of that profit is invested in the form of seed to reproduce corn in the future, while part is paid as rent to landlords owning intramarginal lands. With wages set at some 'natural' level, the central distributional conflict is then between landlords and capitalists. What landlords capture as rent, capitalists necessarily lose in profits. Ricardo's point is that, while capitalists perform the 'useful' function of reinvesting their profit, thereby permitting reproduction of the economy, the landlords are a fetter on accumulation because they only consume. This was an important contribution to debates surrounding the rising importance of capitalists and the resistance of the important landowning class during the onset of industrial capitalism in Britain.

Sixty years later when Marx was writing, history had turned a corner and the central conflict was no longer between landlords and capitalists but between capitalists and workers. In emphasizing this new antagonism on the factory floor, Marx moved away from a 'naturally' defined measure of value such as corn (with which Ricardo flirted), and anchored it unambiguously in the sweat of the labourer's brow. Specifically, commodity prices are determined by the socially necessary labour time contained within them, where 'socially necessary' means the labour time required to produce a good using the average, median or modal technique of production at any given historical moment (for further discussion see section 3.3.1). Although landlords are certainly still important for Marx, they play a negligible role in the core of his scheme (Figure 1.2). None the less, Marx's scheme still bears a close relationship to Ricardo's earlier corn model. Specifically, for Marx combining capital and labour (termed by him constant and variable capital) enables production to occur, resulting in a given level of output. Part of that produced output goes to capitalists as profit (surplus value) and is then reinvested, enabling the reproduction of constant capital. The remainder goes to workers in the form of wages, enabling the social reproduction of labour. It is the so-called rate of exploitation at the workplace that settles the division of the output between wages and profits. Exploitation is formally defined in Chapter 3, but the gist of the idea is that it occurs when the worker produces more value during a day than s/he is paid in wages for that same period.

From these admittedly crude and brief portrayals, the hallmarks of political economic perspective are a concern with value, production and reproduction, and distribution. Of these it is distribution that accounts for the 'political' part of political economy, because within Marx's and Ricardo's frameworks it is this issue that pushed their enquiry beyond the realm of the purely economic, and into the sphere of social and political analysis. For Marx, distribution meant examining questions of property ownership and political power expressed in terms of exploitation; while for Ricardo it meant examining the political power relationships between an embryonic class, the capitalists, and a historically entrenched one, the landlords. Finally, we should also note that neither Marx nor Ricardo presented their work in formal mathematical terms (admittedly Marx made use of some algebra and numerical examples, but in general his forays into mathematics were not successful – witness his difficulties in providing a consistent analytical solution to the problem of transforming labour values into prices; see Steedman, 1977). This does not mean, though, that Ricardo and Marx were against mathematics, or that their work cannot be expressed in mathematical terms. Joan Robinson (1964) argues that Ricardo's work represents the first abstract economic model ever constructed, one portraying causal relationships between dependent and independent variables (see Gudeman, 1986, ch. 3), while Lafargue attributes to Marx the statement that 'a science becomes developed only when it has reached the point where it can make use of mathematics' (Smolinski, 1973, p. 1201).

The neoclassical economics that subsequently arose in the 1870s is a misnomer, for very few of the concerns of the classical economists were retained. In fact, in many ways neoclassical economics represents a rejection of political economy. Neoclassical value theory thus begins not with produced goods and their conditions of production and reproduction, but with scarce non-produced goods and their conditions of exchange (Figure 1.3). In particular, the neoclassical general equilibrium model presumes the existence of a set of rational individuals who are endowed with a given allocation of natural resources. These individuals own such resources, and possess well-defined utility preferences; that is, they know what they like and dislike and can judge the satisfaction from making a

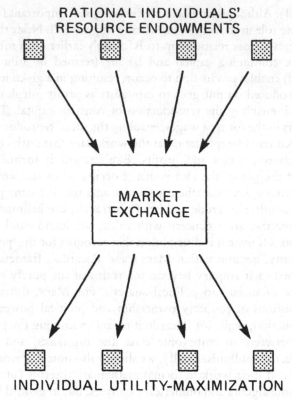

Figure 1.3   The neoclassical model

given choice. The economic problem is then to find 'those prices (equilibrium prices) which bring about, through exchange, an optimum allocation of these given resources; this optimum allocation being defined as that situation at which the individuals maximize their utilities, relative to the original distribution of resources among them' (Pasinetti, 1981, p. 9). In short, the economic problem is one of rational choice. Providing individuals are rational they will exchange the commodities they own for different commodities owned by others until utilities are maximized. Income distribution among individuals is determined solely through exchange relationships. An agent's income is a result of how much his/her resources fetch on the market.

Comparing the political economic approach with this admittedly baldly and tersely stated neoclassical one, we see three dramatic differences. First, in the neoclassical scheme value is created not within the production process, but by the mental desires and preferences of individuals as they exchange. Second, there is no sense of reproduction in the neoclassical scheme. For a given slice of time, individual agents rationally exchange their set of exogenously given resources with another set owned by different individuals in order to maximize utility.

Once exchange occurs, the slate is wiped clean and we start with a new set of resource endowments and preferences. We never know, though, how factors of production are reproduced for the next slice of time. Marshall's solution to the problem was to conceptualize capital as meteoric stones that conveniently fall ready for the next round of production, while later neoclassical economists theorized capital as putty, Meccano sets or, even more fancifully, ectoplasm, which are all readily reusable and malleable (see Chapter 2 for a detailed account). More generally, these assumptions reflect a more fundamental drawback of neoclassical economics, namely, the absence of a clear theory of production. In fact, Pasinetti (1981, p. 16) argues that neoclassical economists 'have eliminated from their analysis all the relevant features of the production process in order to maintain intact the model they have built for the problem of optimum allocation of scarce goods'. Third, unlike the political economic view, neoclassical economics locates the process of income distribution entirely within the realm of the economic. As Nell (1972, p. 21) writes, for neoclassical economics it is only exchange relationships within the market, 'not man [sic], not god, least of all politics that has decreed the shares of labour and capital in the total product'. In this view, the political aspect of political economy has no part to play.

In addition to these differences of substance, there is yet a fourth difference between traditional political economy and neoclassical economics turning on language. In contrast to Marx and Ricardo, neoclassical economists presented their work in rigorous mathematical terms right from the outset. As Jevons (1970, p. 70), one of the founders of neoclassical economics, wrote: 'It is clear that economics if it is to be a science at all, must be a mathematical science.' Through the course of the twentieth century neoclassical economics subsequently amassed an impressive array of increasingly sophisticated mathematical concepts and theories. In so doing, neoclassical economics both legitimated itself within the academy (as 'queen of the social sciences') and fended off radical and institutionalist critics who employed no such mathematical analysis.

In contrast, it was not until 1960, with the publication of Piero Sraffa's book (1960) *Production of Commodities by Means of Commodities* and the neo-Ricardian school it spawned, that a sustained and systematic application of mathematics was undertaken within political economy (thereby also establishing an analytical political economy). The result, according to Wolff (1982, p. 234), was that 'Marxian economics, almost overnight, went from being a semimoribund branch of secular theology to being a lively, developing, controversial, and innovative branch of theoretical economics'. Initially Sraffa's work was used critically in the so-called 'capital controversy' to attack the neoclassical position on its own mathematical terms (see Chapter 2). In fact, the irony of the debate was that the neoclassical school, represented by MIT, conceded only when it was demonstrated that they, the bastions of rigour and logic, had made the mathematical error (this was over the issue of a pseudo production function – Samuelson, 1966).

In the 1970s Sraffa's work once again was a critical foil when Ian Steedman, in 'possibly the angriest bit of mathematical economics ever written' (Wolff, 1982, p. 235), attacked Marx's labour theory of value. The subsequent so-called value controversy continues, but one of the consequences of Steedman's (1977) book, *Marx after Sraffa*, was that it led to the formation of a second group of analytical political economists, termed here fundamental Marxists, who began countering Steedman's arguments against the labour theory of value by employing the same type of mathematical logic that he had used to attack it. In addition, the fundamental Marxists subsequently carried out a rigorous empirical testing of Marx's labour theory of value using formal statistical techniques.

In addition to the neo–Ricardians and fundamental Marxists, yet a third group of analytical political economists emerged in the early 1980s under the rubric of analytical Marxism. Unlike the other two schools, the analytical Marxists were not defined in terms of critique. Rather, from the outset they saw themselves as offering a constructive alternative, albeit shaped by a critical engagement with that which already existed. In particular, the task of analytical Marxism is 'to examine and develop the theory pioneered by Marx, in the light of the intervening history, and with the [analytical] tools of non-Marxist social science and philosophy' (editorial statement to the Cambridge University Press series *Studies in Marxism and Social Theory*, edited by G. Cohen, J. Elster and J. Roemer).

1.2   Analytical political economy: three views

The preceding thumbnail sketch gives some impression of the way in which recent analytical political economy is analytical, and also the ends to which that analysis is put. For the neo–Ricardians, mathematics is used to criticize both the neoclassical and Marxist value theories. In fact, Sraffa subtitled his book 'Prelude to a critique of economic theory'. For fundamental Marxists, mathematics represents a means of, first, defence and counter-attack, and, second, empirically verifying their position. Finally, for the analytical Marxists, mathematics is both a clarifying device and a tool for the further development of theory. These five justifications of an analytical neoclassical economy (critique, defence, empirical verification, clarification, and a tool for further theoretical development) are now explored in greater detail as we turn to a more focused examination of neo–Ricardianism, fundamental Marxism and analytical Marxism.

1.2.1   Neo–Ricardianism

Sraffa's *Production of Commodities by Means of Commodities* certainly attracted mixed reviews. Harry Johnson (1974, pp. 21–2) argued that it is nothing but 'an apparently purely scientific rallying point for ... anti-Americanism', while Robert Wolff (1982, pp. 230–1) likened it to 'a Gregorian plainsong of the middle ages', one engendering 'a deeply moving experience'. Certainly, there are few books quite like it in economics. There is no introduction, or conclusion; no institutions, or social actors. All we are given are the coefficients of production and the conditions for exchange under different sets of circumstances. As Joan Robinson (1965, p. 7) wrote: 'addicts of pure economic logic have here a double-distilled elixir that they can enjoy drop by drop, for many a day.'

In effect, Sraffa's model is Ricardo's corn model extended to many commodities. That model is made up of two main components, which are termed the conditions of production and conditions of distribution. First, the conditions of production are represented by a series of input–output coefficients expressing how much of one good is required to produce a unit of another. Furthermore, production is conceived as circular and interdependent where the output in one economic sector and production period is used as an input in some sector in the next period. Second, the conditions of distribution come into effect when outputs exceed inputs, thereby creating a surplus. This surplus then forms a 'pool' from which each social class draws its respective share of national income. With production coefficients and one of the income shares given exogenously, Sraffa demonstrated that if profit rates are equalized then prices and other income shares are uniquely determined. Because one of the income shares is given exogenously, Sraffa was implicitly suggesting that to close the system one must refer to institutions and social forces that lie outside the purely economic. As Dobb (1973, p. 261) wrote: Sraffa made 'no attempt ... to derive a theory of distribution from *within* the circle of exchange ... [and] thus the boundaries of economics as a subject ... are drawn so as to include social, and moreover institutional and historically relative changing and changeable conditions that were excluded from Economics as viewed in the post-Jevonian tradition.'

Although Sraffa presented a model of the economy, it is a minimalist one. Missing is any kind of economic, social, historical or cultural context in which to interpret the conditions of production and distribution. For this reason some argue that the methodological position offered by Sraffa is a contextual one (Napoleoni, 1978; Hausman, 1981; Gudeman, 1986; Barnes, 1989). In order to fill in the details that Sraffa presumably deliberately left out of his work, we must embed his model into the particulars of the geographical and historical context. In this sense, what Sraffa excluded from his analysis is as important, if not more so, than what he included.

If so, what exactly did Sraffa leave us with? We argue that it is precisely his 'double-distilled elixir of logic'. In his book Sraffa outlined what is logically necessary for the continued reproduction of an economy. But he said nothing more than that. That logic, however, is exceedingly powerful. In particular, the neoclassical concept of capital and the traditional relationship between labour values and prices posited by Marxists (if Steedman's 1977 arguments are accepted) both unravel when scrutinized against such logic.

In our book we make much use of neo-Ricardian analysis in Chapters 2, 3 and 5, where we discuss the capital and value controversies and their implications for a spatial economy. We also discuss Sraffa's work to a lesser extent when we

examine both the issues of land, natural resources and rent in Chapter 6, and the
urban economy in Chapter 7. Yet, if our argument above is correct, to make use
of Sraffa's work in a constructive way (as in Chapters 6 and 7) it is necessary to
embed it within a broader framework that discusses such issues as class and
exploitation, which he excluded. It is here that the other two variants of
analytical political economy, fundamental and analytical Marxism, are central.

1.2.2   Fundamental Marxists

Fundamental Marxists are fundamental in the sense that they defend the labour
theory of value. Although previous analytical justifications of the labour theory
of value existed (for example, in Sweezy, 1942; Seton, 1957; Morishima, 1973),
the most sophisticated and certainly the most numerous appeared in the decade
following Steedman's (1977) *Marx after Sraffa*. Initially, such literature was
concerned with simply rebutting Steedman's claim (1977, p. 202) 'that value
magnitudes are, at best, redundant in the determination of the rate of profit (and
prices of production)' (for example, Wolfstetter, 1976; Shaikh, 1977). More
recently, however, fundamental Marxists are turning to a rigorous critique of the
neo–Ricardian position itself (Shaikh, 1981; Mandel and Freeman, 1984; Fine,
1986). In addition, the same researchers are undertaking empirical work supportive of the labour theory of value; research demonstrating that, first, labour values
correlate highly with prices, and, second, once labour values are empirically
operationalized they very deftly explain various facets of capitalism. For
example, in introducing *Ricardo, Marx, Sraffa*, perhaps the most mature statement of a number of fundamental Marxists, Mandel (1984, pp. xii–xiii) writes:

> all the authors ... apply the best of all scientific tests, ... [and demonstrate]
> that ... Marx's basic hypotheses and his analysis is [*sic*] confirmed ... by all
> the available empirical evidence. ... This is not to say that Marxism has
> closed the book of empirical study. On the contrary, the new statistical
> methods that have become available ... can be used to examine empirical
> issues within a Marxist theoretical framework in a degree of detail not
> previously possible.

A focal point for much of the recent work by fundamental Marxists is Farjoun
and Machover's *Laws of Chaos* (1983). This provides both a critique of the
neo–Ricardian position as well as a foundation for a probabilistic approach to
political economy, thereby meshing well with the use of statistical techniques in
empirical analysis. It is impossible to explore all the nuances of Farjoun and
Machover's work here (see Webber's, 1986, excellent review), but their central
claim is that both profits and prices are stochastic variables, and are analysable by
the methods of statistical mechanics. Farjoun and Machover make two fundamental points: first, neo–Ricardian production prices are just as unobservable as
labour values because there is no single price for the same good – there is only

stochastic variation about some expected production price; second, the labour theory of value is entirely consistent with Farjoun and Machover's analysis. In fact, they demonstrate that prices are proportional to labour values in an aggregate analysis, and that for various reasons labour values are the best standard of measure when comparing average prices of *different* commodity types (Farjoun and Machover, 1983, ch. 4). Furthermore, they provide empirical evidence of a strong correlation between labour values and market prices, and also demonstrate that labour values are calculated under less restrictive assumptions than those required to determine prices of production (Farjoun and Machover, 1983, ch. 8; Semmler, 1984; also see sections 3.5 and 3.6 below).

Our use of fundamental Marxist theories is not confined to a single section of the book, but pervades most of the chapters. Specifically, we make use of Farjoun and Machover's ideas to address the determinants of the average tendencies around which the stochastic variations of prices and profits fluctuate.

1.2.3 Analytical Marxism

Analytical Marxism is the most recent of the three movements. Originating in the early 1980s, it now attracts a number of proponents across many disciplines (economics, sociology, political science and philosophy). Although the antecedents of the movement are in Gerald Cohen's (1978) *Karl Marx's Theory of History: A Defence*, it is John Roemer's (1981, 1982a, 1988) work that provides the analytical foundation for the school.

For both Cohen and Roemer a prime objective is to avoid the naive functionalism of some Marxists who believe that merely pointing to the beneficial consequences of an action/event serves to explain it (Roemer, 1981, Introduction; Elster, 1985). Cohen's solution is to provide a more refined version of functionalism resting on consequence laws. Roemer (1986, p. 192), in contrast, avoids functionalism altogether by arguing that 'Marxists must provide ... explanations of *mechanisms*, at the micro level, for the phenomena they claim come about for teleological reasons' (see also Roemer, 1981, pp. 7–9). What Roemer then offers is the microeconomics of Marxian analysis, one based on 'rational choice models: general equilibrium theory, game theory, and the arsenal of modelling techniques developed by neo-classical economics' (Roemer, 1986, p. 192). In this sense, as Roemer (1986, p. 191) argues, 'with respect to method, I think, Marxian economics has much to learn from neoclassical economics, [but] with respect to the substantive research agenda ... it is the other way around.'

Like neo-Ricardians, the analytical Marxists reject Marx's labour theory of value on analytical grounds (although none have addressed Farjoun and Machover's reinterpretation of Marx's value theory). Unlike the neo-Ricardians, however, the analytical Marxists do offer a complete theory of social and economic life. It is a theory that begins with rational individuals who are endowed with a given share of society's resources, and who are intent on maximizing utility. This, of course, is the same starting point as neoclassical economics. However, the end result for analytical Marxists is not equilibrium and harmony; it is exploitation and class formation. The details of the scheme are presented in Chapter 10, but their central conclusion is worth stating here: for analytical Marxists, individuals rationally choose both the class to which they belong and their status as exploiters or exploited. Of course, this is not a choice that individuals necessarily like making. But they do so because of both the constraints that they live under (their share of society's resources) and their desire to maximize.

In carrying out their theoretical agenda, analytical Marxists employ analytical methods in two different ways. First, they are a means of clarification and critical assessment. As Carling (1986, p. 25) writes, for analytical Marxists 'Marxism was to be sorted into a list of distinct claims: each one deserving its own interrogation for meaning, coherence, plausibility and truth. The logical relation between claims was an explicit topic of the theory, so that it became more open to judgement which parts of a complex Marxist corpus stood and fell together.' The result of this 'interrogation' is the rejection of such concepts as the labour theory of value, the falling rate of profit thesis, and the traditional view of Marxian dialectics (Elster, 1985). But what remains, according to Roemer (1986, pp. 1–2), is a non-dogmatic 'valid core' that has been 'clarified and elucidated' using 'state-of-the-art methods of analytical philosophy and positivist social science'. Second, the use of analytical methods produces conclusions that are unlikely to be derived from a non-analytical approach. For example, Roemer (1986, p. 111) writes: 'formal modelling is also useful in producing ideas. On a number of occasions, formalism led me to conclusions of which I had no prior inkling.'

In terms of our book, we make explicit use of analytical Marxism when we discuss classes, and the conflicts that exist within and among them (Chapters 10 – 12). None the less, we argue that the analytical Marxist general approach to class requires some considerable modification to take into account space and place.

In summary, the analytical turn in political economy is clearly more than a passing whim or fancy. It is already shaping debate and setting a theoretical agenda. Clearly, though, there exist sharp disagreements among the three variants that make up this wider genus. Rather than taking sides, our approach is a pragmatic one; we simply use those ideas that are helpful in accomplishing our particular task, regardless of their origin. Some may well levy the charge of eclecticism at this strategy. But providing that the *specific ideas* we employ are logically consistent with one another, then we do not see the difficulties; after all, in the last instance the three variants are all part of the same tradition of political economy.

1.3 Questions of method

As noted at the start of the chapter, in adopting an analytical approach we employ a methodology that differs from those of many geographers writing on political economy. The strong tradition of radical thought in human geography since the late 1960s is for the most part non-analytical – in fact, radical geography arose in part to counter the widespread use of quantitative methods and abstract theory by positivists. Our use of an analytical approach, however, is not intended to replace non-analytical work. We do not see mathematics as the arbiter of 'truth', nor do we claim that rigorous statistical techniques are necessary to develop empirically accurate theories. Rather, we regard mathematics, as Ruccio (1988, p. 57) argues, 'as merely one discursive strategy among others'.

We do not claim, therefore, that our approach is necessarily superior to, or more scientific than, others. We recognize that many aspects of society and economy are not subject to analytical treatment, and even those aspects that are may well be more sensitively treated by non-analytical methods. As such, we think it crucial to delineate carefully the limits of analytical methods. We should also note here that we do not subscribe to the methodological individualism of analytical Marxists, which is the view that the analytical beginning point in understanding society is with a set of rational individuals (Lebowitz, 1988). While we emphasize that individual behaviour must be accounted for in political economy, we reject the view that it necessarily conforms to rational choice. In this light, we argue that the reality of space and place are central keys to understanding why human actions are more than the sum of the individuals egotistically seeking to maximize welfare. In fact, in Chapter 10 we suggest that it is precisely the existence of space and place that ensures that class formation and collective action are the rule rather than the exception.

Finally, as will become clearer in the course of the book, the analytical approach does not deny the importance of contextual case studies of evolving places in a capitalist space economy; in fact, it suggests that we emphasize their critical role. An analytical model shows only what is logically possible, not what will actually occur. As Ruccio (1988, p. 57) argues, the importance of analytical theory is 'not in terms of representation, but as a form of "illustration". [That is,] mathematical concepts can be understood as metaphors or heuristic devices.' To establish their heuristic value requires studies on the ground.

In summary, we see our analytical approach as clarifying the underpinnings, critically re-evaluating the theoretical claims, and underwriting the research strategies of non-analytical radical geography. It is certainly not its replacement.

1.4 Outline of the book

Our book is divided into two major sections. The first discusses how a capitalist economy would be organized if it were in economic equilibrium, and the second examines the reasons why such an equilibrium cannot be sustained, and the theoretical implications that stem therefrom.

In the first section we present a model of general spatial equilibrium, one in which there is full competition among capitalists, implying that profit rates are equalized across all sectors. Furthermore, we also assume that individuals are rational, and that there is no state action, no corporate monopolies, and no unions. We make these assumptions not because we think this is the way the world works, but because, as we show, even within this limiting set of postulates many neoclassical conclusions do not logically stand up. Specifically, this first section of the book is subdivided into two main parts.

Part I provides reviews of both the capital controversy (Chapter 2), and the value controversy (Chapter 3) in economics. Both debates laid the foundation for a logical critique of the neoclassical school, and staked out the territory of analytical political economy. They must be understood in order to appreciate the central concerns of this book. We summarize both debates here, partly because we realize that many of our readers will be unfamiliar with them, and partly because it enables us to develop concepts that we later employ within a spatial context.

In Part II we examine (following Desai, 1979) the three fundamental circuits of capitalist spatial commodity production: prices, labour values and physical quantities. Four chapters are devoted to the price circuit. Chapter 4 provides the basic spatial general equilibrium model in terms of *prices*. Under simplified assumptions we define a profit-maximizing set of locations, one in which prices and trading patterns are endogenously determined. We also discuss the effects of circulation time on production and location, an issue that is intimately related to the geographical problem of transportation. Chapter 5 then uses the framework of the preceding chapter to develop examples of 'spatial reswitching'. This is the geographical counterpart to capital reswitching, the issue on which neoclassical economists conceded the capital controversy. Likewise we conclude that spatial reswitching has a number of important critical implications for spatial neoclassical theorizing. The two remaining chapters couched in terms of the price circuit respectively examine two specific topics within economic geography: natural resources and rent, and the urban economy. In particular, after discussing the idea of resource scarcity, Chapter 6 reviews both fundamental Marxist and neo-Ricardian theories of differential rent. We conclude that, rather than being viewed as two distinct approaches, the best parts of both the Marxist and neo-Ricardian rent theories can and should be combined. In addition we investigate the applicability of Marx's theory of absolute rent, and the various geographical interpretations made of it. Chapter 7 applies the general equilibrium framework of Chapter 4 to the urban economy. It is an application, however, that explicitly allows for the built environment as fixed capital. Using the so-called Garin–Lowry model of urban land use and population, the model is further extended to cover issues of urban services, residential location and workers' commuting. Chapter 8 examines the second circuit of commodity production, *labour values*. We argue that by explicitly including a geographical dimension it is possible to derive negative rates of exploitation. We also examine in this chapter the work on unequal exchange. Lastly, Chapter 9 turns to the final circuit, which is expressed in *physical quantities* (use values). Here we are necessarily concerned with issues of accumulation, and we define the conditions necessary for unrestricted capital accumulation under spatial equilibrium.

Having presented a general equilibrium model of the economy in the first section of the book, we turn in the second section to an investigation of those forces that undermine such equilibrium. On the one hand, many believe in economic equilibrium because they think it is a reasonable approximation of the actual economy; it is a stylized fact of capitalism. On the other hand, it represents a norm that, if achieved, describes a configuration that is readily reproduced – some would even say a socially desirable configuration. For both these reasons equilibrium is worthy of attention. Here, however, economic equilibrium is rejected because, as we argue, there are endogenous forces within capitalism, such as class struggle, technical change and institutional dynamics, that continually prevent its achievement.

The second section is divided into two parts. In Part III, we examine the implications of introducing classes, and the conflict that exists both within and among them. In particular, Chapter 10 introduces the idea of class using as a framework the concepts of 'class-in-itself' and 'class-for-itself'. Chapter 11 then examines the consequences of three types of inter-class conflict: capitalist–worker, capitalist–landlord, and worker–landlord. Finally, Chapter 12, using the notions of direct and indirect conflict, investigates the potential strife that exists within a class.

In Part IV we examine two types of change that inherently disturb equilibrium: Chapter 13 investigates technical change, including changes in production technology, the built environment and transportation technology; and Chapter 14 focuses on issues of organizational structure. In particular, using the work of Kalecki to develop a model of production and distribution, we explicitly recognize the role of large multi-locational corporations.

In our concluding chapter we summarize our results, and, in a cursory manner, indicate the potential effects on our model of influences that we have not taken into account, such as the state, financial institutions and population change.

Note

1 Some of the central contributions within the political economic approach include: Scott (1980, 1988a,b); Massey and Meegan (1982); Harvey (1982, 1985a,b); Massey (1984a); Smith (1984); Clark, Gertler and Whiteman (1986); Scott and Storper (1986); Storper and Walker (1989).

PART I

The basics

'[T]he usefulness of a theory will only be fully apparent once theories clash and compete.' (Hard, 1988)

2 The capital controversies

Introduction

The theoretical debate between the neoclassical and neo-Ricardian schools of economics became known as the capital controversy because the debate falls squarely within what is known as capital theory – the theory of how capital is valued and measured. This debate has made it increasingly difficult for the neoclassical and political economic paradigms to live in mutual ignorance and disdain of one another, because it established that there are areas of overlap where the logic of both approaches can be directly compared to one another. The origin of this debate lies in the pioneering monograph of Piero Sraffa (1960), who exploited this common ground to scrutinize critically some central theoretical foundations of the neoclassical school. Economic geographers concerned with developing a theoretically consistent framework for their own analysis should be aware of these debates. In the spirit of increasing this awareness, and in order to provide a common foundation for subsequent chapters, this chapter will review this debate over the neoclassical conception of capital (for a fuller treatment see Harcourt, 1972, 1975, 1976, 1986; Dobb, 1973; Blaug, 1974; Harris, 1980; Moss, 1980). In Chapter 3 we shall turn our attention to labour values and the common ground between neo-Ricardian and Marxist economic theory. Readers already familiar with these debates are encouraged to skip sections 2.1 and 3.1–3.4.

2.1 The reswitching debate

During the 1960s a seemingly arcane debate raged in economic theory, dubbed the capital or Cambridge controversies (because most of the protagonists were resident in either Cambridge, Massachusetts, or Cambridge University, England). Although fought on the high plains of mathematical abstraction, the debate in fact dealt with some fundamental social and political questions. At stake was identifying what determines the distribution of income among social classes in capitalist economies.

2.1.1 The aggregate neoclassical model of production

In the view of the dominant neoclassical paradigm, profits are an index of the physical productivity (and scarcity) of capital, and wages are similarly defined by the productivity of labour. As J. B. Clark (1891, p. 313) had put it: 'What a social class gets is, under natural law, what it contributes to the general output of industry.' Such a position represents an ethical justification for the origin and appropriation of profits under capitalism, because profit is thus defined as the proper reward for owning the scarce and productive resource, capital.

Examining the logic behind this argument, the Cambridge (UK) economists showed that this proposition depends on a deliberate simplification introduced early on by neoclassical theorists. The foundation of neoclassical macroeconomic theory was the insight that, if each input used in commodity production is treated as a homogeneous entity that is brought to the market for exchange by its owner, then the process of production can be approximated by a market transaction. In this view inputs are bought from their owners at a price reflecting their marginal productivity to the producer (which is the measure of their utility to him/her). They are then combined, via a mathematical 'production function', to produce some quantity of output. In macroeconomic theory, where the actions of individual producers in a region are aggregated, the output (Y) of an entire regional economy is also conceived as a single commodity that is mathematically related to the availability of production factors, notably labour (L), a single malleable commodity named capital (K) and technology (t) (see Borts and Stein, 1964). The relation between the availability of inputs and output is defined by a continuous production function representing the range of techniques available to producers: $Y = f(K,L,t)$.

In the neoclassical view, this strategy had three advantages. First, an immensely complex system of interdependent producers is reduced to a simple model of the purchase and transformation of 'capital', labour and technical knowhow. Second, it can be shown that the prices of production factors in this scheme (i.e. the wage paid for labour and the rate of interest or profit paid on capital) are equal to their marginal productivity, defined as the increase in output obtainable by employing one extra unit of the factor concerned (or the mathematical derivative of the production function with respect to the production factor). This can be deduced from a mathematical analysis of the above model, under additional assumptions of no scale economies, diminishing marginal returns to labour and capital, and perfect competition. Specifically, when firms employ the profit-maximizing production technique, profits necessarily equal the marginal productivity of capital ($\partial Y/\partial K$) and wages equal the marginal productivity of labour ($\partial Y/\partial L$). Third, a simple relationship between the market value of a production factor and its scarcity can be established. If the productivity of a production factor declines as its abundance (relative to that of other factors) increases, then its price must fall relative to that of other factors. Thus a surplus of labour would imply wages that are low relative to the rate of profit, and a labour shortage would imply the converse. It is this equivalence between prices and scarcity that underlies the often quoted notion that economics (and thus economic geography) is the science of optimally allocating scarce resources.

To explore neoclassical economics further, let us examine the relation between wages and profits, which may be graphically represented as a factor price or *wage–profit frontier* (Figure 2.1a). In the neoclassical model, when only one technique is available for commodity production, and assuming that only a single commodity such as corn is produced, then the relationship between profits and wages is linear (Harris, 1980, p. 45). Each straight line in Figure 2.1a thus represents a single production technique, and shows the possible combinations of wage and profit rates that allow the single good to be produced by this method at its given price. Furthermore, the slope of the line is proportional to the capital intensity of the production technique that it represents. In neoclassical theory there is always hypothesized to be a continuous range of techniques, of varying degrees of capital intensity, available for production. The one chosen by producers depends on the relative cost (and thus the relative scarcity) of production factors. Graphically, the technique chosen will be the one that maximizes profits for a given wage rate. The outer envelope (the thick line) on Figure 2.1a shows the profit-maximizing technology for each possible wage rate, i.e. the technique chosen by profit-maximizing producers. A continuous production function such as that described above represents an infinite family of possible production techniques between which a firm is choosing. This would be depicted graphically by an infinite number of straight lines on Figure 2.1a (one for each production technique), whose outer envelope would be described by a smooth curve (Figure 2.1b; Garegnani, 1970).

Under neoclassical assumptions, the quantity of capital and labour available in an economy, together with knowledge about the production function, determine two things – the marginal productivity of labour and capital (i.e. the wage and profit rates), and the particular technology selected from the production function that maximizes profits. This is represented by wage rate w^*, profit rate

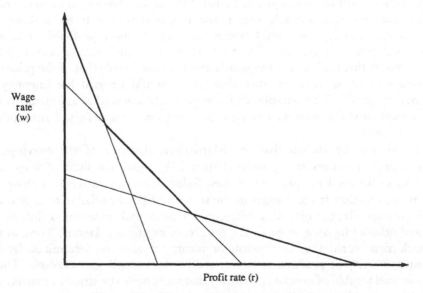

Figure 2.1a Neoclassical wage–profit frontiers

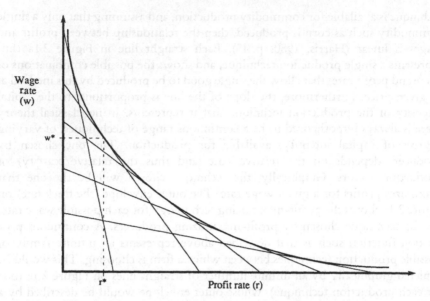

Figure 2.1b   A continuum of neoclassical wage–profit frontiers

r^*, and the point (w^*,r^*) on the curve in Figure 2.1b. By definition, the particular technique chosen is the straight line that is tangential to the curve at point (w^*,r^*), and the slope of that line (or of the curve at that point) equals the aggregate capital–labour ratio of the profit-maximizing technique. The shape of the envelope of optimal techniques in Figure 2.1b implies that any shift to the right along the curve results in a decrease in the slope of the curve. It then follows that the capital intensity, or capital–labour ratio, of the most profitable technique always decreases as the rate of profit increases in a neoclassical world. It is this relationship that leads to the frequently cited economic rule that, if the price of a production factor increases, then that factor should be used less intensely to maximize profits. This parable, which is generally accepted as axiomatic, is not necessarily true for macroeconomic relationships when neoclassical assumptions are abandoned.

It can also be shown that the Marshallian elasticity of the envelope of wage–profit frontiers at any point (Figure 2.1b) equals the share of wages and profits in the total net product. It then follows that any movement along the frontier, resulting from changes in the scarcity of labour relative to capital and the corresponding choice of a different technique, will represent a shift in the distribution of income amongst the owners of capital and labour. Thus, in this neoclassical world the distribution of income is entirely determined by the relative scarcity of production factors and by the technologies available. This is the second parable of neoclassical theory that turns out to be strictly incorrect in a multi-commodity world.

2.1.2 The impossibility of measuring capital

It was Joan Robinson (1953–4) who first raised doubts about the general applicability of the neoclassical simplification. She argued that, if we relax the assumption that the economy produces a single commodity, replacing it by the assumption that many commodities are produced, then the aggregate neoclassical theory of production and distribution founders on the impossibility of developing an independent measure of the quantity of capital (K). In a one-commodity world this is unproblematic because the one commodity must also be the capital good. Because it is the only commodity its price can be expressed in physical terms, making the monetary value of capital equal to the quantity of capital. When a number of distinct capital goods are produced, however, a common unit of measurement is needed. The convention is to use price as the common yardstick. The quantity of capital is ostensibly equal to the total monetary value of capital divided by its unit price, which is the rate of profit. Such a calculation, however, is incorrect. According to this approach, the quantity of capital must depend on the rate of profit (which measures its value). But we saw that in neoclassical theory it is the rate of profit that depends on the quantity of capital available (employing the theory of marginal productivity). The neoclassical definition of capital is therefore circular, undermining any prospect of calculating the rate of profit using this neoclassical approach – except in a single-commodity economy. The initial neoclassical response to this was to argue that Robinson misunderstood the nature of simultaneous equations (Von Weizsäcker, 1971, pp. 97–8), but it was eventually conceded that this is not the case. If 'the relative supplies of capital and labour [are] regarded [in neoclassical theory] as *ultimate* determinants of the shares of profits and wages ... the technical conditions of production ... must exist *before* the equilibrium solutions are derived' (Harcourt, 1982, p. 230).

The paradox noted by Joan Robinson was superseded by a more systematic critique of aggregate neoclassical theory by Piero Sraffa in his book *The Production of Commodities by Means of Commodities* (Sraffa, 1960). Sraffa set up a model of capitalist competition that differs in only one respect from the neoclassical model. Like the neoclassicists he assumed that full capitalist competition exists, implying that the rate of profit is equal for all. However, he replaced the assumption of a homogeneous capital good by a recognition that capital goods are a heterogeneous bundle of commodities produced by other industries. Using the mathematical language of neoclassical theory, Sraffa then showed that such an apparently simple change has a devastating effect on the neoclassical thesis about the distribution of income. The kernel of this critique is the issue of capital 'reswitching' and 'reversal' in the realm of wage–profit frontiers.

2.1.3 Denying the neoclassical parables: Reswitching

Capital reswitching 'is the possibility that the same technique of production may be the most profitable of all techniques at two or more *separated* values of the rate

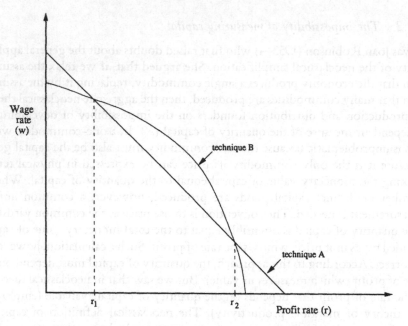

Figure 2.2 Reswitching with heterogeneous capital goods

of profit even though other techniques are the most profitable ones at rates of profit in between' (Harcourt, 1972, p. 232). In the neoclassical world of Figure 2.1, such reswitching is impossible, because the wage–profit frontier associated with a single technology is a straight line and straight lines cross only once. What Sraffa shows, however, is that in a multi-commodity model wage–profit frontiers are normally non-linear (indeed they are described by an N th order polynomial equation, where N is the number of commodities produced in the economy). With non-linear wage–profit frontiers, reswitching is very plausible (Dobb, 1970; Garegnani, 1966, 1970; Pasinetti, 1966). Consider for example Figure 2.2, which shows two such frontiers for two different production techniques. Once again the outer envelope, indicated by the thicker line, identifies the technique yielding the highest profits for each wage level. Technique A is most profitable if the rate of profit is less than r_1 or greater than r_2, whereas technique B is most profitable in between, corresponding to the above definition of reswitching. Such situations are perverse in the neoclassical view of things because neoclassical assumptions allow only for linear frontiers. What is perverse in a neoclassical world, however, is the norm in a more realistic multi-commodity world.[1]

Capital reversing is the possibility of 'a positive relationship between the value of capital and the rate of profits when the switch from one technique to another is considered' (Harcourt, 1972, p. 232). By plotting the values of the capital – labour ratios of the more profitable technique, it is possible to obtain the situation represented in Figure 2.3. When profits rise beyond r_2, the value of capital per worker of the more profitable technology *increases*. In the neoclassical scheme this is impossible because by definition techniques with a steeper wage–profit frontier have a higher capital–labour ratio. Since a move to the right along the wage–profit frontier in a neoclassical world implies choosing a technique with a flatter slope (Figure 2.1), the capital – labour ratio can decrease only as the rate of profit increases. In a multi-commodity world this is no longer axiomatic; higher profits can dictate employment of a more *capital*-intensive technique. This result may seem counterintuitive because neoclassical parables are so broadly taught that we have internalized them as being correct; but it is none the less true for all that.

Clearly there is a close relationship between reswitching and reversing. By definition when there is reswitching one of the switches is an example of capital reversing. Only with more than two techniques can reversing occur in the

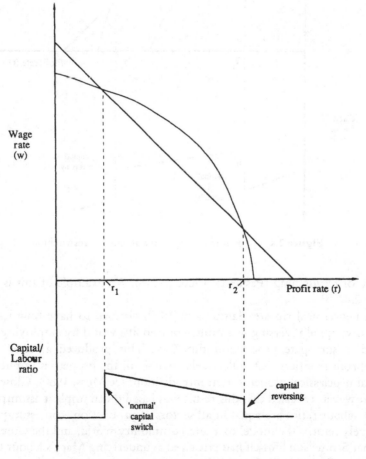

Figure 2.3 Captial reversing and reswitching
Based on Harcourt (1972)

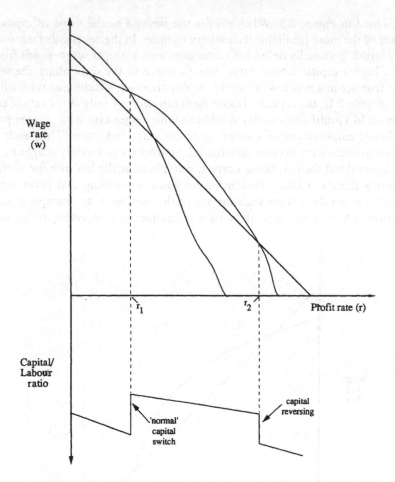

Figure 2.4   Capital reversing in the absence of reswitching

absence of reswitching (Pasinetti, 1966, p. 516). An example of this is given in Figure 2.4.

In a neoclassical riposte, Samuelson (1962) claimed to have found a way to avoid any capital reversing in a multi-commodity world by employing what he termed a 'surrogate production function'. This produced a series of linear wage–profit frontiers that collectively exhibit all the properties implied by the original neoclassical model (for an introduction, see Moss, 1980). Closer inspection, however, revealed that this result was based on an implicit assumption that capital–labour ratios are equal in all sectors of production – an assumption that effectively returns the model to a one-commodity world, and the same assumption that Samuelson himself had criticized as underlying Marx's labour theory of value (see Chapter 3). In addition, the various points on the wage–profit envelope created by the surrogate production function are equilibrium points. As such Samuelson had created what became termed the 'pseudo' production function. Each point on the envelope is a separate equilibrium, and Samuelson provided no mechanism by which the economy could move from one to the other.

It is now generally accepted that the Sraffian criticism of the neoclassical model is valid (Samuelson, 1966). The neoclassical reponse has been either to claim that a neoclassical model can be expressed in such a way that it has all the properties of the Sraffian alternative (Hahn, 1982), or simply to ignore the criticism – *assuming* that capital reversing does not exist by restricting analysis to what are revealingly referred to as 'regular' economies (Burmeister, 1980). The general conclusions are unavoidable, however. In a multi-commodity world profits and wages are not technically determined by the marginal productivity of labour and capital, and there is no simple relationship between profit rates and the degree of capital intensity of the most profitable technique. This makes it impossible to use economic logic to sustain the view that profit is the just reward to the owner of a scarce factor of production (capital), measuring the contribution of that factor to economic productivity; the rate of profit may be high even when capital is abundant.

2.1.4 Politics and the distribution of income

By contrast, the approach to be adopted in this book treats the issue of distribution separately from that of price determination. Wages and profits cannot both be determined from the production relations governing the economy, but require additional knowledge about social and political forces for their determination. Yet once wages, for example, are determined through social struggle, then prices of commodities and the equalized money rate of profit can be calculated without any of the inconsistencies and circularities of the neoclassical account (Bhaduri, 1969; Dobb, 1970). This maintains the position of the classical economists from David Ricardo to Karl Marx that we must specify the institutional framework of the economy in order to understand income distribution. As Joan Robinson continually insisted, it was the *meaning* of capital and not its *measurement* that was at stake in the capital controversy.

2.2 Implications for economic geography

While the Sraffian critique of neoclassical theory was based on introducing an element of realism into macroeconomics by recognizing the multiple commodities characteristic of real economies, the entire debate was carried out under another unrealistic assumption – that economies have no spatial dimension. It remains to determine, therefore, whether there is reswitching in a space economy, and if so what it implies for economic geography. Although some economic geographers have discussed the possibility of reswitching in a space economy, it has not been tackled in a rigorous or systematic way. Gertler (1984) speculated that reswitching is an element in regional capital investment patterns, but did not discuss it in detail. Scott (1979) argued that reswitching in a von Thünen landscape implies that the same land use will be profitable in two distinct land-use bands at different distances from the market. This case, however, is only one possibility.

In order to examine whether there is reswitching in a space economy, space must be incorporated into a multi-commodity economic model of capitalist commodity production. It is not sufficient simply to disaggregate the model by attaching subscripts for each region; interactions between locations must be specified in order to have a properly geographical analysis (Sheppard, 1978). One of the principal tasks of this book, to be explained in Chapter 4, is to illustrate how this may be done. Based on this, Chapter 5 will then show that reswitching is indeed a characteristic of a spatially extensive capitalist economy. This has major implications for the conventional wisdom that neoclassical thinking has introduced into economic geography. We will discuss these implications in detail in section 5.3. In order to emphasize the importance of these conclusions, however, we will summarize them here.

First, if a region is blessed with a relative abundance of labour, it will not necessarily be most profitable for that region to specialize in the production of commodities with a high labour content, nor will a capital-rich region necessarily gain from specializing in the production of commodities that require capital-intensive production methods. Thus free-trade decisions made on the basis of the availability of production factors need not be the most advantageous strategy. Indeed, as we will suggest in Chapter 12, there may be no rule at all for identifying *a priori* those commodities that a region should specialize in, in order to increase the productivity and profitability of an inter-regional system. If this is true, it calls into question the very principle of free trade.

Second, contrary to neoclassical wisdom (see Borts and Stein, 1964; Siebert, 1969), when one region has a relative abundance of capital and is growing rapidly, whereas another region has an abundance of labour and is growing slowly, it need not be the case that the profit rate is higher in the latter region, attracting capital away from the capital-rich region, with labour flowing in the opposite direction. Thus market mechanisms will not automatically ensure some equalization of geographical imbalances in the availability of production factors, as neoclassical growth theory has suggested.

Third, it is not necessarily the case that the highest-paying land use for a plot of land is the most productive way of using that land. This suggests that the use of a land market to allocate land to the highest bidder need not lead to an efficient land-use pattern. Fourth, Adam Smith's principle of the hidden hand, whereby it is argued that when individuals pursue their own economic interests the market mechanism will ensure a location pattern that maximizes overall efficiency and social welfare, is no longer necessarily true. These are certainly controversial claims, since much is written in economic geography that assumes the neoclassical conclusions about trade, labour and capital flows and land productivity. Indeed our conclusions call into question the whole idea of markets as ensuring an efficient and welfare-maximizing space economy. If true, they suggest that we have to radically rethink much of what is taken for granted as conventional wisdom in the field. Yet they follow from the logic of the framework to be developed in this book.

Summary

The explicit recognition that capital is not some homogeneous entity that falls like manna from heaven, but a heterogeneous group of commodities produced by other capitalists in order to make a profit, causes serious problems for neoclassical macroeconomic theory. This is seen most clearly in the fact that reswitching is possible whenever it is allowed that capital is composed of heterogeneous produced commodities. In conventional aggregate neoclassical thought, production methods can be ranked according to their capital intensity. When the profit rate falls, more capital–intensive techniques will become more profitable and more efficient. Indeed the rate of profit equals the marginal productivity of capital, whereas the wage rate equals the marginal productivity of labour. Thus wages and profits seem to be technological measures of the relative contribution of labour and capital. All of these parables, which are so often taken for granted, need no longer be true when reswitching occurs, i.e. when a single technique may be most profitable at both high *and* low rates of profit. Techniques can no longer be unambiguously ranked by their relative capital intensity; it is not predictable which technique is most profitable under which conditions; and, most crucially, wage rates and profit rates no longer measure the relative productivity of capital and labour. In fact, it is not even possible to determine both the wage and the profit rate without knowledge of the relative social power of capitalists and workers, as determined by the political and social history of a society.

Making this apparently innocent assumption of a homogeneous capital good more realistic has, then, critical ramifications for conventional economic thought. The neoclassical vision of a harmonious capitalist society where market mechanisms ensure maximization of efficiency and social welfare must be replaced by the classical political economic vision of capitalism as a society of classes with opposed economic interests, where wages and profits depend on the state of struggle between those classes over their share of the economic surplus. Economics can no longer be separated from society, and social welfare can no longer be separated from class interests and treated as soluble by social engineering. The implication for economic geography is that some widely accepted ideas about the desirability of free trade, and the efficacy of unrestricted labour and capital flows and of a freely operating land market are at best questionable and perhaps seriously misleading.

Note

1 We do not wish to imply that neoclassical economics has ignored multi-commodity models; indeed they have been widely studied. Yet neoclassical theorists generally study only the subset of multi-commodity cases in which the neoclassical parables hold, cases that have been revealingly dubbed by the neoclassicists as 'regular economies' (Burmeister, 1980, pp. 118–34). They provide no evidence in support of the empirical existence of regular economies.

3 The value controversies

Introduction

The idea of valuing commodities in terms of the labour used to manufacture them, rather than in terms of their monetary value, was initially popularized by Adam Smith. David Ricardo determined that, in a complex economy where different commodities require different amounts of capital good input, the relative value of two commodities measured in labour hours needed for their production is not equal to their relative monetary value as measured by the general cost of production. Karl Marx argued, however, that despite this difference there exists a determinate relationship between labour values and money prices. Furthermore, he was able to show that an analysis carried out in labour value terms reveals that the labour market under capitalism is fundamentally distinct from the markets for all other commodities. It was from this analysis that Marx's theory of the nature of exploitation under capitalism evolved.

Ever since Marx wrote about this, three fundamental issues have been at the focus of discussion about the relevance of labour values to economic analysis. These are: the question of the existence of a determinate and logically consistent relationship between labour values and prices; the question of how to measure labour values empirically; and the question of what insights a labour theory of value brings to economic analysis that are missed by a price-based theory.

With respect to the last issue, some argue that an analysis based on labour has greater historical and geographical breadth than one based on price, because there are more types of societies in which labour is necessary for the production and distribution of these goods than there are societies that require prices as signals to regulate the production and distribution of economic artefacts. In order to provide a broader context for our analysis of the particular case of a capitalist space economy we require some conception of what is generally required for a society to support itself economically. The remainder of the introduction will briefly outline such a conceptualization, before devoting the rest of the chapter to an analysis of the relationship between labour and price in capitalism.

Every society requires some system of production and exchange, where production is the process whereby people labour to make products deemed necessary to support the ways of life of that society, and exchange is the process whereby these products are made available to members of society. Such a system of production and exchange has been defined as a set of production relations. Associated with this is a system of social relations representing the interpersonal interaction and social hierarchy within which these production relations operate.

Taken together this entire complex may be referred to as a mode of production. Now it is clearly the case that one cannot weld together any arbitrarily chosen combination of a system of production and exchange and a system of social relations and expect to have a successful society. A successful society is by definition a society that has been able to reproduce itself for a significant period of time, and successful reproduction requires some coherence between the economic, social and political subsystems.

In saying this we do not wish to adopt a functionalist explanation, which would boil down to the argument that the requirement of successful reproduction forces a society to adopt some particular combination of production and social relations, juggling individuals into those slots that ensure reproduction. Nor do we argue that the actions taken by individuals will guarantee societal reproduction. Particular modes of production have experienced long periods of success, but this is not because of some pan-historical determinism in social development. Rather, the constant struggles by people to both improve their lot and respond to the circumstances in which they find themselves affect the structures of society in many potential ways. In those cases where these often unintended consequences feed back to ensure some regularity and permanence in social reproduction then a mode of production may solidify and persist (Lipietz, 1986). This success is predicated on the robustness of a mode of production, since it must be able to absorb and adapt to changing circumstances – whether these stem from external destabilizing influences or from the strategic intentions adopted by individuals within that society (Storper, 1985).

Armed with this general conception, we now go on to discuss commodities, as the particular form taken by objects produced under capitalism, and their valuation (section 3.1). A method for calculating the labour hours embodied in the production of a commodity is given in section 3.2. We argue, however, that such a calculation is consistent with Marx's labour theory of value only if we add the extra condition of 'social necessity'. If this is done, it then becomes possible to define the existence of exploitation in labour value terms – as well as a number of other basic concepts in Marxian value theory (sections 3.3, 3.4). We then review the ongoing debate about the relationship between labour values and exchange values (section 3.5), and show how a particularly well-known critique (an assertion that labour values can be negative) is in fact based on a misconception about the definition of Marxian labour values (section 3.6). Other views about the logical status of labour values are also reviewed (section 3.7). We conclude by drawing implications about the geographical units to be used in our analysis of the space economy (section 3.8).

3.1 Commodities and their value

The distinctive feature of production and circulation in a capitalist system is that the objects produced and exchanged are *commodities*. Commodities are defined as objects that are produced for the purpose of exchanging them. This is not the case in all social systems. Subsistence societies, for example, produce objects because individuals require them in order to pursue their daily lives. It may well be that, in doing this, more is produced than is required, and any surplus may be traded with others. In this case, however, trade is not the driving force behind production, but is more incidental to the purpose of production. In a system of commodity production, the selection of both the commodity produced and the method of production is based on whether the entrepreneur expects that the commodity will sell at a price high enough to realize an adequate profit on the money invested in production. It is left to the market mechanism governing exchange to try and match the search by commodity producers for a profit with the real and perceived needs of society.

Although successful exchange is the purpose of commodity production, this does not imply that production is reducible to exchange, as neoclassical theory claims. It is important to realize that there is a considerable difference between production and exchange in capitalism. Production is the arena of systematic planning. Using the principles formalized by management science and operations research, entrepreneurs work to design a maximally efficient system of factory production – choosing those locations, techniques and labour practices that maximize profits. Control over circulation is far more difficult, however. While consumer demand may be influenced in many ways, it cannot be controlled by capitalists in the same way as a method of production on the factory floor. Market prices fluctuate stochastically, and no capitalist can be certain that the price that recoups expected profits on production will actually be the price at which the product is eventually sold. In short, whereas the realm of production is one of optimal decision making and control, that of exchange is chaotic and uncertain.

This difference in the nature of the forces governing production and exchange, and the fact that successful production of commodities occurs only if the effort put into production is more than recouped in the marketplace, poses one of the most difficult theoretical problems for the economic analysis of competitive capitalism: how can the value created in production be reconciled with the market value obtained from the sale of the commodity? In order to address this, we start by distinguishing between the *use value*, *exchange value* and *labour value* of a commodity.

The use value is the usefulness of a product to an individual. Marx argued persuasively that use values are subjective. There may be little in common between the perceived usefulness of a particular commodity for different individuals, and between the relative value attached to two different commodities by two different individuals. As a consequence, Marx argued that use values must be differentiated from the other two forms of value because no consistent and meaningful scale of measurement for use value can be derived that applies to more than one person in more than one context. He did argue, however, that if an individual decides to exchange one commodity in order to purchase another, then the perceived use value of the product traded must be less than that of the product received at the time and place of exchange (with the other individual engaged in the exchange having an opposite ranking of use values; Marx, 1867, ch. 1).

The exchange value of a commodity is defined as the number of units of another commodity for which it is generally exchanged in the marketplace. By 'generally exchanged' we mean the average exchange value, abstracting from day-to-day fluctuations that reflect temporary imbalances between supply and demand. By definition, if two products are exchanged for one another then they have the same exchange value. By notionally comparing all pair-wise exchanges of one commodity with another it is possible to deduce a common measure of exchange. In practice we usually measure this by the money price of the commodity, because historically money has been the means by which it has become possible to generalize market exchange from a series of spatially and temporally isolated transactions to a global market (Polanyi et al, 1957; Braudel, 1982). Strictly speaking, we shall define the *production price* of a commodity as its exchange value measured in units of the money commodity.

Labour value is defined as a general measure of the value created in the process of production. For reasons detailed in section 3.3.1, we shall define this value as the socially necessary labour value of a commodity. By definition, if a product has no use value it also has no labour value – even if labour is used to produce it. In such a case, the labour involved is not socially necessary because it was devoted to making something that is useless. Similarly, labour that is devoted to making something that is not a commodity has by definition created an object with no exchange value, because the object created was not intended to be exchanged in a market. By the same token, the labour invested in this latter product is not invested in commodity production, and would therefore be excluded from any computation of the total labour value created through commodity production in society.

3.2 Embodied labour values

In his later years, Ricardo puzzled over the relationship between the labour value of a commodity and the production price that that commodity generally fetches in the marketplace. He was convinced that the value of a product is objectively measured by the total amount of labour directly used to make that product, plus the labour invested in non-labour inputs (i.e. the quantity of each input required to make a product multiplied by the labour required to produce that input). Neglecting location, this is expressed as the sum of the quantity of all non-labour inputs used in production multiplied by their labour value, plus the amount of labour directly used in production. Mathematically:

$$\lambda_i = \hat{a}_{1i}\lambda_1 + \hat{a}_{2i}\lambda_2 + \hat{a}_{3i}\lambda_3 + \ldots + \hat{a}_{Ni}\lambda_N + l_i, \qquad (3.1)$$

where λ_i is the labour value of commodity i, $\hat{a}_{1i}$ is the amount of commodity 1 used in the production of one unit of commodity i (effectively an input–output coefficient measured in physical units) and l_i is the hours of labour used in the production of one unit of commodity i. Since an equation such as (3.1) can be written for each commodity i, these may be combined together into a matrix equation. Suppose that there are N different commodities being produced, and define $\boldsymbol{\lambda}'$ as the 1 by N vector $[\lambda_1 \lambda_2 \lambda_3 \ldots \lambda_N]$ including the labour value of each commodity. Similarly, define $\mathbf{l}'$ as the 1 by N vector $[l_1 l_2 l_3 \ldots l_N]$ of direct labour inputs, and $\mathbf{A}^*$ as the N by N matrix of input–output coefficients $\hat{a}_{ij}$ (the amount of commodity i used to produce a unit of commodity j). Then (3.1) becomes:

$$\boldsymbol{\lambda}' = \boldsymbol{\lambda}' \mathbf{A}^* + \mathbf{l}'$$
$$= \mathbf{l}'[\mathbf{I} - \mathbf{A}^*]^{-1} \qquad (3.2)$$

where $\mathbf{I}$ is an identity matrix.

Let us clarify the intuitive meaning of equation (3.2). Assume that the capitalist economy is *productive*. By this we mean that the production techniques are efficient enough that the economy is able to produce a greater quantity of commodities at the end of the year than were required as inputs at the beginning of the year; an eminently reasonable assumption since without it there could be no economic surplus at all. In this case the matrix $[\mathbf{I} - \mathbf{A}^*]^{-1}$ equals $\mathbf{I} + \mathbf{A}^* + \mathbf{A}^{*2} + \mathbf{A}^{*3} + \mathbf{A}^{*4} + \ldots$, or more formally,

$$[\mathbf{I} - \mathbf{A}^*]^{-1} = \lim \sum_{m=0}^{\infty} \mathbf{A}^{*m}.$$

We also know that, since $\mathbf{A}^*$ is the matrix of inputs, $\mathbf{A}^{*2}$ may be interpreted as the matrix of inputs necessary for the production of these inputs, $\mathbf{A}^{*3}$ is the matrix of third-order inputs, and so on (see Chapter 4 for details). Thus, as in standard input–output analysis, $[\mathbf{I} - \mathbf{A}^*]^{-1}$ is interpreted as the sum of all direct and indirect inputs to production. Therefore, equation (3.2) states that the labour value of a commodity is equal to the sum of all the labour hours invested directly and indirectly in its production. For this reason, λ_i is also known as the *embodied labour content* of commodity i, since it equals the labour directly and indirectly used to produce it.

3.3   The labour theory of value

Karl Marx took Ricardo's notion of labour values as the sum of direct and indirect labour inputs and added two crucial insights that revolutionized classical economic theory. First, he pointed out some critical problems in Ricardo's embodied labour theory of value, and replaced it with a subtly but vitally different definition of value. Second, he applied this theory of value and exchange value to the labour market, whereas Ricardo had restricted its application to the markets for produced commodities.

3.3.1 Social necessity

With respect to the first point, note that Ricardo's accounting mechanism of equation (3.1) does not fully explain the quantity of value created in a production process because it does not explain the size of the $\hat{a}_{ij}$ coefficients. Marx felt that a labour theory of value should be able to account for this. Furthermore, the definition of labour values as the sum of direct and indirect inputs leads to nonsensical results. In particular, a strict application of this definition would imply that producers could increase the value of their output by using more labour rather than less to manufacture one unit of a commodity; i.e. the value of a commodity could be increased by producing it in a less efficient manner.

Marx resolved these problems by defining the value of a commodity as the labour *socially necessary* to produce that commodity. In using this term Marx had two kinds of social necessity in mind (Shaikh, 1981). First, for any commodity produced, there are usually several techniques being employed by different capitalists at any point in time for its production. If the value of the product in each factory were equal to the labour embodied in production there, then this would lead to the paradoxical result that less productive producers (i.e. those using more embodied labour) are producing a more valuable product. Yet, very inefficient producers are clearly using production methods that, for that stage of development of society, are unnecessarily wasteful; i.e. they are not the socially necessary production method. It is more reasonable to define the socially necessary production method as either the dominant production method of those currently in use (i.e. the one used most often), or the average production method, or the best production method (Fine and Harris, 1979, p.44; Shaikh, 1981, p.277; Morishima and Catephores, 1978, p.57; Webber, 1988). Throughout this book we will refer to these respectively as the *modal production method*, the *mean production method* and the *best production method*. There is no general agreement as to which of these is the most appropriate definition of social necessity, although the second one is the easiest to apply empirically.

Secondly, the qualifier 'socially necessary' should mean not only that the techniques of production are socially necessary, but also that the commodities themselves are socially necessary (Webber, 1988). Any society producing a large and heterogeneous set of commodities faces a difficult coordination problem of ensuring that the amount of each commodity that is available meets the demand for that commodity made both by all other producers (using it as an input) as well as by households (for consumption). If no way can be found for even approximately reconciling supply with demand, then the long-run viability of the particular mix of commodities and methods of production that exists at a point in time is threatened. The problem of determining the mix of commodities and production methods necessary in order to match supply with demand is the problem of determining the conditions under which individual actors in a capitalist economy behave in such a way that the products of each are demanded by others. This is simply the problem of ensuring that entrepreneurs can realize a profit on their investments so that the economy can continue to reproduce itself year after year. This aspect of coordinating structure and agency might be readily solved through central planning (in theory!). Under capitalism, where exchange is governed by the anarchy of the market, there is no clear reason as to how this should occur. Yet in practice there is none the less a crude if ever-changing correspondence between supply and demand.

Marx argued that the process of exchange plays a crucial role in coordinating the activities of individual capitalists by setting exchange values in such a way as to ensure that the appropriate quantities of each commodity are made available. If this occurs then we can define the corresponding allocation of society's labour to the production of the different commodities as a *socially necessary division of labour*, because it represents the division of labour that ensures that supply approximately matches demand. In this case, capitalists can be sure that there is a demand for their product, and thus that the labour that they employ is performing a socially useful function. This case, where the individual activities of many workers are confirmed as being socially useful because they produce commodities that are in demand, may be regarded as a confirmation via exchange that the labour employed in production was useful (De Vroey, 1981). This is a central proposition in Marx's economic theory:

> Since the producers do not come into social contact until they exchange the products of their labour, the specific social characteristics of their private labours appear only within this exchange. In other words, the labour of the private individual manifests itself as an element of the total labour of society only through the relations which the act of exchange establishes between the products and, through their mediation, between the producers. (Marx, 1867 [1972], p.165)

If such a socially necessary division of labour exists, then it means that, even though each individual worker is carrying out an activity whose concrete characteristics are different from, and difficult to compare with, the other activities of other workers, everyone is contributing *one hour of socially useful labour* to society when he or she performs *one hour of work*. Thus validation through exchange allows us to assert that different concrete labours are equated with one another as being equal units of abstract (i.e. socially useful) labour through the process of exchange. Using this proposition that each unit of labour value is equivalent to one hour of abstract labour, Rubin (1972) argued that the labour values of commodities must be such as to ensure that the socially necessary division of labour is achieved. He argued that money prices are the signals to which capitalists respond in matching supply with demand and thus bringing about a socially necessary division of labour. Because labour values also represent the equal valuing of all socially necessary labour, however, he concluded that when a socially necessary division of labour exists then labour values must be closely related to the exchange values that hold in this equilibrium. This is known as the *law of value* whereby 'social labour values [are] the intrinsic regulators of prices and hence of reproduction' (Shaikh, 1984, p.46).

We should note one final aspect of the definition of socially necessary labour. The socially necessary labour value of a commodity is the labour socially necessary to produce that commodity at the time when it is sold, not necessarily at the time when it is produced. Suppose, for example, that a commodity is made by a technology that was the socially necessary technique at the time of production, but has been superseded by a more efficient technology by the time it is sold. In this situation, the value of the commodity must be governed by the lower labour content of the more efficient production method, not by the labour employed when this unit of the commodity was originally produced.

Important logical implications for the labour theory of value follow from this, since it implies that labour is not the sole source of labour value (Cohen, 1981), i.e that we cannot define the labour value of a commodity as equal to the actual labour expended in the production of that commodity, and the inputs used by that commodity. This, of course, is the reason why the concept of socially necessary labour is important. It is through the adjectives 'socially necessary' that the labour theory of value of Marxist theory differs from the Ricardian definition that the labour value is simply the labour directly or indirectly used in the production of a commodity. Yet, while there are good theoretical reasons for replacing this labour-embodied definition of value with the definition based on social necessity, this does not imply that the mathematical definition in equations (3.1) and (3.2) has to be discarded. If these equations are used, however, their terms must be carefully defined. The $\hat{a}_{ij}$ and l_i coefficients must represent the socially necessary method of production, defined at the time when exchange occurs.

3.3.2   Empirical measurement

If we define the socially necessary production technique for a commodity as the mean production method, then it is straightforward to calculate Marxian labour values empirically from economic data, which include an input–output table. Input–output coefficients measure the average quantity of an input used per unit of output, which is equivalent to the mean production method. In this case Marxian values are easier to calculate empirically than the Ricardian calculation of embodied labour. This is because the latter would require knowledge of quantities of labour employed in the past at those points of time when the inputs to current commodity production were themselves produced (Gibson et al., 1986). The principal difficulty is that input–output coefficients are generally measured in monetary units rather than in physical units. Thus an empirical input–output coefficient measures the dollar value of that input required to produce one dollar's worth of output, and the labour input is measured as the wage bill per dollar of output.

It has been shown, however, that this causes very few problems. The wage bill is easily converted to labour hours by dividing the wage bill by the hourly wage, and input–output coefficients can be transformed into physical units by multiplying each coefficient by the price of the input and dividing by the price of the output. Then, as Marelli (1983) has shown, Marxian labour values can be measured rigorously from data that include an input–output matrix, information on the wage bill and estimates of market prices – all data that are widely available. It is sufficient to substitute the input–output matrix and wage bill data into equation (3.2), giving a calculation of labour values measured in hours per dollar of output, and then to divide these labour values by the market price of the commodity in order to obtain Marxian labour values measured in hours per unit of output.

Indeed, this approach is being used with increasing frequency to measure labour values empirically as well as all the other categories of Marxian labour value theory discussed below (see Marelli, 1983; Shaikh, 1984; Gibson *et al.*, 1986; Webber and Rigby, 1986; *Review of Radical Political Economics*, 1986). In fact, in a wide-ranging review of econometric models of market prices, Semmler (1984) presents evidence that Marxian theories provide a more empirically accurate prediction of prices than do econometric models based on neoclassical or oligopolistic theories of pricing.

3.3.3 Marxian exploitation

Marx's second vital insight in labour value theory was to apply the concepts of labour value and exchange value to the market for labour. He argued that workers in capitalist society have only one resource that they can bring to the market to exchange for other commodities, their capacity for work. He reasoned that the same logic of commodity exchange should apply in the sale and purchase of labour as in any other market. Applying this logic to labour, however, has a startling impact on the classical theory. To see this, let us define the nature of the commodity traded in the labour market, and its value, exchange value and use value.

The commodity that workers seek to sell in the labour market is their capacity to work, defined as 'labour power'. There is a demand for this labour power on the part of capitalists, who wish to combine it with the means of production that they own in order to manufacture other commodities. If labour power is a commodity, its value must be defined in the same way as for other commodities. The value of labour power is therefore the value of the labour that is socially necessary in order to produce that labour power. In practice, labour power will continue to be available only if workers remain capable of doing their job and if other workers are also available in the future once the current labour force becomes too aged or infirm to continue. If we define the bundle of commodities that a worker's family requires to maintain itself, according to current social and cultural standards, as the 'real wage' received by the workers in that family ('real' in the sense that it is not a wage measured in dollars but the actual objects purchased with those dollars), then the value of labour power must be the labour socially necesssary to produce the commodities that make up the real wage. The exchange value of labour power, however, is the quantity of the money commodity for which it is exchanged in the market or, in short, the money wage.

The use value of labour power has a different meaning for the seller and the buyer. For the seller, the use value of labour power is its ability to provide the wages necessary to support the worker's household and thus reproduce his or her capacity to work. For the purchaser, the use value of labour power is the work that it contributes to commodity production, which we can define as the 'labour contributed'.

Marx argued that, under capitalism, the value of a worker's labour power must be less than the amount of labour contributed by that worker to commodity production for the capitalist. He argued that the historical capacity of humans to produce more than they need even in primitive economic systems has been appropriated by entrepreneurs in the capitalist mode of production. Although a worker's labour is exchanged for a wage equal to its monetary value, the labour worked for the capitalist is greater than the labour received by the worker as measured by the value of his or her wage. In short, if λ_L is the labour value of one hour of labour power, then

$$\lambda_L < 1. \tag{3.3}$$

Here the right-hand side is by definition the labour value and use value of one hour of labour contributed to production from the viewpoint of the capitalist. This use value received must exceed the use value traded away (the value of the real wage, λ_L) if the capitalist is to engage in exchange.

Thus an application of the definitions of value to the labour market shows that, beneath the appearance of equality that characterizes the process of exchange in the labour market (where the exchange value sold always equals the exchange value purchased), is a fundamental inequality. The labour value of labour power sold as a commodity is less than the labour contributed by that labour power to commodity production. Because under capitalism both the means of production and the commodities produced are owned by capitalists, then it makes sense to talk of the capitalists as a class receiving more labour value than they trade to workers in the form of the real wage, with the opposite holding for workers as a class. This Marx defined as the exploitation of labour, with the rate of exploitation of labour, e, defined as the ratio of the surplus labour value obtained by the capitalist to the value of labour power:

$$e = [1 - \lambda_L]/\lambda_L. \tag{3.4}$$

It has been shown that capitalists as a group will receive a positive monetary profit on their investment in commodity production only if the rate of exploitation as defined here is positive – thus supporting Marx's thesis (Morishima and Seton, 1961; Okishio, 1963; Morishima, 1973). We shall examine this proposition, known as the Fundamental Marxian Theorem, in more detail in Chapter 8 where we investigate its validity in a space economy.

3.4 Further concepts

There are a number of basic definitions in Marxist value theory that need to be established. These include constant capital, variable capital, surplus value, the rate of surplus value, the value rate of profit, and the organic composition of capital. The purpose of this section is briefly to rehearse both the verbal and the mathematical definitions of these terms. Readers already familiar with them may proceed directly to the next section.

Suppose that we can define λ_n as the labour value of commodity n. The total labour value of all commodity production (V^*) is then the sum of the quantity of each commodity produced multiplied by its labour value. Marx divided this total into three parts: constant capital (C; the labour value of all capital goods used in production), variable capital (V; the labour value of the labour power used in production), and surplus value (S; the labour value of the surplus labour that is appropriated into commodity production due to the difference between the labour value of labour power and the labour value of labour contributed). By definition, $V^* = C + V + S$. Using our definitions, we may write this formally as (Morishima, 1973; Harris, 1977):

$$\sum_{n=1}^{N} x_n.\lambda_n = \sum_{n=1}^{N} x_n \left(\sum_{m=1}^{N} \hat{a}_{mn}.\lambda_m \right) + \sum_{n=1}^{N} x_n l_n \lambda_L + \sum_{n=1}^{N} x_n l_n (1 - \lambda_L), \tag{3.5}$$

where x_n is the quantity of commodity n produced in the economy.

On the right-hand side of equation (3.5) the first term equals the sum of all commodities used as capital good inputs multiplied by their labour values; the second term is the sum of all labour power used multiplied by the labour value of labour power; and the third term represents the difference between the value of labour contributed and the value of labour power. From the definition of the rate of exploitation (equation 3.4), the accounting relationship of (3.5) can also be written as:

$$\sum_{n=1}^{N} x_n.\lambda_n = C + V + S = \sum_{n=1}^{N} x_n \left(\sum_{m=1}^{N} \hat{a}_{mn}.\lambda_m \right) + \lambda_L \sum_{n=1}^{N} x_n l_n + e.\lambda_L \sum_{n=1}^{N} x_n l_n, \tag{3.6}$$

showing that surplus value equals the value of labour power multiplied by the rate of exploitation.

We briefly summarize some other key concepts in Marxist economic theory. The *rate of surplus value*, s, is the ratio of surplus value to the labour value of labour power:

$$s = S/V = \sum_{n=1}^{N} x_n l_n (1 - \lambda_L) \Big/ \sum_{n=1}^{N} x_n l_n \lambda_L. \tag{3.7}$$

When, as we have assumed thus far, the value of labour power is the same in all economic sectors and regions, then the rate of surplus value will equal the rate of exploitation. The rate of profit measured in labour values, or the *value rate of profit*, ρ, is the surplus value divided by the sum of the labour value of all constant and variable capital inputs:

$$\rho = S/(C + V) = e. \left[\sum_{n=1}^{N} x_n l_n \lambda_L\right] \Big/ \left[\sum_{n=1}^{N} x_n \left(\sum_{m=1}^{N} \hat{a}_{mn}.\lambda_m\right) + \sum_{n=1}^{N} x_n l_n \lambda_L\right]. \tag{3.8}$$

Finally, the ratio of the labour value of capital good inputs to the labour value of labour power used, or *the organic composition of capital*, q, is the ratio:

$$q = C/V = \left[\sum_{n=1}^{N} x_n \left(\sum_{m=1}^{N} \hat{a}_{mn}.\lambda_m\right)\right] \Big/ \left[\sum_{n=1}^{N} x_n l_n \lambda_L\right]. \tag{3.9}$$

It is also possible to show that there is a simple equation relating the last three ratios together:

$$\rho = s/(1 + q). \tag{3.10}$$

All of these expressions are directly analogous to well-known measures in monetary terms. The accounting identity is equivalent to saying that the total monetary value of a product equals the cost of inputs, plus the wage bill, plus value added. The rate of surplus value is equivalent to value added per dollar of wages; the value rate of profit is conceptually identical to the money rate of profit; and the organic composition is equivalent to the capital–labour ratio. For a fuller exploration of these equivalences, and parallel concepts for use values, see Table 3.1. This underlines the fact that socially necessary labour time is an alternative to valuing economic activities in monetary units. It remains to investigate the relationship between the two systems of value.

3.5   Labour values vis-à-vis exchange values

The controversy over the labour theory of value is first and foremost a controversy about the nature of the relationship between labour values and exchange values. Recall the different nature of the processes that each of these

Table 3.1 Economic magnitudes in exchange, labour and use value realms

Concept	Exchange value sphere	Labour value sphere	Use value sphere
Unit of value	production price (money, p)	socially necessary hours of labour (l)	quantities (e.g. tons: x)
Labour input	wage bill (W)	variable capital (V)	hours worked
Capital input	cost of capital (K)	constant capital (C)	quantities of input
Surplus	total profits (π)	surplus value (S)	excess produced beyond required inputs
Surplus as proportion of input	money rate of profit (r)	value rate of profit (ρ)	rate of capital accumulation (g)
Surplus as proportion of labour	π/W	rate of exploitation (e)	
Capital input as proportion of labour input	capital–labour ratio	organic composition (q)	technical composition of capital

value systems is supposed to capture (section 3.1): labour values as a measure of the value created in the systematic and efficiency-seeking process of production; and exchange values (production prices) as the measure of the worth of one commodity compared with another when they enter into the often chaotic and unpredictable market of exchange. Because the two value measures reflect very different processes, it is not surprising that it is difficult to reconcile the one with the other. In this sense, we can agree with Himmelweit and Mohun (1981, p.241) that the difficulty of reconciling labour values and exchange values 'is a contradiction in reality, and not at all a problem with Marx's theory'. Yet there must be some resolution, otherwise the value created in production would not be realized in exchange, and the individualistic institutional structure of production that characterizes capitalism would not be easily reproduced.

Essentially a spectrum of opinions exists about the relationship between values and exchange values, ranging from the fundamental Marxist position that exchange values must be calculated directly from labour values to the neo-Ricardian position that exchange values are independent of labour values, which are inconsistent and redundant to theoretical analysis. Before examining this debate, it is necessary to define exchange value.

3.5.1 Prices of production

In general it is agreed by the participants in this debate that there is some 'centre of gravity' around which prices fluctuate in a chaotic manner. This is generally defined as the *production price*, equal to the capital advanced to pay for production incremented by the rate of profit. It is generally assumed that the rate of profit is equal for all capitalists (although this profit rate is better regarded as a mean around which profit rates vary stochastically; Farjoun and Machover, 1983). This definition may be formally written as:

$$p_n = (1 + r)\left[\sum_{m=1}^{N} \hat{a}_{mn} \cdot p_m + w.l_n \right], \qquad (3.11)$$

where p_m is the price of production of commodity m, $\hat{a}_{mn}$ is the quantity of commodity m used as a capital good input in order to produce a unit of commodity n, w is the money wage, l_n is the hours of labour necessary to produce a unit of n, and r is the money rate of profit on capital advanced.[1]

The hourly money wage, w, is used to purchase those consumer goods deemed necessary to reproduce labour power, or the real wage. Following Marx in treating labour as a commodity subject to the same rule as other commodities, the price of labour power must equal its cost of production, the cost of the real wage:

$$w = \left[\sum_{m=1}^{N} b_m \cdot p_m \right]/H, \qquad (3.12)$$

where b_m is the quantity of commodity m that is consumed per week as a wage good in order to maintain a worker and his or her family, and H is the number of hours worked per week. Note two features of equation (3.12). First, the cost of household labour is excluded from this analysis because such labour generally is not a commodity offered for sale in the market and thus has no exchange value. A very important implication of this stems from the fact that under the social relations of capitalism the vast bulk of labour expended in reproducing the worker's family is female labour. Much of this female labour is not compensated because it is not wage labour, implying that female labour is exploited – albeit in a different sense from that referred to above (for a detailed discussion see Carling, 1986, 1987). Secondly, the worker cannot charge a rate of profit on the production costs of his or her labour because no capital is advanced. From the worker's viewpoint, wages (production costs) are paid after the work is done; i.e. after the product (labour power) has been produced.

Now equation (3.12) can be substituted into (3.11). Suppose this is done, and let us define the total amount of commodity m necessary to produce a unit of commodity n, a_{mn}, as the sum of the amount required as a capital good plus the amount consumed as a wage good by workers in order to provide the labour power for production:

$$a_{mn} = \hat{a}_{mn} + b_m l_n / H.$$

It then follows that:

$$p_n = (1 + r)\left[\sum_{m=1}^{N} a_{mn} \cdot p_m\right].$$ (3.13)

3.5.2   The transformation problem

The question of the relation between exchange values and labour values may now be rephrased as an investigation of the relation between prices of production (equation 3.13) and labour values (equation 3.1, which for convenience we reproduce here):

$$\lambda_n = \sum_{m=1}^{N} \hat{a}_{mn} \cdot \lambda_m + l_n.$$ (3.1)

The first solution to this problem was proposed by Karl Marx in the third volume of *Das Kapital* – although it was a problem that Ricardo also worked on. Marx phrased the problem as one of the transformation of values into prices; and it has become known as the 'transformation problem'. The details of Marx's example and the controversy around it have been widely discussed (see Morishima, 1973; Shaikh, 1977). In essence, Marx tried to tackle the following issue. If there is a constant rate of exploitation in all sectors of an economy, those sectors that employ a greater ratio of labour to capital (i.e. those sectors with a low organic composition) stand to make more surplus value because there is relatively more labour being exploited. Referring to equation (3.10), if s is constant (S is proportional to V), then ρ would be greater when q is smaller. If this were the case, Marx further reasoned that capitalists would tend to disinvest from sectors where the rate of profit was low and reinvest in sectors where the rate of profit was high. In equilibrium, where no such progressive disinvestment out of or investment into particular sectors is occurring, all capitalists should be making an equal (value) rate of profit. Only then is there no incentive to change current investment patterns. In effect Marx was looking for a solution to the following problem:

$$p_n = (1 + \rho)\left[\sum_{m=1}^{N} \hat{a}_{mn} \cdot \lambda_m + \lambda_L \cdot l_n\right]$$ (3.14)

where ρ is the value rate of profit (equation 3.8).

Equation (3.14), is, however, inconsistent. First, it entails the highly problematic assumption that capitalists value their inputs in terms of labour values, λ_m, but charge a price of production, p_n, for their output. Because one firm's inputs are the output of another firm, however, capitalists purchasing inputs will in fact purchase them at the *exchange* value or price of production charged by the manufacturer, not at their labour value. In short, the labour values, λ_m, on the right-hand side of equation (3.14) should be replaced by prices of production, p_m, and similarly the labour should be valued by its wage, w, not its labour value, λ_L. If we were to make such changes, the result would be equation (3.11). Furthermore, implicit in the scenario envisaged by Marx is that capitalists respond to differences in the value rate of profit, whereas in fact their direct concern is money profits, as equation (3.11) suggests.

In addition, Marx argued that the sum of the exchange value of all commodities equals the sum of their labour value, and the sum of profits equals the sum of surplus value. These assertions are consistent with the notion that all value is created by labour, and that the process of exchange simply redistributes total value in order to equalize rates of profit in all sectors. These two equivalences cannot hold simultaneously, however, except under very unrealistic conditions (for a full discussion, see Morishima and Catephores, 1978, pp.147–60).

While Marx realized that his solution was only an approximation that 'does not necessitate a closer examination' (Marx, 1867 [1972], p.165), later writers went to considerable lengths to identify the source of the error (see Von Bortkiewicz, 1906[1952]; Seton, 1957). It is now generally agreed that the correct model for determining prices of production is given by equation (3.11), or its equivalent, equation (3.13).

It also has been recognized, however, that there are certain circumstances under which Marx's approximation *is* correct. This is the case if the average structure of production for each commodity is such that exactly the same relative quantities of each input (including labour) are used; an unrealistic assumption, but one that corresponds to the condition under which neoclassical aggregate production theory is correct. Marx's approximation is also correct if the money rate of profit is zero. Marx's two assertions about the equality of labour value and exchange value are true when the quantities of each commodity produced are in exactly the amounts that would ensure that enough of each commodity is produced in one production period to meet the demands of the next production period (Morishima, 1973). When this is the case the socially necessary division of labour has been achieved, ensuring that current supply equals future demand and therefore that value created in production is realized in market exchange.

3.5.3 Controversies over the transformation problem

The mathematical fact that the equation for labour values (3.1) includes no information about prices of production, whereas the equation for prices of production (3.11) includes no information about labour values, has led to the argument that the two systems of value cannot generally be reconciled with one another. The most extreme version of this assertion is Samuelson's (1971, p.400): 'Contemplate two alternative and discordant systems. Write down one. Now transform by taking an eraser and rubbing it out. Then fill in the other one. Voilà.' However, a similar position has been adopted by a number of researchers who are otherwise very sympathetic to the questions posed by Marx:

> Rejection of any kind of 'labour theory of value' can ... be rooted firmly within *the surplus approach itself* ... [N]o quantities of embedded labour play any necessary role in the determination of either the rate of profit or prices of production: embodied labour quantities are entirely redundant ... (Steedman, 1981, pp.13–14; emphasis in original)

The argument of these more sympathetic critics can be summarized as follows (Steedman, 1977; Hodgson, 1982). Labour values and prices of production both depend on the technology used in production, as represented in the capital good inputs, the wage, and the labour inputs (see equations 3.1 and 3.11). Each value system, however, is calculated using this information in different ways, and does not depend on the other. Furthermore, capitalists and workers clearly do not observe labour values but instead react to prices when they make their day-to-day decisions. Finally, many of the conclusions of Marx's theory are arrived at without invoking a labour theory of value. Taking all of these reasons together, labour values are redundant to the analysis. There is a further criticism that, under certain conditions, labour values turn out to be negative. We shall discuss this claim in the next section because it requires us to introduce an additional concept – joint production.

The response to this criticism has taken several forms. First, there have been attempts to modify the transformation problem. It has been shown that it is possible to start with labour values and, through a series of transformations, to arrive eventually at prices of production (Shaikh, 1977; Morishima and Catephores, 1978). A reinterpretation of the transformation problem as referring to net product rather than to gross product also leads to a logically consistent version of the transformation problem (Lipietz, 1982). These results, however, generally have been regarded as curiosa, rather than accepted as a convincing argument. In the former case, it is not clear that the iterative processes described have any grounding in actual price formation, whereas the latter depends on accepting a particular, and novel, interpretation of Marx's intentions. What is clear is that Samuelson's interpretation is wrong: labour values and prices of production are not independent of one another – if only because they both are related to the technical conditions of production. But this does not tackle the core of Steedman's critique, which turns on the notion that labour values, however calculated, are redundant.

Second, it is claimed that the argument against labour values can be applied with equal force to prices of production. The crux of this rejoinder is the observation that prices of production are not the market prices observed every day in the market, to which economic actors respond, but are theoretical prices that would be observed if economic competition allowed all capitalists to make the same rate of profit on capital invested. But are observed market prices approximated by the prices of production? While many have suggested that prices of production do indeed represent an approximation to actual market prices, representing a mean value around which the latter fluctuate, this assertion has come under increasing theoretical and empirical attack (Farjoun and Machover, 1983; Webber 1987c). In fact, we have little idea whether prices of production really are a good approximation of market prices. If they are not, then it is as illogical to argue that capitalists respond to prices of production as it is to argue that they respond to labour values. In short, if prices of production are not readily observed, then it is as questionable to assume that capitalists respond to them as it is to assume that they react to labour values.

This rejoinder might be countered by the argument that prices of production effectively summarize those relationships in the economy to which capitalists effectively respond, even when they believe themselves to be responding to market prices. Exactly this argument, however, traditionally has been used to justify the utility of the labour theory of value: 'From a Marxist standpoint, this, in and of itself, is not a weakness or flaw in the theory – rather the contrary' (Sweezy, 1981, p. 24). Indeed, a major methodological contribution of Marx was that an analysis of the economy that restricts itself, as naive empiricists would like, to 'the surface of appearances' can give rise to misleading interpretations. Seen in this light, the methodological critique mounted by Steedman in many ways undermines his own conclusions. If it were true that prices of production are theoretically a determinant of capitalists' actions, even though they are not strictly observed by capitalists, then surely this defence applies with equal validity to labour values. The only justification remaining for the superiority of prices of production over values is that a monetary measure is theoretically better than one denominated in labour hours; but this is not a convincing argument.

From a purely logical point of view, then, labour values are no more redundant than prices of production. Indeed, from this point of view labour values have one advantage over prices of production; they can be calculated under less restrictive assumptions about the nature of the economy than those required to determine prices of production. The calculation of production prices requires assuming full capitalist competition and an equal rate of profit, neither of which is required to calculate labour values (Farjoun, 1984).

Third, there has been an attempt to rehabilitate Marx's assertions about the existence of a law of value, according to which social labour times are intrinsic regulators of prices and hence of production. This stems from the often-stated, in our opinion over-stated, argument that neo-Ricardians following the critique of labour values popularized by Steedman are technological determinists, who believe that the parameters within which the economy operate are determined by the techniques of production used (see Rowthorn, 1974; Bandyopadhyay, 1981;

Wright, 1981, for a sample of this debate). Shaikh (1981, 1984) has argued that, while the neo-Ricardians state that labour values and prices of production both depend on the triad of capital good inputs, the wage and the labour inputs used in production, they do not tell us how the particular mix of inputs comes into existence that prevails in a particular place and point in time.

At first sight, it seems to be simply up to the individual capitalists to decide what they wish to produce and how, based on current prices. Shaikh argues, however, that there must be more to it than this. He notes that, while any arbitrary set of coefficients can in principle be entered into the right-hand sides of equations (3.1) and (3.11), the success of any such production system will depend on its ability to: match supply with demand, realize profits from sales, and find the funds to invest in expanded production. In other words, while any set of coefficients is mathematically possible, only certain particular combinations will be observed in practice. These must be combinations where labour is allocated among the sectors in such a way as to ensure that appropriate quantities of the various goods are produced. Such a division of labour is the socially necessary division of labour in the second sense outlined in section 3.3. He concludes from this that 'the results of production on which the so-called physical data are based are themselves given only through the actual materialization of social labour time, and hence only because value has actually been created. Values are, so to speak, built into the very fabric of this physical data' (Shaikh, 1984, p.51). In short, it is not possible for any arbitrarily chosen production system to persist because it meets the additional constraint that supply approximates demand. Because this physical constraint is equivalent to there being a socially necessary division of labour, production must be implicitly guided by socially necessary labour time and prices are guided by values.

He assembles both theoretical and empirical evidence to support his argument. Theoretically, he cites as evidence Morishima's observation that total labour value equals total exchange value when the socially necessary division of labour is achieved (see above). He argues that, if all profits are not reinvested in expanded production to meet future demand, then the excess must be spent by capitalists on luxuries; and he further shows that in this case the degree to which total surplus value deviates from total profit depends on the degree to which capitalists consume luxuries for which labour values are not proportional to prices of production (Shaikh, 1984). The deviation of individual labour values from exchange values can also be calculated (Parys, 1982). A related conclusion, arrived at by Farjoun and Machover (1983), is that, in a large economy where profit rates are not uniformly equal but vary stochastically around some average, it can be reasonably concluded that the average labour value in a sector will be approximately proportional to the average price of production.

A number of empirical studies into the relation between values and prices of production have been also carried out. These typically involve taking the input–output coefficients estimated for national economies and using them to calculate values using equation (3.1) and prices of production using equation

Table 3.2 Empirical correlations of labour values and exchange values

Study	Correlation between l_n and p_n	Correlation between l_n and p_n[1]
USA, 1947–72 (Ochoa, 1984)		between 0.963 and 0.987
USA, 1947–77 (Gibson et al., 1986)		between 0.793 and 0.941
USA, 1947, 1963 (Shaikh, 1984)	0.958, 0.949	
Italy, 1959, 1967 (Marzi and Varri, 1977)	0.920, 0.866	

[1]p_n is the observed national market price of commodity n.

(3.11). These two sets of measures are then correlated with one another, and occasionally with market prices. Some results of this analysis, as reported in Semmler (1984), Shaikh (1984) and Gibson et al. (1986), are presented in Table 3.2. The results vary depending on differences in the definitions of what sectors of the economy should be classified as productive, but by and large the correlations are very high, supporting the thesis that the relationship between values and prices of production is meaningful rather than coincidental.

It would be premature, however, to accept the conclusion of Gibson et al. (1986) that such empirical tests constitute a confirmation of the law of value. It is not valid to infer from any empirical correspondence that the theory predicting such a correspondence is correct, unless it is also demonstrated that the processes described by that theory are actually in operation. After all, the same empirical regularity can be consistent with many theories.

What is the mechanism by means of which values intrinsically regulate prices? Shaikh argues that because values represent the socially necessary allocation of labour time, and because a socially necessary division of labour must be present to ensure that there is a market for produced commodities and that the economy can therefore reproduce itself, then values must regulate prices. On the other hand, it can be shown that for an economy with a given set of methods of production there exist many different social divisions of labour that are socially necessary in the sense that they guarantee that supply matches demand and profits are realized. Indeed, there is a different socially necessary division of labour for each different profit rate on the wage–profit frontier (Sheppard and Barnes, 1986; see Chapter 9 for further details). In fact, Roemer (1981, p.201) proposes that the law of value be reformulated as follows: 'Corresponding to any prior specification of the distribution of aggregate social labor time between production for workers' consumption and capitalists' profits ... there is a set of prices and individual demands (by workers) that will realize that distribution of labor time.' Some of these sets of prices of production may approximate labour values, but others may not.

Consequently, the high correlations observed between labour values and prices that seem to support Shaikh's position are not sufficient. We need to understand how and whether an economy narrows its set of possible social distributions of labour from the variety that is analytically possible to a unique distribution determined by labour values in the way that Shaikh argues. In short, the process by means of which labour values influence the structure of production is not detailed, other than arguing that it follows from the definition of socially necessary labour values, and is functionally necessary for the reproduction of the economy. The definition of labour values that is a key to this argument is also ambiguous. Social necessity of the second kind (section 3.3 above) is invoked in order to argue that values regulate exchange values. The calculation of labour values in the empirical analyses quoted above is based, however, on the first definition of social necessity (equation 3.1; labour values as defined by the socially necessary production method). Some equivalence between the two definitions of social necessity is required in order to use the empirical results to confirm the law of value as described by Shaikh. This could be forthcoming if it could be shown that commodities with higher values are also commodities with higher quantities of social labour allocated to their production under the socially necessary division of labour. As yet, however, this has not been shown.

3.6 Does joint production imply negative labour values?

The other major controversy about the possibility of quantitatively calculating labour values stems from an analysis by Steedman showing a production system that seems to result in negative labour values (Steedman, 1975). Since negative labour time is a nonsensical concept, he argued that labour values themselves must therefore be logically inconsistent. In order to evaluate this debate it is necessary to introduce the concept of joint production. *Joint production* occurs when one production process leads to the production of more than one good, none of which can be produced in isolation from the others. For example, the processing of sheep necessarily leads to the production of mutton and sheepskins; oil refining leads to the production of several different kinds of oils and gases simultaneously. Since more than one product can be manufactured during one production process, it is no longer possible to discuss the production of each commodity separately.

We shall repeat here the example that Steedman examined, both to illustrate the concept of joint production and to assess his claim about negative labour values. Suppose we examine just two production processes. In the first one, 1 hour of labour power and 5 tons of commodity 1 are used to produce 6 tons of commodity 1 and 1 ton of commodity 2. In the second process, 1 hour of labour and 10 tons of commodity 2 are used to produce 3 tons of commodity 1 and 12 tons of commodity 2. Schematically (using an asterisk for multiplication):

$$5^\star(\text{good 1}) + 1 \text{ labour} \rightarrow 6^\star(\text{good 1}) + 1^\star(\text{good 2})$$
$$10^\star(\text{good 2}) + 1 \text{ labour} \rightarrow 3^\star(\text{good 1}) + 12^\star(\text{good 2}).$$

Note that, unlike earlier examples in this chapter, it is no longer the case that one production process produces one product. Thus, instead of one production process producing 1 unit of commodity 1, with the other process producing 1 unit of commodity 2, a matrix of products, known as the output matrix, **B**, is defined. In the case examined here, matrix **B** is

$$\mathbf{B} = \begin{bmatrix} 6 & 1 \\ 3 & 12 \end{bmatrix}$$

The entry in the i'th row and j'th column of **B** is the amount of commodity j produced in the i'th production process. Thus, in the above example, $b^{22} = 12$, whereas $b^{12} = 1$. In Steedman's example, only one non-labour input is used in each process. However, this is just an artefact of his example, and in principle all commodities could be inputs into each process. We may therefore characterize joint production as a set of production methods that use a matrix of inputs, $\mathbf{A}^\star$, to produce a matrix of outputs, **B**.

What are the labour values of the two commodities in Steedmans' example? Since labour values are defined as the sum of direct and indirect labour, we may write the following equations to determine the labour values of the two commodities:

$$6\lambda_1 + 1.\lambda_2 = 5\lambda_1 + 0.\lambda_2 + 1$$
$$3\lambda_1 + 12\lambda_2 = 0\lambda_1 + 10\lambda_2 + 1,$$

or, in matrix algebra:

$$\lambda' \mathbf{B} = \lambda' \mathbf{A}^\star + 1',$$

implying that labour values are given by

$$\lambda' = 1'.[\mathbf{B} - \mathbf{A}^\star]^{-1}. \tag{3.15}$$

Now compare equation (3.15) with equation (3.2). In equation (3.2) the identity matrix is substituted for the general output matrix, **B**. Thus, from a mathematical viewpoint, the identity matrix in this case simply means that there is no joint production; that one process produces one product. Replacing it by the general output matrix **B** allows for the possibility of joint production – that each process may produce one or more products.

Steedman shows that the solution to the above example using equation (3.15) is that $\lambda_1 = -1$, whereas $\lambda_2 = 2$. He argues that, since it is meaningless to say that any commodity has a negative labour value, there must be a clear logical inconsistency in the conceptualization of the labour theory of value. Let us, however, re-examine the two processes of Steedman's example by analysing their technical productivity. The technical productivity of a process of production is defined in terms of the net product achieved, i.e. the amount of commodities produced minus the amount of commodities used up in production. Process 1 uses up 5 units of commodity 1 and none of commodity 2 to produce 6 units of commodity 1 and 1 unit of commodity 2. The technical productivity of process 1, therefore, is 1 unit of each commodity $(6-5=1; 1-0=1)$. Applying the same calculation to the second process, its technical productivity equals 3 units of commodity 1 $(3-0=3)$ and 2 units of commodity 2 $(12-10=2)$. We can conclude from this that the second process is absolutely more productive than the first process. As a result, from the point of view of social welfare, it would not be rational to invest any labour at all into the first process, because if that labour were invested instead into the second process it could be used to produce more of both commodities (Morishima and Catephores, 1978, p.33; Kurz, 1979, p.67).

From a societal point of view any labour invested in process 1 is wasteful (i.e. socially unnecessary). Therefore it is perfectly consistent with Marx's labour theory of value to conclude that it would have a negative value. Thus the example itself demonstrates the consistency, not the inconsistency, of this theory. This conclusion can also be generalized. Farjoun (1984) has shown that, whenever there is joint production, then negative labour values will occur if and only if the set of production methods, as described by the matrices $\mathbf{A}^*$ and **B**, includes some processes that are socially unnecessary in the sense that labour invested in them is being wastefully used (see also Webber, 1990). In fact, he points out that there is nothing novel about Steedman's conclusion because, as Marx realized, the same happens if there is no joint production. When two different processes are used to produce the same single commodity and one is more productive than the other, then labour values calculated using the less productive process are negative.

We know that in any real economy several different processes are used at any one time to produce each commodity. Yet this does not mean that labour values are often negative, as long as we consistently adopt a definition of labour values that is based on socially necessary labour time rather than on labour embodied. We have already come across this problem once in section 3.3. There we noted that, if we define the socially necessary production method as being either the best, the modal or the mean production method, then the range of production methods extant at any point in time is reduced to a single socially necessary method. Negative labour values will then not be possible because only one (socially necessary) process is defined for each commodity, or each combination of jointly produced commodities. A second situation where this comes up is of central importance once we inject geography into our analysis. If society's demand for a commodity cannot be met by the output of one plant, then other plants will be constructed at other less productive locations. They may be further from the market or the raw materials, or (in the case of agriculture) they may be using less productive land. These inferior production methods are themselves socially necesssary because sufficient amounts of the commodity in question cannot be produced at the more productive sites. In this situation Marx argued that the difference in productivity is captured as differential rent (Marx, 1896). The least productive site is the one that determines labour values for all sites, because it is socially necessary. The surplus profit made at the better sites is then regarded as a rent, paid out of profits, to use the better land. This provides us with a fourth possible definition of the socially necessary production method – the *marginal production method*. We will investigate this in detail in Chapter 6, where we will see that such a definition of social necessity applies to the use of all non-reproducible resources.

We can conclude, therefore, that the case of joint production poses no logical problems for the internal logical consistency of the labour theory of value as long as we are careful to define values in terms of socially necessary labour. Rather, as we saw in Steedman's example, negative labour values are a useful sign that labour values have been mis-specified, being defined without taking account of social necessity. Steedman's example reminds us that we must think carefully about how to treat different production methods that may be simultaneously in use to produce the same commodity, but these problems are no different from those faced in an analysis where there is no joint production.

3.7   Qualitative arguments for a labour theory of value

The debates summarized in the previous sections indicate how attempts to show that labour values are redundant (section 3.5) or inconsistent (section 3.6) are not conclusive. Furthermore, it turns out that the definition of labour values as socially necessary labour is internally consistent, and that the charge of redundancy is also applicable to prices of production themselves (see also Freeman, 1984). None the less, the question remains as to the relative merits of a labour value based analysis of capitalism compared with an analysis using prices of production as advocated by neo-Ricardians. Several arguments have been put forward to support a value-based approach.

The first is what we might call the 'no difference' argument. In this view, if values and prices of production are closely correlated, then it does not matter which is used as the basis for analysis. This leads to one of two conclusions: that conventional Marxist economic theory posed in value categories can be applied *ipso facto* to describe the inequalities, conflicts and crises that we observe in the exchange value sphere of capitalist economies; or that all analysis can be done in price terms, and that all of Marx's essential theoretical results can be derived in a model that concentrates on prices of production (Steedman, 1977; Hodgson, 1982). Of course the word 'essential' is vital here. Adherents of a price-based approach define those aspects of Marx's theory that are inconsistent with their approach as inessential and dispensable, whereas adherents of a labour value analysis continue to provide counter-examples where Marx's theory leads to conclusions different from those suggested by a price-based theory. A good example of this is the debate over fixed capital and the possibility of a falling rate of profit (see Roemer, 1979; Shaikh, 1980).

Second, it is argued that the whole interpretation of economic relations under capitalism is inappropriate if it is not based on the insight into exploitation and the source of profits provided by the labour theory of value. The interpretation of exploitation that the labour theory of value provides is clearly essential to that theory, because it is the theory of exploitation that allowed Marx to argue that surplus in capitalism stems from the existence of a labour market. Cohen (1981) has cogently argued that an accurate interpretation of socially necessary labour requires that we reject the classical ethical argument of Marxism – that activities of labour are the sole source of labour value and thus that workers are exploited if any part of their labour time is appropriated by capitalists. Yet, he argues that it is still the case that the class structure of capitalism engenders exploitation, since workers are the only ones who create commodities that have labour value whereas capitalists receive part of the value of these products (Cohen, 1981, p. 219). Roemer (1982a) argues that labour values (appropriately modified) can be a measure of exploitation under capitalism. In his view, each economic agent starts with an equal allocation of social labour power and notionally attempts to optimize his or her situation in a capitalist economy. Those who optimize by selling some labour power constitute themselves into one social class, whereas those who optimize by hiring labour power constitute themselves into another social class. 'The Class Exploitation Correspondence Principle ... states that all producers who optimize by selling labor power are exploited at the economy's equilibrium, and all agents who optimize by hiring labor power are exploiters' (Roemer, 1982a, p.111). Both Cohen and Roemer are critical of the internal logical structure of many of the analyses of exploitation, yet both agree that some notion of exploitation exists, based on the worker not receiving all of the value of the objects that he or she has created.

Third, Elson (1979) argues that the adoption of a labour value viewpoint provides an essential insight into capitalism, regardless of the issues of how to calculate labour values. In her view, the labour theory of value is not a theory of pricing but a theory of how and why labour takes on the form that it does under capitalism. This 'value theory of labour' gives insight into the process whereby the many different concrete activities are converted into abstract labour and provides an understanding of how working conditions evolve under capitalism. This is also Harvey's position: 'Value theory deals with the concatenation of forces and constraints that discipline labour as if they are an externally imposed necessity' (Harvey, 1982, p.37).

Finally, there has been some controversy revolving around the mathematical demonstration that in principle any commodity can be taken as the starting point for calculating value quantities, leading to such ideas as an energy theory of value (since all commodities contain some form of energy) or a corn theory of value. Why then choose labour as the key to value? Several reasons have been given: that labour in some form must be invested in the production of every commodity, whereas other vital inputs such as coal or steel can always be replaced by substitutes (Marx, 1867); that the object of study is the historical evolution of people struggling against people and the effect of this on economic relations; that everyone starts with an equal endowment of labour power, making it the most suitable yardstick by which to measure how much they end up with (Roemer, 1982b; but see Roemer, 1986, for a re-evaluation); that when the economy is conceived as a series of sectors reproducing the demand for consumption goods then their production prices equal values (Pasinetti, 1981, 1988); and that historical or geographical comparisons of the value created in an economy cannot be made in price terms. In this last case Farjoun and Machover (1983) argue that there is no common basket of commodities consumed in all times and places that can be used to translate observed prices into real prices, whereas labour values are better for this purpose because labour is a universally available commodity.

In the remainder of this book we will continue to include an analysis of both value and exchange value. On the one hand, it is clearly the case that the actions taken by individual members of the space economy are taken in response to the prices of commodities that they purchase for consumption or for commodity production; prices that, we must emphasize, may or may not coincide with prices of production. These actions may reproduce or alter ongoing production methods, commodity shipments or the location of economic activities. One way of tracing the impact of these changes, and the potential contradictions and conflicts to which they lead, is by examining their effect on socially necessary labour time. This makes it possible to evaluate the generalizations about change in the economy as a whole that Marx proposed as part of his value analysis, and to determine whether these indeed represent a convenient shorthand for explaining the aggregate results of individual action under the social and economic relations of capitalism.

3.8 Geographical implications: Defining regions

We have defined the labour value of a commodity as the labour socially necessary for its production. Once we confront the fact that economies are spatially extensive and not constructed on the head of a pin, then this must include the labour socially necessary to assemble the commodities used up in production at the place of production. There is no question that transportation is a productive activity, since it manufactures the accessibility necessary for any kind of commodity production. It then follows that the value of a commodity depends on the location where it is produced. As we shall show in Chapter 8, this spatial dimension provides some new insights into the complexity of class relations in a capitalist society.

This immediately poses a methodological problem for a labour value based economic analysis. It is often argued that value analysis is carried out at a high level of abstraction – abstracting from the details of observable reality in order to get behind superficial appearances. Thus Marxist scholars set up an abstract model of capitalism designed to capture its essential features. Once an understanding of this abstraction is obtained, it should then be capable of explaining the observable aspect of real capitalist economies. How does the method of abstraction treat space? At one level, space is abstracted from entirely; the analysis is of the capitalist mode of production irrespective of where it may be found. In this context the value of labour power, for example, is seen as its value abstracting from any spatial variations in this that might occur. In practice, however, as soon as value analysis is applied to any actual economy it is given a spatial specificity. Thus an analysis of the United States necessarily implies looking at the value of labour power within that territory, abstracting from any sub-national spatial variations. Social necessity also has an implicit spatial dimension in any application of abstract analysis. In this case, it defines what is socially necessary within the territory of the United States.

It is necessary to define the socially necessary production method to calculate values. For the purposes of argument, define the socially necessary production method as the mean production method (section 3.3). From a geographical viewpoint, defining the socially necessary production method implies taking some kind of spatial average. Thus in the above case we would define the mean production method as the average of production methods in use within the USA weighted by their importance. It is unclear, however, over what territory this averaging should be taken. If we examine a sub-national region, is the socially necessary production method the mean for that region, or for the nation as a whole? If we examine a nation, is the socially necessary production method the mean for that nation, or for the global capitalist system as a whole?

For the purpose of this book, where we wish to concentrate on the regional (sub-national) location of economic activity, we shall take the urban labour market as our unit of analysis (Harvey, 1985b, p.127). This refers to a major city and to the functional region surrounding that city that operates as a market for the supply of and demand for labour. Empirically, this can be thought of as the region that lies within commuting range of the city. We find this definition useful for several reasons. First, manufacturing activities traditionally concentrate in cities, making urban labour markets a logical focus for our analysis of the geography of production. Second, the urban labour market contains the immediate conditions of labour to which production strategies in that place respond and which they seek to shape. Compared with capital and commodity flows, labour remains the least flexible component in the production process.

Third, we shall argue that there is a significant difference in the forces guiding the location of production within, as opposed to between, urban labour markets. Within urban labour markets the relative profitability of different locations depends on rents; between them it depends on some notion of locational advantage. For the remainder of this book we will therefore use 'region' as a synonym for an urban labour market.

Note

1   In some neo-Ricardian analyses the price of production is defined such that no rate of profit is charged on wages, because it is argued that labour is paid after the work is done rather than before (Sraffa, 1960, p.10). Thus

$$p_n = (1 + r)[\Sigma \hat{a}_{mn} . p_m] + w . l_n.$$

However, even though wages are paid *ex post*, typically they still have to be paid before any revenue is made from selling the commodity. From the viewpoint of capitalists, then, money must be advanced for the wage bill before any revenue is made, and the rate of profit expected must be calculated relative to the sum of all money advanced – both for purchasing inputs for and hiring labour. For this reason we shall use some version of equation (3.11) as our definition of wages in this book. All indications are that the choice between the two is not crucial to the value controversies (Steedman, 1977, p.21).

PART II

The profit-maximizing space economy

'The equilibrium position is a target toward which all of the economic variables tend. But it is never actually reached. Instead, prices and outputs "gravitate" ... around this target.' (Duménil and Lévy, 1987, p. 136)

'the equilibrium notion ... is not dependent on a golden age concept [in the sense that all economic agents have perfect knowledge of all that is relevant to them] – though this might hardly be inferred from the formal equations.... Rather, it leaves sufficient room for fierce struggles ... for victory and defeats, risk and uncertainty.' (Franke, 1988, pp. 258, 260)

4 Production prices in a competitive space economy

Introduction

We begin our analysis of the geography of production and exchange under capitalism with an examination of the simplest possible case. Imagine that the space economy is composed of a series of J regions (urban labour markets). Suppose that we know for each region the commodities produced there, the mean (socially necessary) production method used for producing each commodity, and the mean consumption bundle consumed by workers there. Suppose in addition that no natural resources are used, and that land plots place no limits on the production capacity of plants.

An understanding of capitalist production entails unravelling the events that occur during a production period, which we define as the time elapsed between when a capitalist advances capital to pay for inputs to production and when capital is recouped from sale of the commodity. For convenience we label the production period as a year. There are a number of ways of looking at the changes that go on during a period of production, each of which gives a different insight. Consider first what happens if we observe changes at the site where production occurs. In the time that passes between advancing capital and recouping it through sales, the capital advanced goes through a series of transformations. Money is advanced and exchanged for the means of production, i.e. the labour power and capital goods necessary for commodity production. These inputs are transformed into the commodity via the process of production. Finally the commodity is exchanged for money. If the production is successful, this exchange will return a larger quantity of money to the capitalist than s/he started with. This is the circulation scheme for capital developed by Marx (1885 [1972]) in the second volume of *Capital*:

$$M - C(MP) \ldots P \ldots C + \Delta C - M + \Delta M.$$

Here, M is the money advanced; $C(MP)$ are the commodities representing the means of production purchased with that money; $\ldots P \ldots$ is the production process through which the means of production are transformed into a commodity $(C + \Delta C)$ of greater labour value than the inputs; and $M + \Delta M$ is the (increased) money obtained from sale of the commodity.

An observer located at the site of production, following this process, in fact will observe only the end points of larger circulation schemes in time and space that link different economic activities and locations together. To take just one example, the exchange of the produced commodity for money observed at the production site is an end point for two closely articulated circulation schemes. The shipment of the commodity out of the factory to the market where it is to be offered for sale is one moment in the geographical circulation of commodities. The inflow of money received from this sale is a similar moment in the circulation of money capital. Other perspectives can be obtained by following the capital as it flows through space and time, or by following the circulation of the commodities themselves.

We agree with Harvey (1982) that understanding this circulation process is essential to understanding the space economy. If the circulation process is successful, then at the end of the production period a profit on investment will have been realized. This surplus is then invested in expanded production levels in order to achieve capital accumulation and economic growth. Yet success is not guaranteed because of the disaggregated and individualistic nature of capitalism. The transformation from money, to means of production, to the commodities produced and back again to money can be interrupted at any point. Furthermore, successful completion of one stage in the circulation process at one location does not guarantee the same thing elsewhere. The geographical circulation of capital, for example, can drain investment out of one region where profits are not being realized into another, more successful, region. The evolution of the capitalist space economy will depend on how this process of circulation fares.

Of the two ways of observing circulation outlined above, we adopt the latter one in this part of the book. In order to examine the overall structure of the capitalist space economy we examine the circulation through space and time of commodities in the various forms in which they are manifest. We examine three such forms: the exchange value of commodities as measured by prices of production (this and the next three chapters); their labour value (Chapter 8); and the quantities of commodities in circulation (Chapter 9). By examining each of these circuits separately and then looking at how they articulate with one another, we seek to provide insight into how they depend on one another, and how the contradictions of capitalist production systems affect the stability and robustness of capitalist production and circulation.

Before proceeding, it is perhaps necessary to remind the reader of our theoretical strategy. In this part of the book we attempt to describe the equilibrium state that could be obtained in a capitalist space economy if full competition among capitalists holds, and if the socially necessary distribution of labour exists. As we will go on to argue, this can be described as a dynamic equilibrium – a set of prices and a social distribution of labour that if undisturbed can perpetuate itself indefinitely in an orgy of capital accumulation and unending economic growth. This has been dubbed the 'golden turnpike' of economic growth. We term this equilibrium an ahistorical equilibrium because it incorporates time but cannot account for historical change (Sheppard and Barnes, 1986). We will frequently also discuss the equilibrium that maximizes the average rate of profits of capitalists. In this sense it is the space economy that capitalists would wish to exist – or the space economy that could fulfil the 'capital logic' function of best meeting the needs of capital.

We emphasize that the choice of this starting point is not because we believe that this equilibrium will be approximated in practice. Indeed, we will argue the converse (see Harvey, 1982, pp. 82–3). We do believe, however, that the evolution of any dynamic system represents a path that the system is taking away from some theoretical equilibrium states and towards other such states, where we would include among such possible 'end points' the collapse of production altogether, a welfare state, social revolution, and even the chaotic states that mathematicians have referred to as limit cycles. This is no more than the assertion that at any instant of time every changing system is on its way to some end point, whether or not this goal is known to, or intended by, the participants in that system. Typically there are many such end points except in the fantasy world of linear systems. As a result, external fluctuations may disrupt the system, changing the end point towards which it is moving in mid-stream. Indeed the evolution of the system itself may change the nature of these end points endogenously. Nevertheless, we believe that an orderly theoretical procedure is first to attempt to identify these end points, in order to have some points of reference with respect to which the evolution of the system can be understood. In short, we view 'the concept of equilibrium as a convenient means to identify the disequilibrium conditions to which capitalist society is prone' (Harvey, 1982, p. 140).

4.1 The price circuit

Suppose that within the space economy a condition of full capitalist competition holds. By this, we mean that on average the rate of profit made on capital advanced during one year is equal for all entrepreneurs, irrespective of the sector or location of their production facility. Denoting $E\{r_i^n\}$ as the mean rate of profit averaged across all firms producing commodity n in region i, we assume that $E\{r_i^n\}$ equals r for all regions i and commodities n.

The assumption that the average rate of profit is equal in all sectors and locations means that there are no areas of the economy that are systematically more or less profitable. Some capitalists will still face below-average profit rates in one plant, or sector or region, and would attempt to invest their money in other places with above-average profit rates. This will always be happening since it is only the average profit rate that is equal. Yet equal average profit rates mean that there will be no major and systematic net flows of investment from one part of the economy to the other. If there were such systematic differences in average profit rates, capital would flow from less to more profitable parts of the economy. This would then affect prices of production, because disinvestment from certain parts of the economy would reduce the supply from those firms relative to demand, driving prices up; whereas the opposite would happen in those locations into which investment flows. Prices of production would therefore stabilize only at certain fixed values when the rate of profit is on the average equal in all firms (Marx, 1896 [1972]; Morishima, 1973; Pasinetti, 1977). While many possible scenarios can be constructed whereby rates of profit would be persistently different between regions and/or sectors, due for example to the nature of investment behaviour or to differing degrees of corporate monopoly (see Kalecki, 1938; Sheppard 1983a; Webber, 1987c), we shall defer consideration of this to later chapters. Here we shall restrict ourselves to examining the geographical differentiation in production that occurs in the absence of these effects.[1]

Second, we will take as axiomatic that the economy as a whole is productive. By this we mean the reasonable assertion that the economy is capable of producing a surplus – a greater quantity of commodities at the end of the production period than existed prior to the production period. As we shall see in Chapter 8, this means that labour is exploited in value terms, and that the general rate of money profit, r, is positive. Third, we shall assume that the labour force is homogeneous and for the time being we will ignore the possibility of joint production. Other authors have argued that some further assumptions are necessary to the following analysis: namely that the production period is the same for all firms, and that all firms start their production periods simultaneously (Abraham–Frois and Berrebi, 1979). Webber (1987b) shows, however, that such extra assumptions are unnecessary. Finally, we also assume that the economy is fully interdependent. By this we mean that there exists no group of locations that are completely isolated from the rest of the economy, in the sense that they neither purchase from, nor sell to, any firms elsewhere in the economy.

Suppose, as stated above, that we know the mean production method employed in each region for each commodity produced, and we know the real wages of workers in each region. In region j, define the mean production method for the production of commodity n by a vector of capital good and labour inputs:

$$\mathbf{\hat{a}_j^n} = [\hat{a}_j^{1n}, \hat{a}_j^{2n}, \ldots, \hat{a}_j^{mn}, \ldots, \hat{a}_j^{Nn}, l_j^n],$$

where $\hat{a}_j^{mn}$ is the average quantity of commodity m needed to produce one unit of commodity n in region j, and l_j^n is the average amount of labour required. In a similar way define the real wage of workers in region j as a vector representing the bundle of goods consumed weekly by the average worker's family:

$$\mathbf{b_j^n} = [b_j^1, b_j^2, \ldots, b_j^m, \ldots, b_j^N],$$

where b_j^m is the average quantity of commodity m consumed by a worker's family per week in region j. It then follows (Chapter 3) that the quantity of good m consumed per hour worked equals b_j^m / H, where H is the length of the working week in hours. We will assume for simplicity that the real wage is in fact the same in all regions, although it is unproblematic for our discussion even if it did vary between regions. Assume also that the length of the work week, H, is the same everywhere.

The total amount of commodity m used up in the production of commodity n in region j will then equal the quantity used as a capital good ($\hat{a}_j^{mn}$) plus the quantity consumed by workers in order to put in the l_j^n hours necessary to manufacture the commodity ($l_j^n . b_j^m / H$). Therefore, the total quantity of commodity m required to produce one unit of commodity n in region j (a_j^{mn}) is:

$$a_j^{mn} = \hat{a}_j^{mn} + b_j^m \, l_j^n / H. \tag{4.1}$$

Let us note immediately two characteristics of the input coefficients a_j^{mn}. First, by definition, the amount of m used in region j to produce a unit of commodity n equals the sum of all inputs shipped in for that purpose from all regions:

$$a_{ij}^{mn} = \sum_{i=1}^{J} a_j^{mn}, \tag{4.2}$$

where a_{ij}^{mn} is the quantity of commodity m, to be used in the production of n in region j, that is shipped from region i. The coefficient a_{ij}^{mn} represents the inter-regional shipment of commodities or the inter-regional input–output coefficient, and plays a key role in this book. Second, in a productive economy the sum total of each commodity used as an input to production is less than or equal to the total quantity produced. Abraham-Frois and Berrebi (1979) have shown that, when discussing the production of one unit of each commodity in each region, as we are doing since we measure inputs per unit produced, then the sum total used from each sector in each region is less than or equal to the one unit produced:

$$\sum_{j=1}^{J} \sum_{n=1}^{N} a_j^{mn} \text{ or } \sum_{j=1}^{J} \sum_{n=1}^{N} a_{ij}^{mn} \leq 1, \tag{4.3}$$

where the strict inequality must hold for at least one commodity in one region.

Given the assumptions made above, we now wish to describe the spatial structure of commodity production under full capitalist competition; the location of commodity manufacturing; the prices charged for the commodities; and the patterns of commodity circulation and trade that exist between regions. We shall proceed in several stages, determining first the prices and trading patterns and then the locations of firms.

4.1.1 Spatial prices and trading patterns

DEFINING PRICES OF PRODUCTION

The price of production charged by producers of commodity n in region j (p_j^n) is, as before, the total cost of production incremented by the rate of profit. The cost of production will be the quantity of each input required from each location multiplied by the price charged from that location:

$$p_j^n = [1 + r]. \sum_{i=1}^{J} \sum_{m=1}^{N} a_{ij}^{mn} p_i^m. \tag{4.4}$$

This is the production price at the factory gate in region j. Because this includes the transportation of inputs between locations, however, it is necessary to consider the transportation commodity in detail in order to show how prices depend on transport costs.

THE TRANSPORTATION COMMODITY

Transportation is required as an input to ship commodities from each region i where inputs are purchased. Suppose that transportation is hired in region i to ship input m from i to j, and that τ_{ij}^m units of the transportation commodity are required to ship one unit of the commodity this distance for this purpose. τ_{ij}^m is a measure of the physical difficulty of shipping the commodity. If f.o.b. prices are charged for commodities, then the cost of inputs obtained from region i is equal to the factory gate price in region i (p_i^m) plus transport costs:

$$q_{ij}^m = (p_i^m + \tau_{ij}^m p_i^t), \tag{4.5}$$

where p_i^t is the price charged by capitalists providing transportation in region i.

In a space economy, transportation may seem like just another input to be purchased, but this is not so because of its unique situation within the relations of production. Transportation is a commodity that is consumed in moving other commodities from producers to consumers. For capitalists, transportation is not a direct input to production but a commodity needed to make the direct inputs accessible. Whereas use of a certain production method dictates what other capital goods are required, the amount of transportation needed will depend on the accessibility to suppliers. Similarly, the wage goods in workers' consumption bundles each require an input of transportation to make them available (in addition to workers' direct demands for transportation), in a quantity that depends on accessibility to capitalists producing consumer goods.

For example, the total amount of transportation produced in region i that is required in the production of one unit of commodity n in region j, a_{ij}^{tn}, is equal to the amount of all inputs for commodity n that are shipped from region i to j, multiplied by the quantity of transportation required to ship them to region j.

$$a_{ij}^{tn} = \sum_{m=1}^{N} a_{ij}^{mn}.\tau_{ij}^m. \tag{4.6}$$

These transportation inputs are a part of the cost of production, implying that the production price at the factory gate for commodity n produced in region j is:

$$p_j^n = [1 + r] \cdot \sum_{i=1}^{J} \sum_{m=1}^{N} \left(a_{ij}^{mn} \cdot p_i^m + \left(\sum_{m=1}^{N} a_{ij}^{mn} \cdot \tau_{ij}^m \right) \cdot p_i^t \right)$$

$$= [1 + r] \cdot \sum_{i=1}^{J} \sum_{m=1}^{N} (a_{ij}^{mn} \cdot p_i^m + a_{ij}^{tn} \cdot p_i^t) \tag{4.7}$$

While the demand for transportation is derived rather than fixed by production and consumption patterns, it is a commodity whose price is determined in the same way as for any other commodity. Therefore in each region there is a sector providing the means of transportation, which requires certain inputs for its production. It is important to note that the transportation commodity, t, is the service of shipping commodities. Thus, it is not necessary that transportation equipment is produced in each region, just that there are shipping companies that, among other things, use large quantities of transportation equipment to produce their commodity.

As for any other commodity, we can define the inputs required to produce transportation in region j, a_j^{mt}. Figure 4.1 then shows the structure of input–output relations in a space economy, the elements of which have been introduced above. Before the various elements of the matrix in Figure 4.1 can be filled in, so that production prices can be calculated using equation (4.7), it is necessary to determine the pattern of commodity trade between regions.

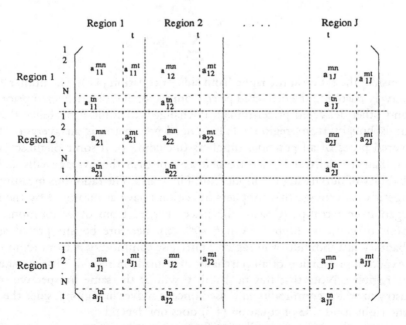

Figure 4.1   Inter-regional input–output matrix

INTER-REGIONAL TRADE

Even if producers are at least temporarily committed to a certain method of production, as defined by an input vector $\hat{a}^n_j$, it is relatively easy for a firm to change suppliers. Indeed, firms would be expected to choose suppliers based on how much they charge to deliver the product. Thus it is unreasonable to assume that shipments are fixed independently of the prices charged by competing potential suppliers. Instead, we expect capitalists to purchase more inputs from those suppliers with cheaper delivered prices. In practice, while producers would like to purchase all their inputs from the cheapest supplier, a number of factors prevent this from happening. Producers may have limited information about prices of competing sellers, or they may be locked into long-term contracts with particular sellers that they are unable or unwilling to break even if that seller becomes more expensive than other sellers. Nevertheless, we expect that the probability that a capitalist purchases an input from region i would be greater if the delivered price from region i is lower relative to delivered prices from other regions. This implies that shipment patterns depend on production prices.

A relatively flexible model describing this dependence is given below. This model states that the average amount of input m purchased from region i by firms producing commodity n in region j equals the total amount of input m required (both as an input to production and for workers' consumption; recall that $a^{mn}_j = \hat{a}^{mn}_j + b^m_j \, l^n_j / H$) multiplied by the probability that this will be purchased from region i. This probability in turn depends on the delivered price from i compared with the average delivered price from all competing suppliers:

$$a^{mn}_{ij} = a^{mn}_j \frac{e^{-\beta\{q_{ij}{}^m\}}}{\sum\limits_{k=1}^{J} e^{-\beta\{q_{kj}{}^m\}}}. \tag{4.8}$$

Consider the ratio on the right-hand side of equation (4.8). The numerator is negatively related to the delivered price from region i; if that delivered price goes up and other delivered prices remain unchanged, then the probability that the input is bought from region i falls. The denominator is an average of the delivered prices of all potential suppliers (including i). If some of those prices, other than the delivered price from i, go up, then this average falls and the probability of purchasing the input from i increases. The ratio thus measures the competitive advantage that suppliers in region i have as measured by the price charged from i compared with the price charged from other locations. The amount of good m shipped from i, a^{mn}_{ij}, can therefore be interpreted as the probability that individuals in region j purchase commodity m from region i, or the expected proportion of all purchases of m made in region j that are shipped from region i. Note that this probability will be the same irrespective of the industry in which commodity m is used once it arrives in region j, since the ratio on the right-hand side of equation (4.8) does not depend on n.

Conceptually, this probability reflects a process of spatial competition.

Capitalists seeking to purchase inputs have some information about the delivered price at which m can be obtained from various sources. The lower the delivered price from i relative to other potential suppliers, the more likely is the capitalist to place an order with a supplier from i. By the same token, workers seeking to consume commodity m are also more likely to obtain it from region i if the delivered price is lower. This occurs either because retailers selling that commodity in region j are more likely to order it from i, or because workers themselves may order it from i or travel to i to purchase it. The degree to which inputs are purchased from the cheapest location will depend both on the price differentials and on the responsiveness of purchasers to those differentials. This responsiveness is captured in the coefficient β of equation (4.8). Mathematical analysis of this equation shows that when β is approximately zero, the probability of buying an input from a location is equal for all competing locations regardless of the price charged. This would be a situation where price plays no role in affecting commodity shipments. When β is extremely large, however, virtually all inputs are bought from the cheapest location even if the price difference is very small, suggesting that there is a very efficient marketing of commodities. This latter case implies that all purchasers have full information and make the economically rational choice. The smaller is β the less this is true (Williams, 1977; Leonardi, 1982). The advantage of equation (4.8), then, is that it expresses the idea that commodity shipments and trade are responsive to price differentials, without requiring the extreme assumption that all purchases are rational cost-minimizing decisions. Indeed, it is not necessary to know what the value of β is in order to continue with our analysis; it can be left to be determined empirically.

The idea that commodity trade depends on production prices considerably complicates the determination of these prices. A casual glance at equation (4.4) would suggest that production prices are calculated given knowledge of the inter-regional input–output coefficients. Indeed, one of the central conclusions of aspatial neo-Ricardian and Marxian models is that prices depend only on production techniques and the workers' consumption bundle (see Pasinetti, 1977; Roemer, 1981). Consider equation (3.13) of Chapter 3, which is used to calculate production prices in a non-spatial economy. The prices depend only on input–output coefficients which are known if the technology for each sector and the size and mix of the consumption bundle are fixed. Certain well-known theorems in matrix algebra, the Perron–Frobenius theorems, guarantee that there is a unique set of positive relative prices and a positive rate of profit that are associated with these coefficients.

In a space economy, however, the situation is far more complex. The coefficients a_{ij}^{mn} on the right-hand side of equation (4.7) include commodity flows that themselves depend on production prices, as can be seen if a_{ij}^{mn} and a_{ij}^{tn} are replaced by their definitions as given in equations (4.6) and (4.8). As long as we know the transportation requirements for each type of shipment (τ_{ij}^{m}) this adds no extra unknown variables to the model, so in principle prices of production can still be calculated. It is necessary, however, for the prices of production and the trading patterns to be consistent with one another. It must be simultaneously the case that the trading patterns accurately reflect spatial price differentials (equation 4.8), and that these prices are consistent with the trading patterns (equation 4.7). In other words, not any trading pattern is possible, but it must be a trading pattern that is consistent with both of these relationships.

We know that, for a given set of technologies in use in each region, real wage consumption bundles in each region, and a particular spatial configuration of regions, there exists a mutually consistent configuration of spatial prices and trading patterns (Sheppard, 1987; and see Appendix to this chapter).[2] In addition, a number of numerical experiments with this kind of space economy have shown that if technologies, workers' consumption and the spatial configuration remain unchanged then, when prices of production and trading patterns differ initially from this mutually consistent configuration, the profit-maximizing actions of capitalists will alter prices and trading patterns until they become consistent with one another in a unique trading and pricing equilibrium (Sheppard 1983b). In short, there exists a spatial trading and pricing equilibrium, which represents a possibly unique pattern of prices and commodity flows enabling all entrepreneurs to gain an equal rate of profit. Furthermore, this is a stable equilibrium as long as all produced commodities can be readily sold (but see Chapter 9).

To summarize, for any fixed geographical distribution of production methods and real wages among a group of urban-centred regions, it is possible to determine three attributes of a fully competitive capitalist space economy: the patterns of inter-urban commodity flows; the spatial distribution of prices of production; and the equalized rate of profit that would prevail. These prices, trading patterns and profit rate are not the prices and trading patterns that each firm would face, because we follow a stochastic form of reasoning; they are simply the expected trading patterns, and the average prices and profit rate expected in the long run (see Farjoun and Machover, 1983; Semmler, 1984). Thus, even if a real space economy could be found where all the assumptions employed thus far were true, the particular set of prices and trading patterns observed there at any one point in time would be in all likelihood somewhat different from these mean values – which simply represent an expectation of an economy stochastically governed by the rules of capitalist competition.

4.1.2 Wage–profit frontiers

Suppose that the real wage rises or falls, while the distances between urban regions, the production methods used in each region, and the industries found in each region remain the same. Each change in the real wage gives rise to a new trading and pricing equilibrium with a new rate of profit. The changes in the rate of profit that result from changes in wage levels can be graphed as a wage–profit frontier, as depicted in Figure 4.2. It can be shown that as real wages increase, for example by reducing the length of the work week or by increases in workers'

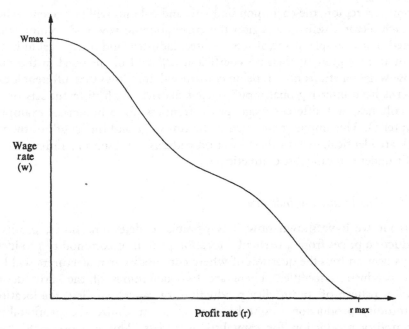

Figure 4.2 A wage–profit frontier

consumption levels, then the rate of profit falls. Thus conflict between capitalists' profits and workers' wages is inherent to capitalist commodity production in a space economy.

This wage–profit frontier is constructed for a particular spatial configuration of production, i.e. for a particular set of production methods (as defined by the capital good input requirements for each sector and each region, $\hat{a}_j^m$) and a particular spatial configuration of regions as defined by the difficulty of shipping goods between those regions (i.e. the transportation coefficients, τ_{ij}^m). Changes in any of these elements would essentially represent a different geographical configuration of production, with a different wage–profit frontier – the geographical version of the different technologies whose wage–profit frontiers were being compared in the capital controversies (cf. Figure 2.2).

Of particular interest are the different wage–profit frontiers that result from different location patterns of industry. By 'different location patterns' we mean a different allocation of economic sectors among regions. For example, a situation where region A produces steel and region B produces grain is a different location pattern from one where both regions produce steel and only region B produces grain. Clearly with many regions and sectors a very large number of such different location patterns are possible. Now each possible location pattern for commodity production represents a different set of input coefficients, a_j^{mn}. To see this, consider Figure 4.1. Each row and column in Figure 4.1 represents a particular sector that is present in a particular region. For each sector that is present in a region, the corresponding row and column will be present, whereas for each sector absent in a region the corresponding row and column will be deleted. For example, if sector n is the steel industry, and if the steel industry is absent from region B, then no coefficients a_{iB}^{mn} will be included in the matrix whose wage–profit frontier is being constructed. It follows that different location patterns have inter-regional input–output matrices with different sets of rows and columns, and different wage–profit frontiers (for a numerical example see Chapter 5). This implies that, even if the real wage and the length of the work week are identical, each of these location patterns can result in a different rate of profit under full capitalist competition.

4.1.3   The location of industry

Thus far we have shown how it is possible to determine flows, profits and production prices from a particular location pattern of commodity production. Let us now address the question of where commodity manufacturers will locate their production facilities. There are two definitions of the most desirable location pattern of commodity production for capitalists. First, the location of commodity production might be such as to maximize the profitability of commodity production for capitalists as a class. This is represented by the geographical pattern that maximizes the average rate of profit, r, which we call the *profit-maximizing location pattern*. Because higher profits entail a faster rate of capital accumulation (see Chapter 9), this location pattern could be interpreted as that to be expected by applying a strict 'capital logic' interpretation of the economy, where it is argued that economic production takes on the form that is best for the purposes of capital accumulation. A second possibility is to recognize that individual capitalists choose those locations that maximize the rate of profit that they realize for themselves as individuals. Such choices by individual capitalists need not result in a location pattern that maximizes the collective benefits for capitalists as a class, if individual self-interest conflicts with class interest.

Considering the first definition, we have seen above that each location pattern is associated with a different wage–profit frontier. Because neoclassical parables are inapplicable (see section 2.3) it is not generally possible to predict *a priori* which location pattern is most profitable, except in the trivial case of one technology being more productive in all respects than all others. It is possible in principle, however, to calculate which pattern yields the highest rate of profit for any given level of real wages. Wage–profit frontiers can be constructed for each possible location pattern (resulting in a very large number of frontiers!) We then simply determine the frontier with the highest profit rate (Sheppard and Barnes, 1986), and conclude that the location pattern associated with this frontier is the profit-maximizing location pattern.

Figure 4.3 shows a simple example of this, where at wage level w_1 location pattern B maximizes the expected profits of capitalists as a class. Notice from the figure that at different wage levels different location patterns may be profit maximizing. This has the important implication that, if capitalists are able to realize profit-maximizing location patterns, then any shift in the balance between wage and profit levels in society will also trigger a change in location patterns. If this were true, changes in the relative power of different social classes, as they struggle to increase their share of the economic surplus, would lead to shifts in the landscape of commodity production.

Little will be said here with respect to the second definition. We shall examine the relationship between the interests of individual capitalists and those of capitalists as a class in Chapter 12, finding that the two are not generally coincident. As Marx has argued: 'individual capitalists – coerced by competition ... make ... adjustments which drive the economy as a whole away from "a 'sound', 'normal' development of the process of capitalist production"' (Capital, vol. 3, p. 255)' (Harvey, 1982, pp. 188–9). We shall also argue that actions that are in the interest of individual capitalists may lead to a reduction in the general rate of profit. If this is the case, then the least we can expect is that the location of commodity production under the second definition will be different from that calculated under the first definition. For the time being we shall simply indicate that it would be possible to construct an argument about how locations, prices and trading patterns might evolve under this second definition, but the details of that process can be postponed to parts III and IV of the book.

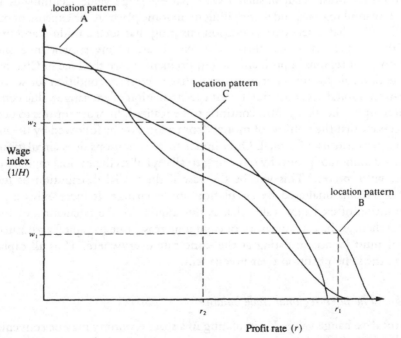

Figure 4.3 Wage–profit frontiers for three location patterns (A, B & C)

4.1.4   Circulation and exchange

There are two different possible interpretations of the geographical pattern of prices and commodity flows that represents trading and pricing equilibrium under full capitalist competition: one more economic, and the other more geographical. The economic one, corresponding to examining the circulation process from the observation point of individual production sites, states that prices everywhere are equal to the cost of inputs incremented by an equal rate of profit. The rationale for this kind of interpretation is that described above, whereby if this were not the case investment capital would flow from one firm to another until profit rate differentials disappeared, supplemented by the neo-Keynesian argument that demand has little influence on prices because firms typically respond to fluctuations in demand by changing production levels rather than by changing prices (Kalecki, 1938).

If we observe the circulation process, however, by following the overall geographical circulation of exchange value, an alternative interpretation is possible. According to this interpretation, there is trading and pricing equilibrium when exchange value is distributed among locations in such a way that the spatial circulation of money will not alter this distribution (for a demonstration of this, see the Appendix to this chapter). While commodities circulate from places where prices are lower to places where prices are higher (i.e. up the spatial price gradient on average), the circulation of money in the opposite direction as payment for these commodities must occur in such a way that exchange value accumulates at the same rate in all regions, and is not piling up in some places at the expense of others. This implies that a necessary condition ensuring that such a trading and pricing equilibrium persists is that there is an equal rate of investment in economic growth in all regions; a problematic requirement, as we shall see in Chapter 9.

The rationale for this interpretation is that a necessary condition for successful long-term capital accumulation is that the circulation of exchange value continue uninterrupted, i.e. that profits continue to be realized. In order for this to occur it is necessary that the outflow of money from a location be followed by the inflow of a larger quantity of capital. Only under these conditions does circulation lead to the realization of profits by returning to all capitalists their initial capital investment, with interest. This will be the case if the spatial distribution of money capital remains unaltered by circulation and exchange. If there exists a spatial distribution of exchange value that is not altered by the circulation of money capital then, as the total quantity of capital increases due to capital accumulation, capital must be accumulating at the same rate everywhere. Thus all capitalists realize the same profit on their investment.

4.1.5   The inter-regional trade balance

The total exchange value, Y, circulating in a space economy may be conveniently divided into three parts: that necessary to purchase capital good inputs, K; that necessary to pay for labour, W; and monetary profits Π. The latter two together are often referred to as 'net income'; net, that is of the payment for capital goods.

$$Y = K + W + \Pi. \tag{4.9a}$$

If we know the quantity of each commodity n produced in each region, j, x_j^n, then we can see exactly what each term includes. Total exchange value at the end of a production period is simply the total cost of production in the economy, added up over all sectors and regions weighted by production levels in each place, incremented by the rate of profit:

$$Y = [1 + r].\sum_{j=1}^{J} \sum_{n=1}^{N} \left(\sum_{i=1}^{J} \sum_{m=1}^{N} (a_{ij}^{mn}.p_i^m).x_j^n \right).$$

Recalling the definition of equation (4.1) and the identity of (4.2):

$$Y = \sum_{j=1}^{J} \sum_{n=1}^{N} \left(\sum_{i=1}^{J} \sum_{m=1}^{N} [(\hat{a}_{ij}^{mn}.p_i^m) + b_{ij}^m\, l_j^n\, p_i^m/H)].x_j^n \right)$$

$$+ r.\sum_{j=1}^{J} \sum_{n=1}^{N} \left(\sum_{i=1}^{J} \sum_{m=1}^{N} (a_{ij}^{mn}.p_i^m).x_j^n \right), \tag{4.9b}$$

where $\hat{a}_{ij}^{mn}$ is the quantity of commodity m bought from i that is used as a capital good input in producing one unit of commodity n in region j, and $b_{ij}^m l_j^n/H$ is the quantity of commodity m bought from i that is used as a wage good in producing one unit of commodity n in region j. Both inputs are defined as those that exist in trading and pricing equilibrium. In matrix algebra this identity is:

$$[1 + r].\mathbf{p}'\mathbf{\hat{A}x} = \mathbf{p}'\mathbf{\hat{A}\star x} + \mathbf{p}'\mathbf{Bx} + r.\mathbf{p}'\mathbf{\hat{A}x}, \tag{4.9c}$$

where $\mathbf{\hat{A}}$ is the matrix of commodity flows in trading and pricing equilibrium (i.e. the inter-regional input–output matrix equalizing capitalists' profits), $\mathbf{\hat{A}\star}$ is that part of $\mathbf{\hat{A}}$ representing capital good flows; $\mathbf{B}$ is the matrix of commodity flows representing wage goods ($\mathbf{B} = \mathbf{\hat{A}} - \mathbf{\hat{A}\star}$), and $\mathbf{x} = [x_1^1, x_1^2, \ldots, x_j^n, \ldots, x_j^N]$ is the vector of production levels in all sectors and regions.

The accounting relationships of equation (4.9b) can be further disaggregated by region in order to examine balance of payments implications of inter-regional commodity circulation (Sheppard, 1987). Denote the total exchange value flowing from region i to region j, which is the total revenue made on sales of products of region j in region i, as Y_{ij}. This is equal to the sum of: the profits made on the sale of commodities manufactured in region j and sold in region i (Π_{ij}); the money needed to purchase capital goods inputs for this production (K_{ij}); and the money needed to pay labour for this production (W_{ij}):

$$Y_{ij} = K_{ij} + W_{ij} + \Pi_{ij}.$$

The monetary balance of trade for region i, y_i, is then:

$$y_i = \sum_{j=1}^{J} (Y_{ji} - Y_{ij}). \tag{4.10}$$

Because the spatial distribution of exchange value remains unchanged in equilibrium (see section 4.1.4), it follows that in the trading and pricing equilibrium representing full capitalist competition monetary trade is balanced in each region ($y_i = 0$). We shall show in Chapter 14 how this kind of accounting formula aids analysis of the spatial reallocation of monetary profits. In general it is a useful way of accounting for the geographical circulation of exchange value. It may also be compared with circulation of labour value in order to evaluate claims of unequal exchange (Chapter 8).

4.2 Incorporating production and circulation time

In the analysis thus far, it has been assumed that the cost of all capital goods and labour used in production equals the amount of capital that capitalists must advance each year. This is because we assumed that production periods last one year and that all inputs are completely paid for at the beginning of the year and completely used up at the end of the year. Furthermore, it has been assumed that at the end of the production period (the year) the produced commodity is sold immediately. Thus all inputs are paid for on 1 January, and the revenue is received on 31 December. The annual profit rate, the extra revenue made by 31 December per dollar of capital advanced on 1 January, is then simply the ratio of net revenues to costs. In fact, the situation is both more flexible and more complex than this assumes.

If the production period is six months then, instead of buying all inputs on 1 January, half of them could be bought then, and the other half purchased on 1 July with the revenue made from sales on 30 June. In that case the capital advanced by the capitalist would be only one-half of the total cost of inputs used during the year. In the same way, less capital is advanced initially if payments for inputs are spread throughout the production period. On the other hand, the time involved in bringing products to market and selling them lengthens the production period, thus reducing the rate of profit per unit of time. This last factor depends in part on the efficiency of the transportation system, and is of particular interest to economic geographers.

Underlying these examples is the principle that the rate of profit of concern to capitalists is the increment in exchange value per dollar advanced per year, whereas the relationships developed in section 4.1 calculate the rate of profit as the increment in exchange value per dollar of production costs per production period. To calculate a rate of profit accurately it is necessary to take into account those various factors making the production period different from a year or making capital advanced different from production costs. In a series of papers Michael Webber has examined the adjustments needed on the production side, and we will summarize his findings in section 4.2.1 (Webber and Rigby, 1986; Webber, 1987b). A complementary analysis of the circulation side will be conducted in section 4.2.2.

4.2.1   Capital advanced in production

There are five factors to be taken into account in calculating the capital advanced per year: fixed capital; the capacity utilization rate; the number of production periods per year (the turnover rate); deferred payment for labour and capital goods; and delayed delivery of inputs.

Fixed capital refers to all capital goods that are not completely used up in the course of production – notably buildings and machinery. Such equipment must be available at the start of the production period, as production is impossible without it. Fixed capital eventually must be replaced in order to continue production, meaning that capitalists must recoup the purchase price of this equipment during its economic lifetime. This is done by charging depreciation on fixed capital as a part of production costs.

Straight-line depreciation, for example, divides the cost of fixed capital by its lifetime to determine annual costs for replacing fixed capital. In this case, the annual cost of fixed capital of type k, with a mean lifetime of θ_k years, will be equal to $a_{ij}^{kn} p_i^k / \theta_k$. Here, a_{ij}^{kn} is equal to the mean total quantity of fixed capital of type k required, divided by the engineering-related capacity of that fixed capital (Eichner, 1976).[3] (For circulating capital θ_k equals 1 by definition.) In general, the economic lifetime of fixed capital and its depreciation schedule depend on how it is produced and on the cost of other commodities, a complication that we will reserve for Chapter 7. The depreciation on fixed capital that must be charged as a production cost per unit produced, in order to recoup the full value of fixed capital before it becomes obsolescent, depends on the *capacity utilization rate* of fixed capital as well as on the length of its economic life. For example, if the amount produced in a year falls by half, the fixed cost charged per unit produced that year must double so that the capitalist can recoup the annual depreciation cost. If C_n^k is the mean capacity utilization rate for fixed capital of type k in industry n, the mean annual production costs associated with fixed capital input k, per unit of commodity n produced in region j, are: $a_{ij}^{kn} p_i^k / \theta_k C_n^k$.

If there are an average of T_j^n production periods per year in industry n in region j, the capital that must be advanced per unit of production to pay for circulating capital and wages equals the mean annual production cost divided by the mean number of production periods per year, T_j^n. This does not apply to fixed capital because those costs are already calculated per year rather than per production period. For the time being, a production period is defined as the time taken to manufacture the commodity; but see section 4.2.2.

The capital to be advanced must all be available prior to the start of the production period, since it must all be paid out before any revenue from sales of the commodities produced during that period is forthcoming. It generally is not all paid out in advance, however. Wages are paid out weekly or monthly, and capital goods are paid for when they are delivered. If labour and circulating capital costs are paid for periodically during the production period, then the capitalist can invest that amount of the bill not yet required as finance capital and earn interest on this investment. Webber (1987b) has suggested that the interest thus earned can be calculated in advance and incorporated in the financial plan, thus reducing the amount of capital that must be invested at the start of the production period. This would seem to be an example of a situation where the timing of wage payments does matter (contra Steedman, 1977).

We suggest, however, that most capitalists would regard such interest income as a supplement to the profits earned from commodity production. Strictly speaking, this is not profit made from commodity production but interest earned on finance capital. The extra profits made from interest on deferred payments for labour and circulating capital will increase if the rate of interest is greater, if the length of time between deferred payments is greater, and if the turnover rate is less (Sheppard, 1990). For example, if labour is paid monthly rather than weekly, then more interest is made per year by investing wage capital prior to using it for wages. A second source of such interest income stems from the use of fixed capital. 'Whereas circulating capital ... is invested gradually but recouped all at once at the end of the [production] period, fixed capital is advanced in a lump and recouped gradually ... The return of fixed capital at the end of each production period can be loaned' (Webber, 1987b, p. 1317). Thus money received as depreciation payments on fixed capital may be invested to earn interest until it is required to replace worn out fixed capital. Since all such extra earnings result from interest on finance capital investments rather than profits from commodity production, we exclude them from our calculation of the rate of profit below.

There are, however, real deductions in production costs that result from the *delayed delivery* of circulating capital goods. If all such goods are purchased and delivered at the start of the production period, there are extra storage and inventory costs incurred that could be avoided if the goods were delivered just prior to when they are required in production. Indeed, one of the principles of the Japanese 'just-in-time' system of production is that storage costs are minimized by delivering capital goods immediately before they are required in production.[4] Such savings on storage costs represent a reduction in costs that must be accounted for.

Since storage is a part of the cost of making an input available at the right place and time, it can be regarded as part of the cost of delivering an input. Previously, under the assumptions that the production period is one year in each region and industry and that inputs are delivered prior to production, this cost could simply be subsumed under transportation. This simplification will now be dropped. Storage costs are directly proportional to the lag between the time when an input arrives at the factory and the time when it is required. Thus even if all inputs are delivered at the start of a production period, storage costs for them are proportional to the length of the production period. In addition, delayed deliveries will reduce storage costs up to a point. Obviously, deliveries that are delayed to the point that they arrive too late imply significant financial penalties. In addition, when there are considerable economies of scale in the costs of shipping commodities the storage costs associated with fewer larger shipments can be smaller than the savings associated with bulk shipments.

The total annual delivery costs for inputs are then equal to the sum of shipping and storage costs. Storage costs equal the quantity of warehouse space required, multiplied by the price of warehouse space. Define warehouse space as fixed capital of type s. Then total delivery costs are:

$$\sum_{i=1}^{J}\left(\sum_{m=1}^{N} a_{ij}^{mn}.\tau_{ij}^{m}\right).p_i^t + \sum_{i=1}^{J} a_{ij}^{sn}.p_i^s,\qquad(4.11)$$

where

$$\sum_{i=1}^{J} a_{ij}^{sn} = \sum_{m=1}^{N} a_j^{mn}.\sigma_m.\delta_m \quad \forall\, m \notin K,$$

where K is the set of fixed capital inputs, σ_m is the mean quantity of warehouse space needed to store one unit of input m, and δ_m is the mean number of days that input m has to be stored before it is employed as an input to production. As for other circulating capital inputs, the capital advanced to pay for transportation and storage of inputs equals total delivery costs divided by the turnover rate (T_j^n).

If adjustments are made to the relationship of equation (4.7) to account for fixed capital, the turnover rate and storage costs, we obtain the following:

$$p_j^n = [1 + r].\left\{\left[\sum_{i=1}^{J}\sum_{m=1}^{N} a_{ij}^{mn}.p_i^m + \sum_{i=1}^{J}\left(\sum_{m=1}^{N} a_{ij}^{mn}.\tau_{ij}^{m}\right).p_i^t\right]/T_j^n\right.$$

$$\left. + \sum_{i=1}^{J} a_{ij}^{sn}.p_i^s/\theta_s C_n^s + \sum_{i=1}^{J}\sum_{k=1}^{K} a_{ij}^{kn}.p_i^k/\theta_k C_n^k + \sum_{i=1}^{J}\left(\sum_{k=1}^{K} a_{ij}^{kn}.\tau_{ij}^{k}\right).p_i^t\right\}.\qquad(4.12)$$

In equation (4.12), r is measured as the rate of profit per unit of capital advanced per year. Capital goods, k, that are fixed capital have been separated from the others. This equation notes that storage costs are reduced by delayed deliveries (third term); that the capital advanced to pay for the production, delivery and storage of circulating capital and for labour (the terms in square brackets) is inversely related to the turnover rate; and that the capital advanced to pay replacement costs for fixed capital is inversely related to the durability of fixed capital and its capacity utilization rate (fourth term).

The implication of this analysis is that there are other important determinants of the rate of profit than production methods and wage levels. An increased turnover rate (T_j^n), achieved for example by increasing the speed of production lines or by increasing the speed at which products are marketed and revenues earned, will increase the rate of profit. By the same token, reduced lifetimes for fixed capital (θ_k) or reduced capacity utilization rates (C_n^k) can put downward pressure on profits. The former occurs, for example, when technological change is fast enough to accelerate the speed at which old machinery and buildings become economically obsolete. The latter occurs in times of falling demand and economic crisis. Finally, increased time lags between the delivery and utilization of circulating constant capital (δ_m) will reduce the rate of profit.

Note that there is also a close relationship between the turnover rate and the rate of capacity utilization. When production periods are shorter (i.e. a high turnover rate) there is greater annual output from the same fixed capital; which implies a higher capacity utilization rate. Thus the turnover time and the capacity utilization rate will increase or decrease together over time, implying a tendency for the capital advanced for both fixed capital and other inputs to increase or fall simultaneously.

The importance of accounting for turnover rates and fixed capital effects in calculating the rate of profit on commodity production has been studied by Webber and Rigby (1986; Webber, 1987a, 1988). They used an aggregate version of the framework developed above to analyse the rate of profit in Canadian manufacturing since the Second World War.

4.2.2   Circulation time

It is frequently remarked in economic geography that the cost of transportation is not just the money spent but also the time involved. In the case of capitalist commodity manufacture there is an extremely close relationship between time and money, through the way in which the length of the production period affects capital advanced. The production period includes: the time taken up by production itself, the time required to deliver the commodity to the market, the length of time it sits on the shelf prior to purchase, and the time taken to return the revenue to the capitalist. Thus far, we have considered only the first of these, as we concentrated on examining processes within the place of production. If capitalists are realizing profits, implying no great imbalance between supply and demand, then the third of these four factors will show no systematic variation from place to place. The high mobility of money means that the fourth one is generally very small. The second factor, however, shows systematic spatial variation and has certainly played an important role historically. The balance of this section is therefore devoted to examining the impact of circulation time on the production and circulation of commodities in a space economy (see Puchinger, 1979).

The circulation time that elapses before a producer of commodity m in region i can earn revenues from production depends on three things: the state of development of the transportation system, the locations to which output is shipped (as defined by the trading coefficients a_{ij}^{mn}), and the relative accessibility of region i to these locations. Suppose that we know the time it takes to reach each region from each other region, given their relative location and the state of development of transportation. Then the mean circulation time for commodity m produced in region i depends on the time taken to reach each destination, weighted by the amount shipped to that destination:

$$\Psi_i^m = \sum_{j=1}^{J} \sum_{n=1}^{N} a_{ij}^{mn}.\psi_{ij}, \tag{4.13}$$

where ψ_{ij} is the time taken to deliver a commodity from region i to region j, and Ψ_i^m is the mean circulation time.

Recall that the turnover rate is the number of production periods per year, which we previously defined as the time required to manufacture the commodity. The length of this period, expressed in fractions of a year, is simply the reciprocal of the turnover rate; $(T_j^n)^{-1}$. Dividing the costs of capital advanced to pay for circulating capital goods and labour in equation (4.12) by the turnover rate is thus equivalent to multiplying by the length of the time required for manufacturing. In other words, this equation has taught us that the capital advanced is proportional to the time required for manufacturing. We must now amend this to define the production period as the sum of the time of manufacture, over which the capitalist has some direct control, and the circulation time, over which s/he has little control; $(T_j^n)^{-1} + \Psi_j^n$.

This affects the rate of profit, since circulation time must be incorporated into equation (4.12). As suggested above, the capital advanced to pay for circulating capital goods and labour equals total production costs for this component multiplied by the sum of manufacturing and circulation time, measured in fractions of a year:

$$p_j^n = [1+r].\left\{\left[\sum_{i=1}^{J}\sum_{m=1}^{N} a_{ij}^{mn}.p_i^m + \sum_{i=1}^{J}\left(\sum_{m=1}^{N} a_{ij}^{mn}.\tau_{ij}^m\right).p_i^t\right].[(T_j^n)^{-1} + \Psi_j^n]\right.$$
$$\left. + \sum_{i=1}^{J} a_{ij}^{sn}.p_i^s/\theta_s C_n^s + \sum_{i=1}^{J}\left(\sum_{k=1}^{K} a_{ij}^{kn}.\tau_{ij}^k\right).p_i^t + \sum_{i=1}^{J}\sum_{k=1}^{K} a_{ij}^{kn}.p_i^k/\theta_k C_n^k\right\}, \tag{4.14}$$

where

$$\sum_{i=1}^{J} a_{ij}^{sn} = \sum_{m=1}^{N} a_j^{mn}.\sigma_m.\delta_m \quad \forall\, m \notin K.$$

This relationship, which defines the rate of profit given the location of production and commodity flows, shows how under capitalist commodity production time is indeed money. Profit rates are calculated per unit of time and they depend on the time taken for manufacturing and circulation. As the time taken for manufacturing and circulation falls, the capital that is advanced to finance commodity production also falls, leading directly to an increased rate of profit. Circulation time also influences the capital advanced to pay for fixed capital – the remaining term on the right-hand side of (4.14). As noted in section 4.2.1, the turnover rate is directly related to the capacity utilization rate. This is not changed by including circulation time. The quantity of output obtainable annually from a certain amount of fixed capital equals the total capacity of that fixed capital at any one time, multiplied by the number of times that the fixed capital is used per year. Thus, as the length of the total production period falls, the rate of capacity utilization increases. It follows that a reduction in circulation time reduces the production period, increases the capacity utilization rate, and thus reduces the capital advanced to pay for fixed capital goods (see the last term in equation 4.14).

4.2.3   Implications for the geography of commodity production

Assuming that capitalists are concerned with the rate of profit they obtain per unit of capital advanced, rather than per unit of production costs, the modifications suggested in sections 4.2.1 and 4.2.2 suggest that the equation for production prices (equation 4.7) must be modified. A mutually consistent configuration of production prices and trading patterns that equalizes the expected rate of profit on capital advanced is defined by a set of equations for each sector and each region of the form of equation (4.14), with trading patterns defined by equation (4.8) and circulation time defined by equation (4.13). This is conceptually more complicated than the case developed in section 4.1, but in principle the pattern of prices and trade can still be determined if circulation times, ψ_{ij}, are known, because no new unknown quantities are introduced. It then also follows that the profit-maximizing location pattern can be found for any real wage level. In short, all the procedures outlined in section 4.1 can also be followed for this more complicated case in order to determine the spatial organization of commodity production in a fully competitive capitalist space economy.

In addition, some important implications for the transportation system can be developed, which will be pursued in more detail in Chapter 13. David Harvey (1985a, pp. 1–35) has argued that one of the imperatives of capitalism is the annihilation of space by time. It is clear from the discussion of section 4.2.2 that the time taken for commodity trade has a broad influence on the rate of profit and, we might add, on the most profitable geography of production. Reduced circulation time increases the rate of profit for all capitalists by reducing the quantity of capital that is advanced for production. For this reason it is understandable why there has been considerable pressure to improve transportation technologies and reduce the time of travel, and why the spread of capitalism has been accompanied by the increased importance of time as a measure of transportation costs (for a brief discussion and further references, see Harvey 1985a, pp. 10–16). It also follows that those locations within an interregional system that have high levels of accessibility will have a comparative advantage for the location of economic activities, because the concentration of commodity production at such locations would be most effective in maximizing capitalists' rates of profit.

Summary

This chapter has concentrated on the geographical circulation of exchange value in a fully competitive space economy. Suppose we know the real wage consumed by workers in each region, the socially necessary (mean) production technology used in each region for each commodity that might be produced there, and the difficulty of transportation within and between regions (i.e. their relative location). If capitalists make decisions about where to purchase the inputs they need based on the relative delivered prices of competing suppliers, then a mutually consistent spatial pattern of factory gate production prices and interregional commodity trade exists that ensures that capitalists get the inputs they need at competitive prices, and that all capitalists make the same expected rate of profit on their investment (4.1.1). This 'trading and pricing equilibrium' is stable in the absence of technological or geographical change.

Transportation plays an essential and unique role in the circulation process necessary to realize profits on commodity production as an intermediate commodity – producing the accessibility that is necessary to move commodities from places of production to places of consumption. Commodity shipping patterns were not given exogenously because they are not fixed by production technologies; the socially necessary technology determines how much of some input is required but not where it is purchased. This is why we took account of the decision of where to purchase inputs, making it dependent on the relative delivered prices of competing suppliers.

The commodity circulation patterns that result may be interpreted geographically as money flowing down the spatial gradient of production prices as payment for the commodities that flow up this gradient (4.1.4), in such a way that the spatial distribution of exchange value remains unchanged. The monetary balance of payments can be calculated for each region, broken down if desired into payments for capital goods, payment for labour and profits. In trading and pricing equilibrium the exchange value flowing into each region equals the exchange value flowing out of it (4.1.5). Associated with each combination of location patterns and socially necessary techniques is a wage–profit frontier showing how profits fall as wages increase (4.1.2), revealing the struggle between capitalists and workers for economic surplus. These frontiers may be compared for different location patterns to determine which pattern maximizes the collective rate of profit for capitalists as a class (4.1.3). Changes in the distribution of income between capitalists and workers may well alter which location pattern is most profitable

This scheme had to be modified to take into account the fact that the capital advanced by capitalists to pay for commodity production at any one point in time need not equal the total cost of production (4.2). Within the sphere of commodity production, fixed capital, different numbers of production periods per year, and the timing of the delivery of inputs can all affect the proportion of the cost of production that must be advanced (4.2.1). Within the sphere of circulation, the time taken to deliver commodities to the market for sale extends the production period and also affects the quantity of capital that must be advanced (4.2.2). These complications must be taken into account in order to determine the real rate of profit of concern to capitalists − the rate of profit per year per dollar of capital advanced. This was done by modifying the relationships developed in section 4.1 (4.2.2), without affecting the main results of that section. It did, however, point to a second way in which transportation plays a central role in profit realization. Increases in circulation time increase the capital that is advanced for all kinds of inputs, pointing to an imperative in capitalism to reduce circulation time in order to increase profit levels (4.2.3).

Notes

1   There has been extended debate about whether capitalists will reinvest in such a way that inequalities in profit rates would disappear over time. Initial scepticism expressed by Nikaido (1983) has been tempered by several models where convergence does occur (see Flaschel and Semmler, 1986).
2   The spatial configuration of regions is defined by the distance regions are apart from one another as measured by the size of the transportation coefficients τ_{ij}^m. Thus, knowledge of what these coefficients are is tantamount to knowing the relative geographical location of the various regions.
3   The engineering-related capacity is the total quantity of commodity n that can be produced per annum by the fixed capital, if it is operating at full capacity. Thus a_{ij}^{kn}/θ_k is the quantity of fixed capital required per unit of commodity n produced.
4   Deferred delivery of inputs is often associated with deferred payment for inputs, but this need not be so.

Appendix: Existence of trading and pricing equilibrium, and potentials

In part 1 of this appendix we prove that a trading and pricing equilibrium exists. In part 2 we show how the geographical interpretation of prices is derived.

Part 1

We demonstrate here that there is always a positive set of spatial prices of production and a positive profit rate that will occur under full capitalist competition for the conceptualization of the space economy developed in this part. Consider equation (4.4), reproduced here for convenience:

$$p_j^n = [1 + r] . \sum_{i=1}^{J} \sum_{m=1}^{N} a_{ij}^{mn} p_i^m. \tag{4.4}$$

This equation can be rewritten as:

$$\mathbf{p}' = (1 + r) . \mathbf{p}' . \mathbf{A}, \tag{4A.1}$$

where $\mathbf{p}'$ is the vector of production prices, $[p_1^1, p_1^2, \ldots, p_1^N, p_2^1, \ldots, p_J^N]$, and $\mathbf{A}$ is the matrix with entries:

$$a_{ij}^{mn} = \begin{cases} a_{ij}^{mn}\{\mathbf{p}\} & \forall\ m \neq t \\ \sum_{m=1}^{N} a_{ij}^{mn}\{\mathbf{p}\}.\tau_{ij}^m & m = t \end{cases}$$

where $a_{ij}^{mn}\{\mathbf{p}\}$ means that a_{ij}^{mn} is a function of the price vector, defined by equations (4.8) and (4.5), and that matrix $\mathbf{A}$ is also a function of $\mathbf{p}$; $\mathbf{A}\{\mathbf{p}\}$. Because of this dependence on $\mathbf{p}$, equation (4A.1) is not a conventional eigenvalue system whose properties may be investigated using the Perron–Frobenius theorems. However, equation (4A.1) may be regarded as a non-linear eigenvalue problem with eigenvalue $(1 + r)^{-1}$, and certain properties of such systems are known (see Nikaido, 1968). If $\mathbf{A}\{\mathbf{p}\}$ is greater than or equal to zero for any non-negative price vector, and if $\mathbf{A}\{\mathbf{p}\}$ is a continuous mapping $\mathbf{A}: R_+^n \rightarrow R_+^n$ except possibly at $\mathbf{p} = 0$ (where n is the order of $\mathbf{A}$), then there is at least one solution to (4A.1) that gives a positive price vector $\mathbf{p}$ and for which $(1 + r)^{-1} \geq 0$ (Nikaido, 1968, theorem 10.1). It is clear from inspection of the definition of $\mathbf{A}$ that it is a non-negative and a continuous function of $\mathbf{p} \geq 0$, implying that this result applies to equation (4.4).

For any positive solution $(\mathbf{A}^*, \mathbf{p}^*, (1 + r^*)^{-1}$ to the non-linear eigenvalue problem (4A.1) it can be shown that $r^* > 0$. This stems from the assumption that the entire space economy is productive. This assumption implies that the entries of $\mathbf{A}^*$ can be scaled in such a way that the row sums of the input–output matrix are less than or equal to one, with at least one row sum being less than one (equation 4.3). This implies that all the eigenvalues of $\mathbf{A}$ are less than one. The eigenvalue associated with $\mathbf{A}^*$, $(1 + r^*)^{-1}$, must then fall between zero and one, implying that $r^* > 0$. These conclusions also apply to the more complex model of equation (4.14) (Sheppard, 1990).

Part 2

In this part we show why it is that, in trading and pricing equilibrium, exchange value is distributed among locations in such a way that the spatial circulation of money will not alter this distribution (for a fuller discussion, see Sheppard, 1987). Consider a geographical distribution of commodity production that is also in trading and pricing equilibrium. In trading and pricing equilibrium:

$$\mathbf{p}' = [1 + r].\mathbf{p}'.\mathring{\mathbf{A}}, \tag{4A.2}$$

where $\mathring{\mathbf{A}}$ is the matrix of inter-regional commodity shipments in trading and pricing equilibrium. We know (equation 4.3) that the rows of matrix $\mathring{\mathbf{A}}$ all sum to less than or equal to one, meaning that $\mathring{\mathbf{A}}$ is a stochastic matrix (see Sheppard, 1979). We also know that $(1 + r)^{-1}$ is the largest eigenvalue of the matrix $\mathring{\mathbf{A}}$, and that $\mathbf{p}'$ is the associated eigenvector. It can therefore be shown, by successively substituting the definition of $\mathbf{p}'$ (the right-hand side of 4A.2) into the right-hand side of 4A.2, that, for any positive integer k:

$$\mathbf{p}' = [1 + r]^k.\mathbf{p}'.\mathring{\mathbf{A}}^k \tag{4A.3}$$

Summing equation (4A.3) over all k, and taking the limit (for proof, see Sheppard, 1987):

$$\mathbf{p}' = r/[1 + r].\mathbf{p}'.[\mathbf{I} - \mathring{\mathbf{A}}]^{-1}. \tag{4A.4}$$

Now equations (4A.2) and (4A.4) are two different expressions defining the geography of production, prices and trade that is consistent with full capitalist competition. Equation (4A.2) represents this relationship from an economic point of view; prices everywhere equal production costs incremented by a constant rate of profit. Equation (4A.4) represents this relationship from a geographical point of view. It describes a spatial potential equation, with the spatial distribution of potentials being given by $\mathbf{p}'$ (Sheppard, 1979; 1987). From potential theory, the expression $r/[1 + r].\mathbf{p}'$ can be interpreted as the spatial distribution of prices at the start of the production period, whereas $\mathbf{p}'$ is the spatial distribution of prices at the end of the production period. Each entry in the matrix $[\mathbf{I}-\mathring{\mathbf{A}}]^{-1}$ represents the probability that a commodity produced within a particular sector at a particular location has reached some other sector at some other location by a direct or indirect route (i.e. it was either sold directly from the one firm to the other, or it arrived there indirectly by being first sold to some intermediate firm, incorporated in its product, and then used by the destination firm).

Because $\mathbf{p}'$ is a vector of potentials, we know (as we would expect) that the average direction of flow of commodity trade is from places with lower spatial prices to places with higher spatial prices. Equation (4A.4) states that prices represent a spatial circulation of money resulting from commodity trade (Sheppard, 1987). It also states that, while the relative distribution of monetary value remains unchanged, the quantity of money in circulation is increasing. We start with $r/[1 + r].\mathbf{p}'$, but at the end of the production period we have the larger amount, $\mathbf{p}'$. This describes the increment in exchange value due to circulation, which Marx (1885) expressed as $M - C(MP) \ldots P \ldots C + \Delta C - M + \Delta M$ (see introduction to this chapter). Under full capitalist competition, then, money circulates and accumulates, but the spatial distribution of exchange value remains unchanged.

5   *Reswitching in a space economy*

Introduction

The purpose of this chapter is to provide some examples to elucidate and substantiate the arguments developed in the previous chapters. Section 5.1, based largely on the work of Claire Pavlik (1990), provides an example of pricing and trading patterns under full capitalist competition for a simple case of two regions. This shows construction of trading and pricing patterns and a wage–profit frontier for a particular location pattern; comparison of wage–profit frontiers for different location patterns; and the existence of reswitching in both a closed inter-regional economy and a single region interacting with a larger economy. Section 5.2, based largely on Barnes and Sheppard (1984), examines the internal structure of a region where rents are important in determining location patterns, and again argues for the existence of reswitching.

5.1   An example of reswitching

5.1.1   Background

Suppose that there are two regions, producing four commodities in all: two circulating capital goods (1 and 2), a single wage good (c) and transportation services (t). To simplify matters, assume that: each production period is one year long, there is no fixed capital, and all inputs are paid for and delivered at the start of production. Thus capital advanced equals production costs – the sum of wage costs and the costs of capital goods. We also assume that full capitalist competition prevails, leading to equal rates of profit for all capitalists, implying that the factory gate price charged by each producer in each region is equal to the cost of production incremented by a constant rate of profit. Thus we are essentially examining a two-region version of the framework developed in section 4.1.

We start with knowledge of: the quantity of each capital good and of labour required per unit of the commodity produced (the input–output coefficients) for each commodity in each region; the quantity of each wage good consumed by a worker's family per week in each region; the number of hours worked per week; and the difficulty of transporting goods between regions. For industry n in region A, this implies that we know the following: the quantity of each capital good required per unit produced (a_A^{1n}, a_A^{2n}); the labour time required (l_A^n); the weekly consumption of the wage good by workers in region A (b_A); the length of the working week (H) and the quantity of transportation necessary to ship each commodity m from each region to the other (τ_{AB}^m, τ_{BA}^m). We would need to know this for all four industries in both regions. This is all information that is empirically available if there exist input–output tables for regions, as well as information on prices so that input–output coefficients are convertible into physical units (see section 3.3.2).

From Chapter 4, the factory gate production price is:

$$p_A^n = (1 + r). \left[\sum_{i=1}^{2} \sum_{m=1}^{2} a_{iA}^{mn} p_i^m + l_A^n . \left(\sum_{i=1}^{2} b_{iA} p_i^C \right) \middle/ H + \sum_{i=1}^{2} a_{iA}^m p_i^t \right], \qquad (5.1)$$

where we know that for each capital good the sum of capital good inputs equals the input–output coefficient ($a_A^{mn} = a_{1A}^{mn} + a_{2A}^{mn}$); total consumption equals the sum of wage goods bought from both regions ($b_A = b_{1A} + b_{2A}$); and transportation purchased from region i is the sum of transportation needed to ship each good required from i in the process of production ($a_{iA}^{tn} = a_{iA}^{1n} . \tau_{iA}^1 + a_{iA}^{2n} . \tau_{iA}^2 + b_{1A} . \tau_{iA}^c$). Equation (5.1) states that the production price equals: the cost of capital good inputs (the first expression inside the square brackets) plus the cost of labour (the second expression) plus transportation costs (the third expression), incremented by the rate of profit (i.e. multiplied by $[1 + r]$). If each commodity is produced in each region there are eight such equations (four for each region).

The proportion of each commodity purchased from region i rather than region j depends on prices, and on the efficiency of the trading system:

$$a_{iA}^{mn} = a_A^{mn} \frac{e^{-\beta\{q_{iA}^m\}}}{\sum_{j=A}^{B} e^{-\beta\{q_{jA}^m\}}}, \qquad (5.2)$$

where q_{iA}^m is the delivered price for inputs obtained from region i; equal to the cost in region i plus transport costs ($p_i^m + t_{iA}^m . p_i^t$). There would be six such trading equations, three for each region (for the two capital goods and the consumption good).

This set of fourteen (eight plus six) equations is then solved under the constraint that the rate of profit is equal in all sectors and all regions. Consider the following example, where the production techniques for region A are represented as a matrix as follows:

$$a_A^{11} \quad a_A^{12} \quad a_A^{1c} \quad a_A^{1t}$$

$$a_A^{21} \quad a_A^{22} \quad a_A^{2c} \quad a_A^{2t}$$

$$a_A^{c1} \quad a_A^{c2} \quad a_A^{cc} \quad a_A^{ct}$$

The first column represents the inputs of the two capital goods and the wage good required per unit of the first capital good produced, the second column represents the technology used to produce the second capital good, and the third and fourth columns represent, respectively, inputs for the wage good and the transportation commodity (t).

For the numerical experiments reported on here there are two combinations of techniques used in a region. Combination a is:

0.089	0.0	0.780	0.010
0.010	0.022	0.018	0.010
0.005	0.018	0.018	0.010

and combination f is:

0.080	0.0	0.780	0.010
0.038	0.022	0.018	0.010
0.009	0.018	0.018	0.010

Notice that, in this case the two differ only in terms of the production method used for the first capital good. More dramatic differences can of course also be used; these particular coefficients are chosen so that they are as similar as possible to a previous numerical example developed for a spaceless economy (Garegnani, 1966). This allows a direct comparison of the spatial and spaceless cases. Readers can compare the results here with Garegnani's example to see the difference that space makes.

The amount of transportation required (τ_{ij}^m) is assumed to be the same for all commodities, and equal to 0.5 for intra-regional, and 5.0 for inter-regional trade. These values are further multiplied by a scalar, t^*, which can be varied to indicate different degrees of transportation difficulty. t^* is varied between 0 (zero distance friction) and 2.5. The parameter β of equation (5.2), the index measuring the efficiency of trading patterns (see Chapter 4), is also varied between 0 and 0.2.

For any given real wage level, combinations a and f are assigned to the two regions, values for β and t^* are chosen, and the trading and pricing equilibrium is determined, giving production prices, trading patterns and the rate of profit. Wage–profit frontiers are then constructed by varying the wage level from 0 to the maximum possible, in each case solving the equations to calculate the new rate of profit. Wage–profit frontiers are then constructed for both a and f, and compared to see whether reswitching occurs.

5.1.2 A two-region economy

We investigate whether there is reswitching; i.e. whether one technique is more profitable for high and low wages whereas the other is more profitable at intermediate wages. If this is the case, it implies that no simple rules can be used to predict which location pattern is the more profitable (see section 4.1.3). Suppose

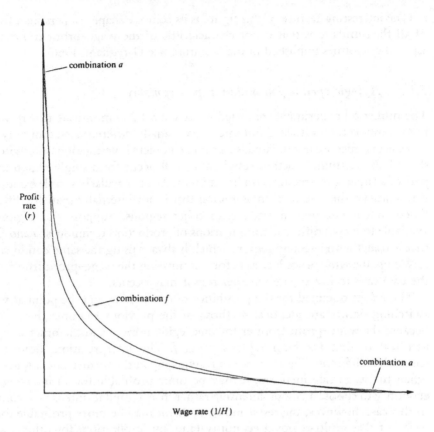

Figure 5.1   Example of reswitching in a space economy

initially that all commodities are produced in both regions, and that for each commodity the same technology (combination *a*) is used in both regions. The wage–profit frontier constructed from solving for trading and pricing equilibrium at different wage levels, with $\beta = 1.0$; $t^* = 0.095$, is shown in Figure 5.1. A second situation was constructed under the assumption that technique *f* is in use in both regions. This led to a second wage–profit frontier, also depicted in Figure 5.1.

Reswitching is indeed present in this case, shown by the fact that the two wage–profit frontiers cross one another twice. Because there is reswitching, one of these shifts in the rank ordering of the profitability of the two alternatives must seem perverse from the neoclassical perspective. This experiment was in fact repeated with different values for β and t^*, and in every case but one ($\beta = 1$; $t^* = 0.2$) there was reswitching. Given the very small difference in the two combinations employed (implying considerable spatial homogeneity), and the simplicity of just using two regions, it is safe to conclude that reswitching and capital reversing are a real possibility in a multi-regional economy.

One interesting feature of this figure is its concave shape – a persistent feature of all the simulations that is not characteristic of the wage–profit frontiers for aspatial economies published in the literature (see Garegnani, 1966).

5.1.3　A single open region within a space economy

The numerical experiments reported in section 5.1.2 demonstrate that reswitching is possible for an isolated but internally spatially differentiated economy. Yet space economies are not isolated, so a more complete investigation of reswitching should also examine whether reswitching will occur for a single region that is part of a larger inter-regional trading system. As a particularly simple example of this, consider the case of a single region that is not internally spatially differentiated, but is engaged in trade with other regions. Suppose this region has available to it two different combinations of production techniques, a and f, and that it trades with a second region, which is always using the same combination, a. We use the same procedure as before, comparing the wage–profit frontiers for the two cases to investigate whether reswitching occurs.

The results obtained for the possibility of reswitching and the point at which switching occurs are identical to those of the previous example (Figure 5.1), because the wage–profit frontier for one region using a and the other using f is identical to that for both regions using f. The interpretation, however, is somewhat different. The existence of reswitching in the former case implies that, when the wage rate increases, it may be more profitable for an inter-regional economy to choose a labour-intensive rather than a 'capital'-intensive technique. In this case, however, the result indicates that it may be more profitable for *one region* of that multi-regional economy (and, by implication, for other regions also) to adopt a more labour-intensive technique as the general wage rate in the economy increases. This second case is important because it provides a counter-example to conventional neoclassical wisdom concerning both the effect of labour and capital migration on aggregate regional economic growth, and the benefits of specializing in producing commodities that intensively use locally abundant production factors (see section 5.3).

5.2　Intra-regional location and reswitching

The examples examined above, while revealing, do not exhaust the different situations in a space economy where reswitching is relevant. Two features in particular are absent. It was not demonstrated that the examples of reswitching represent cases where commodity production is occurring at the most profitable locations, because we did not exhaustively compare different location patterns in the way suggested in section 4.1.3. Thus it might be argued that reswitching occurred only because industries were not optimally located in space. Second, the internal spatial structure of the regions was ignored. In order to extend our analysis to take partial account of these issues, we will consider in this section a simplified intra-regional space economy, highly analogous to a von Thünen model. This example should also be seen as a prelude to more detailed discussion of the role of land rent in Chapter 6 and urban land-use patterns in Chapter 7.

Suppose there is a single marketplace at the centre of an isolated urban-centred region, with land-consuming production activities distributed across a uniform plain surrounding this point. At this intra-regional scale, land rent plays an important role in determining which activities are to be found at which locations. Land rents are paid to the owner of a site by the capitalist located there, and represent an additional payment over and above those made to purchase capital good inputs and hire labour. Therefore if we define the *net income* accruing to a capitalist as the difference between total revenue and the cost of capital good inputs, this net income must be divided three ways; into land rent, wage costs and profits (taxes, dividends and other drains on profits are ignored).

Suppose that there are two kinds of land-consuming economic activities, each producing a different commodity – land-use 1 located within distance D of the market, and land-use 2 found between distance D and distance D^* from the market (Figure 5.2). Depending on the scale of analysis, this might represent agricultural production around a market town, or the intra-urban location of industrial activities about a central business district. In either case, however, we assume that all inputs required are purchased from the centre, and that all commodities produced must be shipped there. This idealized landscape is chosen in order to simplify the spatial structure of production and consumption in our region to a measure of distance – distance both from the marketplace and from

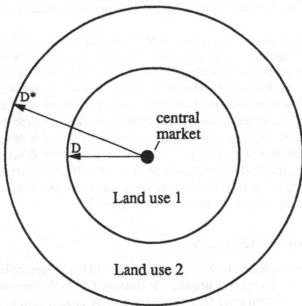

Figure 5.2 A single centred region

other producers located in the same direction from the marketplace. In short, we simplify from a fully two-dimensional spatial structure to a spatial structure that can be represented as a one-dimensional line. We contend that if reswitching exists in this simple situation then it will also occur in more complicated intra-regional spatial configurations of markets, production facilities and transport routes.

A full analysis of this situation is provided by Barnes and Sheppard (1984), and we repeat only the bare bones of the argument here. As before, we assume that full competition among capitalists ensures that the rate of profit is equal for all producers of both types of commodity in the region. For capitalists to be competitive they have to be able to deliver the product to the market at the same price as other producers of the same commodity. Clearly those located further from the market face a potential locational disadvantage because transport costs for inputs purchased from, and commodities sold to, the market are higher. If the nearby producers can manufacture an unlimited quantity of the product, then more remote producers could never compete, and thus never would produce the commodity. If there is a limit on the quantity that can be produced per acre with the given technology, however, then, when demand exceeds the capacity of nearby producers, more remote locations are required. The price in the central market must then rise so that the more remote producers can also obtain the going rate of profit on their investment. Such a price rise gives closer producers a chance for excess profits, but then landowners will charge rent for superior locations. As in von Thünen's framework, the rent charged in a fully competitive land market will reduce profit rates for nearby producers to the same rate of profit as is made by the farthest producer whose output is required to meet market demand.

Land rent plays an important role in this intra-regional case for two reasons. First, there are capacity constraints, meaning that a single location in a region cannot produce unlimited quantities of the product on a fixed quantity of land, allowing other disadvantaged locations to begin production. Second, the market is so localized that all producers of the same commodity must sell their product at the identical price in order to obtain a share of the market. Implicit in this analysis is that, at the inter-regional scale, capacity constraints are of little relevance and markets need not operate perfectly. In order to simplify the following argument, we assume that capital advanced equals production costs, that both commodities produced are capital goods, that wages are fixed at w per hour, and that no profits are paid on wage costs (see note 1, Chapter 3).

5.2.1 Fixed transportation costs

Let us define the price at which commodity n is sold in the central market as p_n^*. For a capitalist producing commodity n at distance i from the central market, the factory gate production price (p_i^n) equals the cost of production, plus the rate of profit, plus land rent. Production costs equal the factory gate cost of commodity

n, plus the cost of any input of another commodity, m, required from the central market, plus the transportation costs in both bringing commodity inputs, m, to where n is produced and shipping n to the market. This sum is then incremented by the rate of profit, to which is added wage and land rent costs. Mathematically:

$$p_n{}^* = [\hat{a}_{nn}p_i^n + \hat{a}_{mn}p_m{}^* + \{\tau_i^n + \hat{a}_{mn}\tau_i^m\}p_t](1 + r) + w.l_n + R_i\phi_n \qquad (5.3)$$

where $\hat{a}_{mn}$ is the quantity of good m required as a capital good per unit of n produced, τ_i^n is the amount of transportation required to ship a unit of commodity n, ϕ_n is the acreage of land required per unit of n produced, and R_i is the land rent per acre at distance i.

The differential rent paid by a producer at distance i, R_i, can be endogenously calculated as the savings in transportation costs relative to those paid by the outermost (marginal) producer of the same commodity (Barnes and Sheppard, equations (6)–(10)). These are differential rents. Land is allocated to the activity paying most rent at each location, implying that each commodity producer is optimally located in the sense that landlords maximize rent payments (Figure 5.3).[1] Rents are proportional to the cost of transportation. If this cost increases (as a result of an increase in either the price of transportation or the effort required to transport goods, τ), the slope of each rent curve in Figure 5.3 increases, making rents higher at every location closer to the market than distance D^* (Figure 5.4).

An analysis of this model shows that there is an inverse relationship between each of the three components of economic surplus; wage rates, profit rates and rent levels. The relationship between the wage rate, w, and rent levels, R, is linear

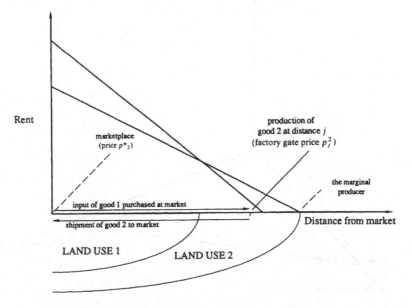

Figure 5.3 Production around a central market

... plus the cost of any input of another commodity as required from the central market, plus the transportation costs in both bringing commodity inputs to ... where it is produced and moving ... to its market. This sum is then divided by the area of production to obtain realized wage and land rent costs. Mathematically,

$$c_{z,d} = [p_z - (w a_z + \sum_i c_i p_i)]/A_z = w(1 - a_z) + Z_z \qquad (5.3)$$

where $c_{z,d}$ is the quantity of good z required at ... area ... per ... product ... produced; A_z is the amount of transportation ... ; a_z is the average ... land rent per ... of a product; and Z is the land rent per acre at distance ...

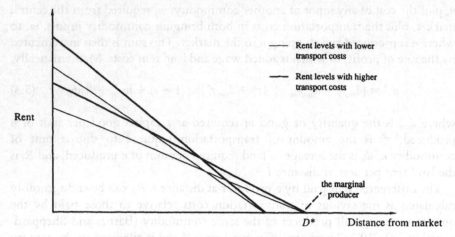

The difference in land type at each distance d can then be directly calculated as ... location ... transportation costs relative to those paid by the ... producer at the ... of the same commodity (Barnes and Sheppard equations (4)–(10)). ... expected to be accruing to every unit of land at each location, supplying that ... commodity, produced is spatially located in the same that land bids maximize rent (reference Figure 5.5). Rents are proportional to the ... of transportation ... this can increase (a ... or an increase in either the price of transportation or the effort required to transport goods. In the slope of each rent curve in Figure 5.4 increases, leaving land at every location closer to the market to obtain ... rent D^* (Figure 5.4).

Sir Thomas More's model applies the ... relationships between each of the three components of cost ... supply, wage rate, profit rate and rent levels. The relationship between the wage rate, w, and rent levels, R, is linear

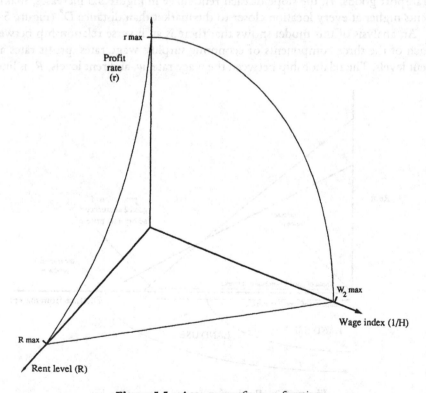

Figure 5.4 Transport costs and differential rents

Figure 5.5 A wage–profit–rent frontier

because the quantity of both of these inputs can be expressed in units of measurement that are given exogenously (i.e. the wage level and the price of transportation). The relationships between profit and wages, and between profit and differential rent, are, however, non-linear because the profit rate depends on commodity prices. An example of the resulting three-dimensional wage–profit–rent frontier is given in Figure 5.5, where for simplicity the non-linear relationships between profits and rents, and between profits and wages, are shown as simple curves. This shows that if landlords increase rents, then wages or profits must fall, implying a conflict between the interests of landlords and those of capitalists and/or workers.

If two methods of production, A and B, are available for one of the commodities, and if Figure 5.5 represents the wage–profit–rent (w-r-R) surface when technique A is employed, then a different wage–profit–rent surface will generally occur if technique B is used. This is in part because a different production technique for one of the commodities may lead to different land–use patterns, and in part because for a given level of wages and rents the rate of profit will differ. It is clearly possible that the w-r-R surfaces for the two techniques might intersect in such a way as to make reswitching possible, because the wage–profit–rent surfaces are non-linear in two of their three dimensions. An example of this is drawn as Figure 5.6. A similar conclusion was reached by Metcalfe and Steedman (1979b) – albeit without examining the relationship between location and rent levels. In Figure 5.6, for low rent levels technique A is more profitable than technique B at both low wage and high wage levels, whereas B is more profitable at intermediate levels.

5.2.2   Endogenous transportation costs

In the above analysis we assumed that the price of transportation was exogenously given. Yet, as suggested in Chapter 4, transportation is a commodity like any other, whose production price equals the cost of producing the transportation commodity incremented by a rate of profit. Profit-seeking freight shipping companies produce the commodity of accessibility. They do this by applying labour to inputs in order to overcome spatial barriers for their customers. This element of realism can be added to the framework by assuming that transportation is produced at the marketplace, using as inputs the commodities m and n available for purchase there. The price of production of transportation then equals the cost of inputs purchased at the central market incremented by the rate of profit, plus wage costs:

$$p^t = [a^{mt}p^{m\star} + a^{nt}p^{n\star}](1 + r) + w.l^t \tag{5.4}$$

where a^{mt} is the quantity of m required to produce a unit of transportation. It is important to recall that what is being produced here is not the *means* of transportation, but shipping services. In the same vein, the price of transportation, p^t, is the price of movement.

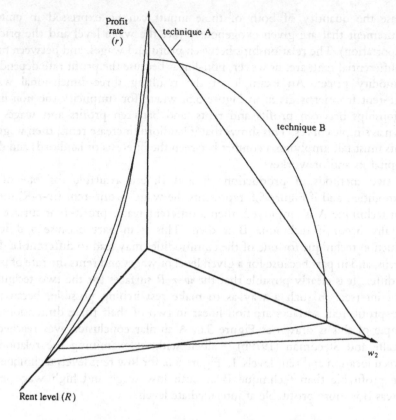

Figure 5.6   Reswitching: transport costs exogenous

Summarizing the analysis of this case from Barnes and Sheppard (1984, section 8), the relationship between differential rent and the wage rate is no longer linear, implying that the wage–profit–rent surface is now non-linear in all directions. This increases the likelihood of reswitching when two techniques of production are available. A hypothetical example of such reswitching is provided in Figure 5.7, but many other possibilities can be envisaged.

In landscapes where the physical separation between producers and markets is greater, the terrain more difficult, or the efficiency of the transportation system lower, then overall differential rent levels must be greater for given profit and wage rates, because the total cost of transportation is higher. This is because it takes more effort to produce the accessibility necessary to make commodities available. Reswitching between techniques can then occur as the friction of distance increases, as suggested by the horizontal plane of Figure 5.7. The possibility of reswitching as the friction of distance increases is superficially similar to the case of spatial reswitching suggested by Scott (1980, p. 52). Scott defines spatial reswitching as a situation where a particular land use is most profitable both near to and far from the market, with another land use being more profitable in between (see also Hartwick, 1976; Schweizer and Varaiya, 1976, 1977). This is also possible in the model developed here, under certain conditions specified in Chapter 6. Yet the type of reswitching discussed in this chapter is conceptually different since it refers to aggregate possibilities faced by a region, which is how reswitching is defined in capital theory. By contrast, in Scott's analysis reswitching refers to decisions taken by individual commodity producers.

5.3 Inconsistencies in neoclassical economic geography

We now use the findings from this chapter to critically re-examine conventional wisdom about aggregate economic relationships in the space economy as they have been imported from the neoclassical tradition. Before doing that, however, we want to summarize these findings briefly. The existence of reswitching in a spatially extensive economy has at least the following implications. First, if wage levels are high at some location, then it is not necessarily more profitable to adopt a capital-intensive method of production there. Secondly, in a space economy

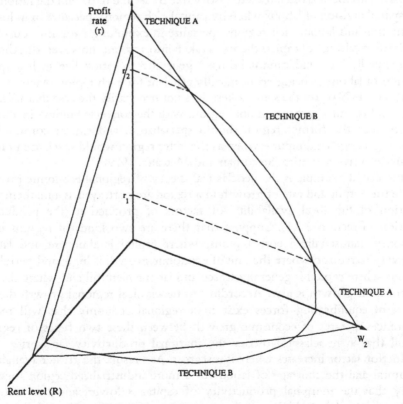

Figure 5.7 Reswitching: transport costs endogenous

with high friction of distance, it is not necessarily most profitable to adopt a method of production requiring a lower transportation input.[2] Thirdly, as a corollary, the level of differential rent at a location is not a measure of the marginal productivity of land at that location, just as wage and profit rates are not equal to the marginal productivity of labour or capital.

Such results have significant implications for some commonly accepted ideas in economic geography. To explore these implications, we will examine four common theses propounded in neoclassical economic geography in the light of reswitching: comparative advantage and trade, regional growth equilibria, the notion of the highest and best use for a plot of land, and the welfare implications of allowing individuals to make self-interested location decisions.

Consider first neoclassical trade theory, which centres on a thesis originating with Ohlin (1933) and subsequently extended into the so-called Hecksher–Ohlin–Samuelson theory of trade and comparative advantage. According to this theory, if there are two regions, one with a low profit rate and high wage levels and the other with high profits and low wages, then the former region should specialize in the production of capital-intensive commodities and the latter region in labour-intensive commodities, and they should trade their surpluses with one another. This thesis is often used to justify free trade, and to defend the rationality of a spatial division of labour whereby capital-rich regions specialize in industrial production and labour-rich regions specialize in exploiting natural resources. If the Sraffian critique of capital theory is taken into account, however, this thesis is not generally true and cannot be used generally to rationalize such a spatial division of labour as being economically efficient for each region (Metcalfe and Steedman, 1979a). By the same token, it is not necessarily the case that if land is cheap and labour expensive in one region, with the converse holding in another region, then the former region should specialize in and export commodities requiring a large land input – whereas the latter region would specialize in more labour-intensive activities (Steedman and Metcalfe, 1979).

The second example is the neoclassical theory of regional economic growth, where the output and rate of growth in a region are portrayed as a mathematical function of the local availability of factors of production (the production function of section 2.1.1). Suppose that there are two kinds of regions in an economy: industrialized core regions, where capital is abundant and labour relatively scarce and where the rate of economic growth is high; and peripheral regions where capital is generally scarce and labour plentiful and where the rate of economic growth is low. According to neoclassical regional growth theory, powerful equilibrating forces exist in a regional economy that will reduce differences in rates of economic growth between these two types of regions. Recall that in neoclassical theory the marginal productivity (and price) of a production factor increases where it is scarcer. As a result, the greater availability of capital and the shortage of labour in a more industrialized region together imply that the marginal productivity of capital is lower, and the marginal productivity of labour higher, in that region than in the periphery. It follows that profit rates must be higher in the periphery than in the core whereas wage rates are higher in the core than the periphery. It is therefore argued that, if factors are allowed to migrate freely between regions by encouraging market forces, capital will flow from the core to the periphery while labour flows from the periphery to the core. Over time this will reduce the initial differences in capital and labour endowments between the regions, bringing about a convergence in growth rates.

One consequence of recognizing the existence of reswitching is that the price of a factor does not equal its marginal productivity. Therefore, it is very possible that a region using more capital and less labour in fact experiences a higher rate of profit on investment than a labour-rich and capital-poor region. When this occurs, capital will flow from the capital-poor to the capital-intensive region. It is possible, therefore, that a freely operating capital market will attract more capital investment to industrialized core regions rather than less, accentuating regional imbalances instead of encouraging inter-regional equilibrium.

The third example entails a critical re-examination of the common thesis in land-use theory that when land uses are allocated so that rents are maximized at each location, then each location will have been given over to its 'highest and best use'. This assertion is directly based on the assumption that rent at any location is a measure of the marginal productivity of that piece of land for the economic activity located there. If this were true, it clearly follows that the activity paying the most for a plot of land must use it most productively. A corollary of this is that an unrestricted land market will allocate land in the most efficient manner by ensuring that it goes to the highest bidder (the most productive activity). Once we accept, however, that rents are not equal to marginal productivity, then this argument is no longer valid. Indeed, as we shall see in Chapter 6, rents reflect a whole series of other factors. This implies that the land market is a problematic device for determining appropriate land uses.

Finally, and perhaps most importantly, it is generally assumed in neoclassical economic geography that, if each economic actor makes those locational decisions that are most beneficial for him/her within a freely operating market, then the aggregate result of all such actions is an optimally organized space economy. We shall see in subsequent chapters how the results of the reswitching debate imply that the actions that seem to be best at one point in time in fact have consequences that may be the opposite of those intended by the actor. Thus a location decision that may appear to increase profits for an individual capitalist may have unintended ramifications that reduce the general profit rate for capitalists as a whole. One message, central to the neo-Ricardian critique of neoclassical theory, is that the interdependent capitalist economy with its anarchic market mechanisms is so complex that simplified rules of behaviour tend to be misleading. It is frequently difficult to tell what the consequences of an action will be, even if the actor is far-sighted enough to wish to consider long-run consequences. Furthermore, this complexity increases when we introduce space (Chapter 12).

While it is not yet clear how generally these counterintuitive conclusions of the neo-Ricardian analysis occur in practice, we clearly cannot just ignore them and continue to draw on accepted wisdom simply because that is how we have been taught to think. Neoclassical macroeconomic theory rests on some unrealistic assumptions that are critical to the theory in the sense that when these assumptions are relaxed the theory loses its coherence. It is ironic to note that, since the dawn of marginalist thinking in economics, Marxist economic theory has been vilified for some logical inconsistencies in Marx's theory of the relation between labour values and prices, known as the transformation problem. By contrast, neoclassical theory was argued to be internally consistent and thus a more scientific theory. Yet one indisputable conclusion of the neo-Ricardian critique is that the assumptions necessary to make aggregate neoclassical theory internally consistent are identical to those necessary to make Marx's transformation of labour values into prices consistent (Harcourt, 1972, p. 145). In an economy where every commodity is produced by exactly the same technology, Marx's solution to the transformation problem is correct, and neoclassical theory is also consistent since reswitching is impossible. In every other situation Marx's solution is inaccurate, but aggregate neoclassical theory is also inconsistent because of the possibility of reswitching. In short, aggregate neoclassical theory has been hoist on its own petard.

Summary

We took two examples of commodity production in a space economy that demonstrate the conflict of interest among capitalists, workers and landlords. The case of a two-region space economy was analysed to show that, when production prices prevail as described in Chapter 4, then a trading and pricing equilibrium can be calculated, with a wage–profit frontier demonstrating conflict between capitalists and workers. It was shown numerically that reswitching is plausible for the entire multi-regional economy as well as for a single region within that economy. It then follows that there is no simple relationship between the type of technique that maximizes profits and wage and profit levels. The second case examined was an internally spatially differentiated but isolated region where differential rents determine the geography of production, in a manner superficially analogous to von Thünen's model of agricultural land use. The presence of landlords means that surplus is in this case divided into three parts, implying a three-dimensional wage–profit–rent frontier and a three-way conflict of interest. The possibility of reswitching in any of the three dimensions of this frontier (Figure 5.7) adds to the difficulty of finding any simple principle by means of which the profit-maximizing technique or location pattern can be determined.

Using the existence of reswitching to re-examine some commonly held beliefs in economic geography, we see that a number of widely accepted ideas are not generally true. These include the arguments that openly competitive markets:

allow the development of maximally efficient specialization and trading patterns by following the procedures of comparative advantage and free trade; bring about convergence in growth rate differentials between regions; enable land to be used most productively; and enable efficiency and profitability to be maximized in the space economy by allowing all economic actors to pursue actions that maximize their own individual economic gain (the hidden hand).

These conclusions have important theoretical implications. Of particular importance is that there is no fully competitive geographical pattern of capitalist commodity production that is in the interest of all social classes. This implies that, if capitalists wish to develop a location pattern and a set of production methods to maximize their profits as a group, then they must be able to agree on what that location pattern might be. Theoretically, agreement might be arrived at in one of two ways: by collusion, whereby the optimum is determined and all capitalists agree to work together to achieve it; or through individual action, whereby the actions taken by individual capitalists in their own interest happen to converge on the configuration that is most profitable for capitalists as a class. The former is implausible, both because simple rules for identifying the best configuration (of the kind sought by neoclassical theorists) are at best difficult to identify, and because capitalists are in competition with one another. The second approach, which is similar in spirit to Adam Smith's hidden hand, depends on whether the interests of individual capitalists are in the long run consonant with the interests of capitalists as a class. We are sceptical that this is the case, for reasons to be developed in Chapter 12. Needless to say, workers and landlords face the same difficulties in realizing a space economy that is optimal in their view, coupled with the additional difficulty that they possess less economic power under capitalist social relations. In short, these properties lay the logical foundation for a view of a capitalist space economy as unstable and under the influence of political conflicts stemming from contrary economic interests.

Notes

1 The rate of profit achieved, however, may not be maximized owing to class conflict between landlords and capitalists. See Chapters 7 and 11.
2 Note that 'method of production' refers here to a method adopted for the production of a single commodity, in the absence of changes in methods of production for all other commodities.

6 *Incorporating natural resources: rent theory*

Introduction

The relationship between nature and society is central to geography in general, and economic geography in particular. A subset of that wider enquiry, and one with which we are concerned in this chapter, is the relationship between natural resources and economy – an aspect of economic geography that is often neglected in theoretical work. We examine such a relationship within the framework of the production-based model constructed in Chapter 4. In so doing we both extend that model and also demonstrate its flexibility.

Historically, there are two theoretical visions of the relationship between 'nature' and the economic process, both originating in economics. The first is the neoclassical view, which begins with a pristine 'nature' and its presumed inherent scarcity. The second, adopted in this book, is a political economic perspective, which starts with nature already transformed (i.e. socially produced from the outset), and scarcity established by sets of broader social relations. The central cleavage between these two approaches is in their conception of nature: for neoclassical economics nature exists independently of society, while for the political economic approach nature is conceivable only within a broader complex of social and economic relationships. For our purposes, the significance of this conceptual difference is in shaping the theoretical framework in which the issues of natural resource scarcity and rent are conceived and analysed. Specifically, we argue below that, by beginning with a view of nature as independent of society, neoclassical economics is led to conceive scarcity; and even rent itself, as a product of 'natural forces'. In contrast, by viewing nature as socially constructed, the political economic approach conceives scarcity and rent as precipitates of a set of social relationships rather than natural ones.

We begin by contrasting neoclassical and political economic theoretical visions of nature and scarcity. This is followed by an analysis of the forms of differential rent respectively proposed by Marx and Sraffa. This discussion is then elaborated by explicitly introducing the effects of transportation costs on rents. After demonstrating that our analysis applies to all natural resources including land, the chapter concludes with an examination of absolute/monopoly rent.[1]

6.1   Nature and scarcity

Neil Smith (1984, ch. 1) argues that philosophers and social scientists historically treat the relationship between society and nature in terms of a duality. Nature and society are viewed as distinct and separate entities, each with their own respective attributes and laws. This dualistic view is clearly evident in neoclassical economics and in the part of economic geography that accepts its precepts. For in the neoclassical conception the central fact about nature is its inherent finiteness, while its comparable claim about society is that individuals possess infinite needs that are met only by the consumption of resources (see Chapter 1). With nature finite, and infinite demands made on it by an insatiable society, neoclassical economics is necessarily driven to make the central problem of the economy one of scarcity; it is the problem to which all economic relations are directed (Rowthorn, 1974; Hodgson, 1982). It follows that the role of prices in the neoclassical scheme, including the price of land, is to serve as particular indices of a presumed generalized dearth. In so doing, prices fulfil a major function within the neoclassical scheme. Although the rent (price) on land and natural resources does not obviate the inherent scarcity of nature, it ensures, through Adam Smith's invisible hand of the market, that nature is used in the way that best satisfies society's infinite needs. In this sense, rent as an index of natural scarcity reconciles the finiteness of nature with the infinite demands placed upon it by different individuals (Matthaei, 1984).

Although this is the general neoclassical vision of nature and scarcity, we need to address the specific issue of rent determination. In neoclassical economics, rent on natural resources is like any other price. Specifically, all prices reflect the scarcity of the commodity: as the scarcity of a good increases, so does its price. By constructing a neoclassical model in which there are only two factors of production, land/natural resources and labour, Barnes (1988) formally demonstrates that for this simple case rental levels indeed vary 'correctly' with different levels of 'natural' scarcity: rents are more as the supply of land is less. But, perhaps more importantly, Barnes (1988) also shows that such a conclusion is *not* generalizable to the case where capital inputs are included, for reasons that parallel the critique of profits as an index of the marginal productivity (scarcity) of capital (see Chapter 2). In particular, because of a logical contradiction first revealed in the capital controversy, there is no general systematic relationship between land/natural resource scarcity and rent once produced means of production are included. Such a conclusion is very damaging, for it clearly raises doubts about both the natural basis of scarcity, and the idea that rent is a faithful index of it (for details, see Barnes, 1988). The naturalism of neoclassical economics is thereby severely compromised.

Neil Smith (1984), however, recognizes a second approach to society and nature originating in Marx's work (see also Burgess, 1976, 1985; Smith and O'Keefe, 1985). This view denies any duality between society and nature. Rather, as Smith and O'Keefe (1985, p. 80) write: '[A] nature separate from society had no meaning for Marx; nature is always related to societal activity. He meant this materially as well as ideally; the entire earth bears on its face the stamp of human activity.' For Marx, nature is not an asocial given, but takes on significance only in so far as it is embedded in a particular set of social and economic imperatives – such as capitalism. The implication is that the separation between society and nature disappears.

As with the case of neoclassical economics, we need to show how this general political economic vision is translated into the determination of specific rental levels on natural resources. To do this, however, we need first to say something about the peculiar characteristics of natural resources within the political economic approach.

In contrast to the neoclassical scheme, the prices (rents) of land and natural resources in the political economic approach are *unlike* other commodity prices. For the general system of production prices in a political economic perspective is based on reproduction costs; one where price levels of produced goods are set so that they cover the costs of production (input costs, wages and profits), thereby enabling the commodity to be *reproduced* in the future. With non-produced goods such as natural resources, however, there are no reproduction costs. Land, mineral deposits and forests cost nothing when first 'produced', and they are not easily reproduced, if at all. But if there are no (re)production costs for natural resources then two questions arise. First, what determines rental levels? (Rents cannot be equal to costs of production because there are none.) Second, and more fundamentally, why should natural resources receive a rent at all? (If there are no reproduction costs there is seemingly no justification for levying a price.)

These two questions are not as intractable as they appear, however. There are good explanations within the political economic approach as to the existence of rent, and the determination of its level, and these explanations lie precisely within Marx's broader conception of nature and society discussed above. For Marx is arguing that to understand natural resources, including the issues of scarcity and rent, we must examine the broader social formation in which such resources are exploited. In that light, we will argue below that, first, the existence of rent is quite understandable once set within the wider context of capitalism's social relations of *production*, and, second, rent levels are readily defined once set within the context of capitalism's class relations of *distribution*.

To address the first question of why rents exist, we begin with the origin of that payment, which is in the sphere of production. As Leitner and Sheppard (1989, p. 67) write:

> rent is first and foremost a result of production. It is through production that a surplus is made in the space economy, and it is from this profit that rents must be paid to owners of land. ... [R]ent theory [within the political economic approach] therefore proceeds by grounding rent within the process of production, and thereby the political and social forces immanent in that production process.

That resource owners are able to appropriate a portion of the total surplus to which they did not contribute is a result precisely of the 'political and social forces' that Leitner and Sheppard note. Through the institution of private property, resource owners have collective control over a necessary means of production, and are therefore able to exact a levy for its use. For this reason, rent is not the gift of nature but stems directly from the imperatives and social relations of the broader production system in which nature is enmeshed.

Second, the levels of rent are established on the basis of scarcity, but, in accordance with Marx's broader view, scarcity is socially created, not natural. As Perelman (1979, p. 84) writes:

> Marx does not treat scarcity as an independent category, but in relation to the mode of production, i.e., to the historically specific set of relations and forces of production, ... [The scarcity] of natural resources must be seen within this context.

In particular, in the next section we will demonstrate that the scarcity of natural resources (which sets rental levels) is determined by class conflict over distributional shares of the surplus. For this reason the very definition of scarcity and rental levels is inseparable from the wider structural issues of class formation, power and struggle. One cannot in any sense deduce levels of scarcity from the laws of an independent nature.

We can now see in what sense the notions of the rent and scarcity of natural resources within the political economic tradition emerge from Marx's insistence that nature is understandable only as part of a broader social system. First, rent is conceivable only as an aliquot part of the surplus originating within the wider production system; and second, the scarcity of natural resources and land is only defined by the conflict among classes within the broader social formation. Furthermore, by drawing upon Marx's conception of nature and society, the political economic tradition is able to incorporate consistently non-reproducible commodities into its analysis. Although such prices are not established on the basis of reproduction costs, the forces that create and determine rent are the very same ones that a system of prices based on reproduction costs makes pivotal, namely, the relations of production and distribution (Chapter 1).

In summary, we tried to link here the issues of scarcity and rent of natural resources to broader geographical concerns about nature and society. We argued that embodied within neoclassical economics is a dualistic view of nature and society manifest in its thesis that scarcity is 'natural' and rent the index of it. Although one may wish to take issue philosophically with this dualistic view, we suggested that a more immediate problem with neoclassical economics is that its vision of scarcity is incompatible with its own formal analysis of rent. For once produced goods with a positive rate of profit are included in the neoclassical model, there is no necessary correspondence between scarcity and rental levels. Because of this formal inconsistency we will not discuss the neoclassical approach further. Rather, the rest of the chapter is concerned with formally establishing some of the broader claims we made above for the political economic approach. In particular, we will show that a penetrating analysis of scarcity and rent can be provided by starting with the social relations of production and distribution. In such a scheme, scarcity is defined within a set of broader class relationships, and rent is defined within a broader system of production. Both these claims are demonstrated analytically in the next section where we focus on two traditions of political economy: Marxism and neo-Ricardianism.

6.2   Rent and scarcity within political economic theory: two views

Although Marxists and neo-Ricardians both approach issues of land and natural resources from the political economic perspective, they propose quite different mechanisms for the actual determination of scarcity and rent. To establish these differences, we need to say something about the formal derivation of prices and labour values presented in Chapter 4. Recall that to determine either prices or labour values there must be a separate equation for each good specifying the physical costs of its production. But when we deal with non-produced goods, such as land and natural resources, 'there is by definition no cost-of-production equation; the system of price or value-determining equations is indeterminant and must be closed by some other means' (Gibson and Esfahani, 1983, p. 86). This is just a formal way of expressing the point already made in the preceding section, namely, that the peculiar characteristic of natural resources is their non-reproducibility. Because there are no cost of production equations for this class of commodities, their prices must be set by other means. What divides Marxists and neo-Ricardians on the question of land and natural resources is precisely the means by which such prices are set. Formally, the schools are divided over how to specify the social conditions creating scarcity, which, in turn, closes the price/value equations and determines rental levels on non-produced goods.

Marxists usually argue that it is the power of landlords as a class in limiting access to land that creates scarcity. Neo-Ricardians, in contrast, argue that it is only the conflict between workers and capitalists, through defining the technique of production, that determines scarcity. We will suggest below that there is room for compromise between these two positions. We will argue that, on the one hand, the neo-Ricardian rent theory is improved by linking it to Marx's conception of social relations – as currently conceptualized the neo-Ricardian theory undertheorizes social class and conflict. On the other hand, Marx's rent theory is made analytically 'tighter' by drawing upon what Dobb (1975–6, p. 468) calls Sraffa's *anti-kritik*. More generally, we suggest that the neo-Ricardian and Marxian positions are not mutually exclusive, but rather mutually supportive.

The discussion of land rent in this section will be carried out with respect to an idealized economy with the following characteristics: (1) the only natural resource discussed for the time being is land;[2] (2) all production is capitalist; (3) capitalist farmers lease land from landlords and hire workers; (4) there is only one market price for each commodity; (5) only one agricultural crop is grown, corn; and (6) profit and money wage rates are equalized for all lines of production.

6.2.1   Marxist rent theory

It is well known that Marx left his theory of land rent in, at best, an incomplete form. This is despite the centrality of rent both as a distributional variable and as a means to understand the spatial organization of capitalism (Harvey, 1982; Katz, 1986). The result has been a number of quite different attempts to reconstruct Marx's rent theory. None the less, a consensus has emerged among recent writers on the preconditions necessary for land rents to be levied, namely, the power of landlords as a social class (Ball, 1977; Fine, 1979; Katz, 1986). In terms of our earlier discussion, this thesis is critical because the landowner's monopoly power represents the means of 'closure' of the set of price/value equations describing the non-produced commodities. The precise way in which monopoly power determines scarcity and thereby rents varies with the particular kind of land rent discussed, of which Marx recognized two main kinds: differential rent (DR) and absolute/monopoly rent. Katz (1986, p. 67) argues that these can be distinguished as follows: absolute/monopoly rent expresses the power embodied in particular individuals, while 'differential rent expresses the monopoly power of capital as a whole'. In this section we focus on the power of landlords as a class in determining differential rents I and II (DR I and II), postponing discussion of absolute/monopoly rent until later (section 6.5).

DIFFERENTIAL RENT I
For Marx (1959, p. 650), DR I occurs because of the 'unequal results of equal quantities of capital applied to different plots of land'. Given that equal capital – labour ratios (organic compositions of capital) imply that prices are proportional to labour values (Dobb, 1973, pp. 155–7), it follows that differences in output among lands where equal quantities of labour and capital are applied can be the result only of differences in fertility, and not of differences in capital intensity. In terms of the wage–profit curves constructed in Chapter 4, each plot of land is associated with a straight-line wage–profit frontier (Figure 6.1; Kurz, 1978, pp. 21–2). These lines are parallel to one another, with the plot of the lowest fertility represented by the innermost frontier.[3] If the price of the product is given by the least efficient land (price is established by the marginal producer), and if the wages are set at w^*, DR I on each land plot i is equal to the difference between w_i and w^*. In more intuitive terms, if the price of corn is set by the least efficient land, it follows that all other intramarginal lands produce corn at lower production prices. The greater efficiency of such intramarginal plots is represented by their wage–profit frontiers all lying above the comparable frontier for

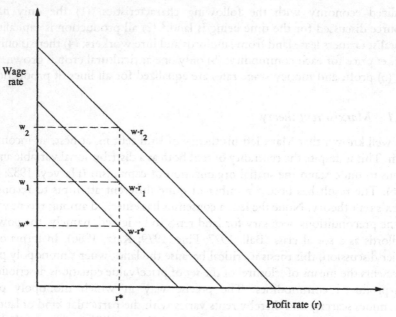

Figure 6.1 Straight-line wage–profit frontiers (price proportional to labour value); each representing a different level of land fertility

the least efficient land. By then setting wages at some arbitrary, but common, level for all plots, it follows that all intramarginal lands will make higher rates of profit than the marginal one: with net output on each land different, and resolvable into only profits and wages, the same level of wages must entail different levels of profit. It is the differential profit rates that are then the source for the different levels of DR I charged on each plot.

Two additional points need to be made about this scheme. First, under the conditions for DR I, relative fertility is defined unambiguously in the sense that the ordering of land plots by fertility will not change even though wages and profits vary (Kurz, 1978, pp. 21–4). We will see later in this section that for the neo-Ricardians this case is the exception not the norm. Second, and a general point applying to all our discussions of differential rent, by suggesting that the marginal plot determines prices we are employing a very different pricing rule from when discussing produced goods. Prices of produced goods, we argued in Chapter 4, are determined by the average, median or modal technique of production. In contrast, prices of non-produced goods here are set by the least efficient technique of production, represented in our case by the marginal plot. The reason for such different pricing rules is the different capacity constraints that apply to the two different types of commodity, produced and non-produced. For non-produced commodities, such as land, each type of plot of a given fertility level has an upper limit on the quantity of output that is possible at a given cost of production. In other words, for non-produced goods each type of technique has a capacity constraint. In contrast, the type of techniques of production used to manufacture produced goods do not have capacity constraints. For example, there can be ten or a hundred car manufacturing plants all using the same technology. Certainly, each individual plant has a limited capacity, but more total units are easily produced using the same technology by simply constructing more identical plants. Given such differences we can now understand the reason for the two different pricing rules. With a given level of demand to satisfy, and with capacity constraints on non-produced goods, prices for such goods must be set to cover the highest-cost (marginal) producer necessary to satisfy this demand. If this were not so, the marginal producer would be out of business and, because of capacity constraints on the other plots/techniques employed, insufficient output is produced. This is not the case for produced commodities. Producers there always have the option of installing the average, median or modal technique of production, thereby ensuring that sufficient output is produced and that their revenues cover costs. In short, the constraint on the output of non-produced commodities implies a fourth definition of socially necessary production method to be added to those of section 3.3.1, the marginal method.

DIFFERENTIAL RENT II

There is far less agreement about DR II than about DR I. In general, DR II arises when capitalists decide to intensify production on a single plot of land, rather than to increase production by expanding the number of plots they own. A consequence of the intensification of production is generally a different capital–labour ratio on each plot of land, resulting in prices not being proportional to labour values (Chapter 3). In discussing DR II we will examine two influential, but different, interpretations, namely, those of Fine (1979) and Ball (1977).

Fine (1979, p. 251) argues that 'DR I is to be distinguished from that of DR II by the latter's dependence upon unequal applications of capitals to lands. ... [Where t]he significance [for Marx] of unequal capitals is their unequal size as a source of productivity increase and surplus-profits.' For Fine, the monopoly power of landed property is seen in its role in setting the level of 'normal' capital, that is, the average amount of capital invested on a plot of land. Only when the level of normal capital is defined, thereby setting the market price, can surplus profit or DR II be calculated on lands with larger than average capitals (Fine, 1977, pp. 256–7). In terms of the $w - r$ curves, Fine's model is one of a series of non-intersecting frontiers, where the frontier closest to the origin has the lowest capital intensity, and the furthest one has the highest capital intensity. (Re-using Figure 6.1, $w - r_2$ is more capital intensive than $w - r_1$, which, in turn, is more capital intensive than $w - r^*$.) In other words, Fine assumes that, with greater capital intensification, the rate of profit necessarily increases, thereby allowing rents to be realized.

Fine's argument is problematic, however, because it fails to recognize that once capital–labour ratios are unequal among plots of land, then $w - r$ frontiers need not be parallel and may cross. This conclusion stems directly from the capital controversy discussed in Chapter 2. If $w - r$ frontiers do cross, however, then at some wage level the land with a relatively low capital–labour ratio will be the one that is more profitable, with its frontier lying outside that of the land with a higher capital–labour ratio, thereby undermining Fine's conclusion. More generally, Fine presumes that a magnitude of capital can be defined independently of the rate of profit. He wishes to use an ostensibly independent entity, the degree of capital intensity, to explain variations in profit and rent levels. But we saw that a central conclusion of the capital controversy was precisely that this cannot be done because variations in the magnitude of capital are in part explained by variations in the rate of profit itself (Chapter 2).

Ball's (1977) argument is more complex than Fine's, and also better articulates the active role of landed property in appropriating rent. His argument is best understood by an example. Suppose that there are two plots of land both producing corn, and that the price of corn at the farm gate equals the cost of production plus a uniform rate of profit. Farm gate prices are given by:

$$y_i p_i = (1 + r) A_i^c . p^c, \qquad (6.1)$$

where A_i^c is the quantity of capital used per acre on plot i; p^c is the price of capital; y_i is the yield per acre on plot i; and p_i is the farm gate price on plot i under an equal rate of profit.

Suppose that the same amount of capital is applied to an acre of land in both places, but that output per acre is greater on plot 1 because of higher fertility. This means that the capital used per unit produced is less on plot 1 than on plot 2, implying that the production costs per unit produced are lower, and thus the farm gate price is less on plot 1 than plot 2. If plot 2 is required in order to meet demand in the market, the market price must equal the farm gate price on plot 2. This is above the farm gate price on plot 1, and the difference is DR I, which is pocketed by the landowner. Specifically, DR I equals the difference between the two farm gate prices multiplied by the yield per acre:

$$DRI = y_1(p_2 - p_1), \qquad (6.2)$$

where DRI is the rent per acre.

Suppose now that further increases in demand occur that require a more intensive use of plot 1, employing a more capital-intensive technology. This leads to an increase in the quantity of capital used per acre on plot 1, but also to an increase in yield per acre. An increase in yield (y) and in capital per acre (A) affect the farm gate price. The direction of this effect is found by dividing both sides of (6.1) by the yield:

$$p_1 = (1 + r)(A_1^c / y_1) . p^c. \qquad (6.3)$$

Ball assumes that the rate of increase of yield is less than the rate of increase in capital inputs (i.e. the ratio A_1^c/y_1 increases), leading in turn to an increase in p_1. Thus, while yields increase on plot 1, the farm gate price also increases, reducing the price differential between the two plots.

These changes have opposite effects on the possible rent per acre that the landowner of plot 1 could receive. Recalling equation (6.2), which represents rent levels before capital intensification, an increase in p_1 reduces DRI, whereas an increase in yield increases DRI. Yet as long as p_1 continues to be lower than p_2 there is positive rent on plot 1. This is DR II because it results from a different production technique.

Crucial to Ball's argument is that, if such an increase in capital intensity were desirable, it could never lead to a fall in rents. In effect, landlords continue to demand rents at least equal to those obtained as DR I. If so, landlords are acting as a barrier to the degree and type of capital-intensive investment that a capitalist farmer can pursue because s/he employs only those forms of capital intensification that make DR II at least as large as DR I on plot 1.

The problem with Ball's analysis, and it is the same one that besets Fine, is that his conception of capital is inconsistent. Ball assumes that a more capital-intensive technique is associated with higher per unit costs, and greater total output. In terms of $w - r$ frontiers, Ball begins with the linear frontiers of Marx discussed earlier where prices are proportional to labour values (Figure 6.2). At some point, demand exceeds current production levels, and a new, more capital-intensive, technique is introduced on land 1. This is represented by the dashed line, 1', in

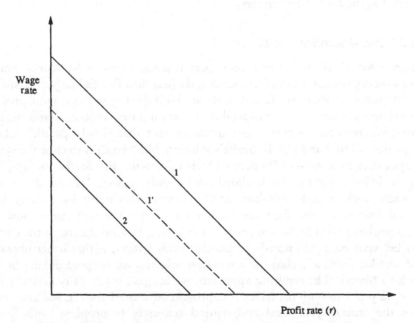

Figure 6.2   Three wage–profit frontiers (illustrating Ball's theory of DR II)

Figure 6.2. By definition, 1' lies inside 1 because the former implies greater costs per unit. Although rent per unit falls, yields increase and hence rent per acre also increases or at least remains the same. If yields per acre do not increase sufficiently to offset declining rents per unit (a result of higher per unit costs), then Ball's 'ratchet effect' comes into operation. That is, investment is pared back until rents per acre are at least as high as they were formerly. The problem with this scheme is that a capital–intensive technique need not be more costly per unit throughout the entire range of wages and profits. Without prices proportional to labour values, the result might well be that some sections of 1' may lie above or below the $w - r$ frontiers for either 1 or 2. If the $w - r$ frontier for 1' lay below that of 2, then DR I and DR II would equal zero because land 1 is now marginal.

In summary, in Marx's analysis of both DR I and DR II the landlord's monopoly power is used to define scarcity and rent levels. Rent is not simply nature's beneficence, but arises only within a particular social formation – one where landlords, because of their power in creating 'scarcity', are able to reap rewards. Furthermore, rent cannot be understood outside of the production system. Along with wages and profits, rent represents a claim on the economic surplus remaining after all inputs have been replaced from total produced output. The problem with the existing Marxist analysis is either its restrictiveness or logical consistency. Under the limiting assumption of equal capital–labour ratios DR I is consistent but unrealistic, while under the assumption of different capital–labour ratios DR II is realistic but inconsistent because of the complexities revealed in the capital controversy.

6.2.2   Neo-Ricardian rent theory

Whereas Marx's theory of land rent sprawls across many of his works, Sraffa's (1960) theory occupies a single, brief chapter (less than five full pages). In spite of the differences in style, Sraffa and Marx are both dealing with the same problem of determining prices for non-produced commodities. Indeed, Sraffa makes a distinction between extensive and intensive rent that closely parallels Marx's categories of DR I and DR II. Sraffa's solution, like Marx's is to close the system by appealing to some socially defined level of scarcity. For Sraffa, though, scarcity is defined not by the landlord's monopoly power, but by the conflict between workers and capitalists. Sraffa achieves this result by making land/ natural resources a so-called non-basic commodity. We will argue, however, that a problem with Sraffa's approach, and by implication the use of the distinction between basic and non-basic goods in this setting, is that it peripheralizes land and landlords as a class. As a tentative solution we propose linking Sraffa's work to Marx's. The resulting approach, we suggest, is generally logically consistent (a problem with the Marxist approaches discussed above), and also recognizes the centrality of land and natural resources (a problem with Sraffa's account).

EXTENSIVE RENT

In providing a formal account of Sraffa's theory of extensive rent (corresponding to Marx's DR I), suppose a region produces N industrial commodities that do not require land inputs for production, together with one agricultural product, corn $(N+1)$, grown on K plots of different land $(k = 1, \ldots, K)$. On each plot of land there is a separate technique of production. Wages and profits are equalized for both industry and agriculture. Production prices are then calculated as:

$$p_n = (1 + r) \sum_{m=1}^{N+1} a_{mn} p_m \quad n = 1, 2, \ldots, N \qquad (6.4)$$

$$p_{N+1} = (1 + r) \sum_{m=1}^{N+1} a_{m,N+1}^K p_m + R^K Q^K. \quad K = 1, 2, \ldots. K,$$

where p_n is the per unit price of industrial good n; p_{N+1} is the per unit price of corn; p_m is the price of the mth input, including inputs that make up the real wage rate; a_{mn} is the amount of good m required to produce a unit of good n, including labour inputs; a_{mN+1}^k is the amount of m required to produce a unit of corn on land k; R^k is the land rent per acre on land k; Q^k is the quantity in acres of land k required to produce a unit of corn; and r is the rate of profit.

Equation (6.4) consists of $N+1$ prices, K rents, and the profit rate, but only $N + K$ equations. To close the system, first set the price of corn to 1 as a numeraire:

$$p_{N+1} = 1. \qquad (6.5)$$

Second, let one land be marginal, paying no rent:

$$\prod_{k=1}^{k} R^k = 0. \qquad (6.6)$$

To solve for rents and prices, it is necessary to determine which of the K lands is marginal. To do this, set the level of rent on any one plot to zero, and draw the corresponding wage–profit frontier (Barnes, 1984). Such a step is equivalent to assuming that the production price of corn grown on this plot sets the market prices. Then repeat this procedure for all K lands, and place the resulting $w - r$ frontiers on a single diagram. For any level of wages, the marginal land will be given by that plot for which the $w - r$ frontier lies inside all other curves. For example, in Figure 6.3, with three plots of land, land 1 is marginal at wage rate w^*.

A central feature of Sraffa's theory of extensive rent is that the marginal land cannot be defined in terms of some absolute measurement of fertility. Rather, the relative fertility of different plots depends upon the distribution of income between wages and profits. Defining the marginal land in terms of the innermost $w - r$ frontier, the example in Figure 6.3 shows how land 1 is marginal at w^*, whereas at w^{***} land 3 is marginal. A corollary is that the order of rentability

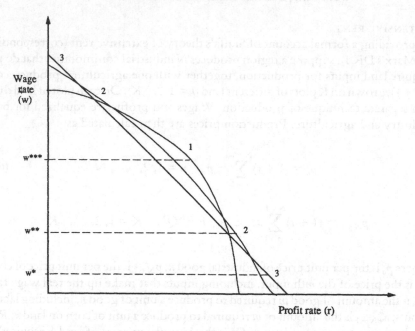

Figure 6.3 Wage–profit frontiers for three different lands

may also change with changes in income distribution. For example, in Figure 6.3 at w^* land 1 is marginal and pays no rent, land 2 pays an intermediate level of rent, and land 3, because it is most profitable, pays the highest level of rent. With wages now equal to w^{***}, land 2 is still intermediate, but now it is land 3 that is marginal and land 1 the most profitable (highest rent-paying) plot. Lands 1 and 3, therefore, have changed their order of rentability even though the 'intrinsic' fertility of soil is the same (Montani, 1975). Within the context of the capital controversies, these 'anomalous' results are easily explained. Because there is a complex relationship between the value of capital goods and the rate of profit, it is possible that, as the rate of profit changes, the value of inputs on the marginal land decreases relative to the value of inputs on an intramarginal land. That change in input value may be so great that a former marginal land becomes an intramarginal one, and vice versa. A similar explanation holds for changes in order of rentability.

Sraffa's strategy for determining the price of land, a non-reproduced commodity, is based on making a distinction between basic and non-basic goods. Basic goods are those that enter directly or indirectly into the production of every other good, while non-basics enter only as inputs into some goods, and possibly none. It can be shown that only the conditions of production of basic goods set prices and the profit rate for the economy as a whole; non-basic goods, in contrast, are peripheral to price and profit determination (Pasinetti, 1977, pp. 104–10; Abraham-Frois and Berrebi, 1979, pp. 39–43). Sraffa makes land a non-basic commodity in his theory by letting only the marginal land producing corn be part of the basic price-determining system. Intramarginal lands, in contrast, are taken as non-basics and therefore have no price-determining role. As a consequence, although the non-produced commodity land has a price, it is a price that in some sense is secondary to the basic price-determining set of equations. For it is only the zero-rent, marginal land that is included in such equations; the level of land rents on intramarginal lands has no influence on either prices or the profit rate. More generally, by making land a non-basic commodity, Sraffa represents extensive rent as price determined rather than price determining (Gibson and MacLeod, 1983). Extensive rent, therefore, is a residual payment. The implication is that landlords are not active in influencing the level of rents charged; like land itself, landlords as social actors are peripheral to the core of the economy.

At this point, we can discuss Sraffa's conception of scarcity. Clearly for Sraffa scarcity is not determined by nature, because, as we have seen, fertility levels depend upon wage and profit levels. Nor is scarcity created by the power of the landlords because they are passive in his scheme. Rather, Sraffa defines scarcity only in terms of the struggle between workers and capitalists in setting the wage rate within the basic sector. Only when the wage rate is established can the marginal land be defined, and only when the marginal land is known are the intramarginal rental levels determined. Land scarcity for Sraffa, then, is determined by two factors that lie outside of the land market: first, the technological conditions of the basic sector; and, second, the conflict between capitalists and workers in determining the wage rate.

Comparing Sraffa's account of extensive rent to Marx's DR I, it is clear that Sraffa's is the more general *in the sense* that its logical consistency does not depend upon equal capital–labour ratios on every plot of land. It does, however, relegate land/natural resources and their owners to 'purely secondary complications, and as such [they are] eliminated from the simplified system that forms the nucleus of ... the theory of prices' (Roncaglia, 1978, p. 6). This feature of Sraffa's acccount becomes even more problematic once intensive rent is considered.

INTENSIVE RENT

Intensive rent arises when two or more different techniques of production are employed on the same quality of land. Initially suppose that for a given type of land all farmers use the same profit-maximizing technique of production. Now let the demand for corn rise, but suppose there is no more readily cultivatable land. To satisfy the extra demand some farmers must employ a more land-intensive, but also more costly, technique of production on the existing cultivated plots. (The technique must be more costly – less profitable – or it would already be in use; for proof, see Abraham-Frois and Berrebi, 1979, p. 97). Formally, the use of two different techniques on a single land plot k is represented as:

$$p_{N+1}^k = (1 + r) \sum_{m=1}^{N+1} a_{mN+1}^1 p_m + Q^1 R \qquad (6.7)$$

$$p_{N+1}^k = (1 + r) \sum_{m=1}^{N+1} a_{mN+1}^2 p_m + Q^2 R$$

where all terms are as before except that superscripts now refer to the technique of production, where technique 1 is less costly but requires more land per unit of output than technique 2.

The broader rationale for intensive rent is in reconciling two different profitable techniques. Farmers always prefer to employ technique 1 because it is less costly. That technique, however, because it is land extensive, is unable to satisfy demand given the available stock of land. To ensure that demand is met, some farmers must employ a more costly (less profitable), but also more land-intensive technique. The intensive rent charged to farmers is the mechanism that allows simultaneous use of

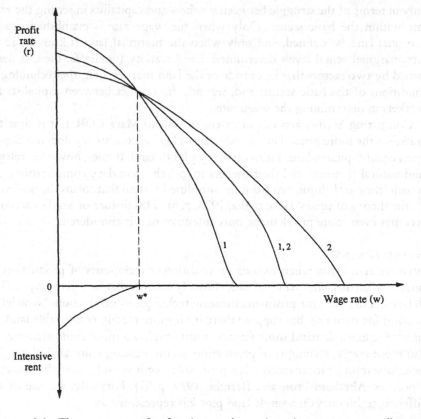

Figure 6.4 Three wage–profit frontiers and an intensive rent curve illustrating Sraffa's theory of intensive rent
Based on Mainwaring (1984)

two different profitable techniques. Technique 1's costs, because of its higher use of land per unit of output (lower yields), rise proportionately more than technique 2's when intensive rent is charged, thereby making both techniques equiprofitable.

Intensive rent is represented in Figure 6.4. At low levels of wages, technique 1 is most profitable. If it is unable to satisfy the current level of demand, however, then both techniques are used together on each plot (technique 1,2), resulting in intensive rent (the difference between the $w - r$ frontiers for technique 1 and 1,2). At wage rates greater than or equal to w^*, technique 2 becomes most profitable. Because by assumption technique 2 is able to meet the high level of demand, there is no reason to employ any other technique and rent levels are again zero.

The definition of non-basics for the case of intensive rent is much more problematic than for the case of extensive rent. This issue is complex and involves the definition of basics and non-basics under conditions of joint production.[4] But the upshot of this discussion is that one cannot exclude land from the central price- and profit-determining system of equations; land and landlords, in other words, are no longer peripheral. Intuitively, we can see why this is so. With intensive rent as defined here, there is no marginal land because, first, all available land is employed, and, second, all lands are of identical quality. As such, there is no cost of production equation for corn where rents are zero. If all lands pay rent, rent becomes price determining rather than price determined. The broader implication is that landlords can no longer be treated as passive; they have a potentially active role in setting rent levels. As such, the neo-Ricardian view of scarcity is also challenged. Rental levels are not only the result of the conflict between workers and capitalists in the basic sector, but also depend upon the active intervention of landlords.

6.2.3   Towards a reconciliation

Both the Marxist and neo-Ricardian approaches to differential rent are problematic. On the one hand, Marx's DR I is logically consistent but limited in application, while DR II is more general but suffers from problems of internal consistency. None the less, both DR I and DR II emphasize the power of landlords as a class in creating scarcity and levying rent. On the other hand, Sraffa's extensive rent is mathematically consistent and general in that there are no restrictions on capital–labour ratios, while his intensive rent, although mathematically consistent on its own terms, does contradict his more general definition of scarcity couched in terms of only worker–capitalist relations. Indeed, more broadly, the problem with both Sraffa's intensive and extensive forms of rent is that they marginalize the position of land and its owners, thereby implicitly derogating the spatial organization of the economy.

There is a clear charter here to combine the mathematical rigour of neo-Ricardianism with the socially penetrating class analysis of Marx. This,

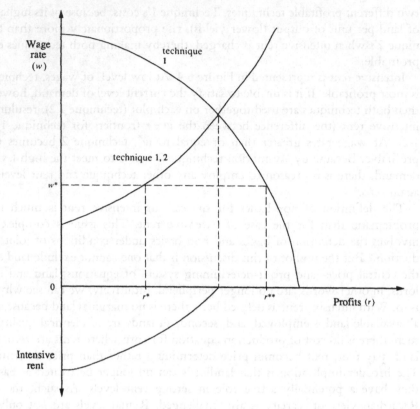

Figure 6.5 Intensive rent with a positively sloping wage–profit frontier
Source: Mainwaring (1984)

however, is not an easy task, if for no other reason than the conflicting ideo-logical freight associated with each position (see Chapter 1). In addition, of course, there are the analytical problems. Because of the complexity of the issue we cannot provide a comprehensive integration of Marxist and neo-Ricardian rent theories here. Rather, we will provide just two examples (both of intensive rent) where such an integration appears fruitful.

To take the first example, a number of writers have noted that the neo-Ricardian theory of intensive rent allows the possibility of positively sloped wage–profit frontiers (Montani, 1975; Gibson and MacLeod, 1983; Mainwaring, 1984, ch. 4). Such an occurrence is ruled out in extensive rent because only basic industries, which by definition exclude land, determine prices and profit rates. In that case the surplus remaining after all replacement costs are accounted for is divided only between profits and wages, thereby ensuring a negative relationship between those two variables. In contrast, in the case of intensive rent, the basic price–distribution system does include a rent term. As a consequence it is possible for wages *and* profits to increase simultaneously providing that intensive rent declines by a sufficient amount. The economic limit of a positively sloped $w - r$ curve is where rents fall to zero. The interesting situation that follows from a positively sloping $w - r$ curve is that of multiple equilibria solutions to price, profit and rent determination (Montani, 1975; Mainwaring, 1984). For example, Figure 6.5 represents two techniques, 1 and 1,2, where the latter is positively sloped. At wage rate w^* there are two solutions: either profit rate r^*, where an intensive rent is levied, or r^{**} where there is no rent. Capitalists will favour using only technique 1 with a solution of r^{**}. If, however, landlords through their social power can restrict capitalists in terms of the technique they use, it is possible for landlords to garner intensive rents. In this case, the potentially active role of landlords is made the focus of the analysis.

The second example is where a time dimension is recognized in the discussion of intensive rent. In Figure 6.6 suppose, for a given wage rate and level of demand, technique 1 is employed. Now let demand increase, and for the reasons discussed above the combination technique 1,2 is used, yielding an intensive rent (traced by the line 1,2 on the lower portion of Figure 6.6). Now let demand

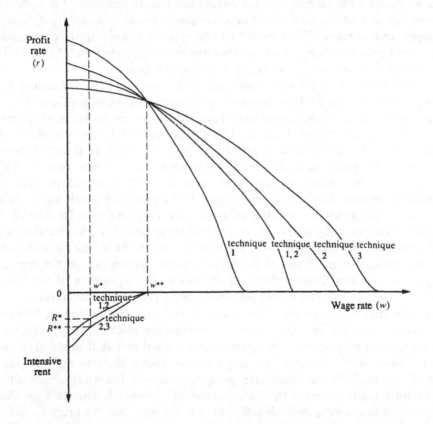

Figure 6.6 Changing levels of intensive rent with changes in the technique of production
Source: Mainwaring (1984).

increase even further, necessitating use of only technique 2 (more costly but more land intensive than both techniques 1 and 1,2). The usual neo–Ricardian interpretation here is that rents will fall to zero because only one technique is used (Quadrio–Curzio, 1980). If this occurs we have a determinate system because rents drop out of the analysis, making the production of corn a produced commodity like any other. In contrast, and following Ball's (1977) suggestion, an alternative scenario is one where rents do not fall once the second, more land-intensive, more costly technique is introduced because landlords will not let rents decline. Mathematically this creates a problem because there are too many unknowns given the number of equations. The strategy of setting the price of corn to 1, and letting wages be given exogenously (section 6.2.2), is inapplicable because we still have an extra unknown. Unlike the case of extensive rent, there is no marginal land paying rent. By letting landlords be active, however, the system can be closed. In Figure 6.6 rent levels are established in the range R^* and R^{**} on the basis of the landlords' power in society. The reason that there is an upper limit to the power of the landlords to levy rent is that at a certain point a new combination technique 2,3 is introduced (where technique 3 is both more land intensive and more costly than technique 2), thereby technically defining an upper limit of rents, R^{**}. It should also be noted that, for wage levels exceeding w^{**} on Figure 6.6, there are no rents because the single technique 2,3 meets all the demands for corn, while also being the most profitable.

Admittedly these two examples are both crude and brief. Lacking is an explicit mechanism to show how the equilibrium position of profits and rents is moved to favour either capitalists or landlords. It is here, however, that there is a potential intersection between David Harvey's (1985a,b) manifesto of a Marxist materialist historical geography and the kind of theory suggested above. The problem with the neo–Ricardian view is not one of logical consistency, but rather its truncated view of social relationships. Only the conflict between workers and capitalists in the produced goods sector defines scarcity. By casting the neo–Ricardian framework into the wider web of class relations discussed by Marx, and the web of geographical and historical relations discussed by Harvey, there is the potential to etch in the mechanisms of change and conflict that are absent in the neo–Ricardian framework. In this sense, we again return to a familiar theme of Harvey's: the importance of the level of abstraction. The formal theory presented above can clarify certain fundamental relationships in capitalism – for example, the antagonisms among landlords, workers and capitalists. But when we examine the mechanisms by which such antagonisms are played out in a particular place and period, then we must move to a lower level of abstraction. In particular, taking the two cases above, we need to establish the particular geographical and historical details of the landlord–capitalist nexus that was portrayed in only an abstract way above. Clearly this is a very difficult task, but it is the only way 'to bring theory and historico-geographical experience together in such a way as to illuminate both' (Harvey, 1985b, p. xv).

6.2.4   Summary

We have suggested here that non–produced goods are dealt with effectively within a political economic approach based upon reproducibility. To accommodate land and natural resources it is necessary, however, to specify a means by which the economic system can be closed and rents calculated. For Marxists that closure stems from the power of the landlords, whereas for neo–Ricardians it is the conflict between workers and capitalists. In formally implementing that 'closure', there are clear differences between the neo–Ricardian and Marxist approaches, differences that give rise to particular problems in each scheme. Specifically, while the neo–Ricardian theory undertheorizes social relations, the Marxist theory has problems in providing a consistent solution. Both Marxist and neo–Ricardian positions are modifiable, however. Moreover, they are modifiable in ways that are mutually supportive: Marxian rent theory is improved by Sraffa's critique of capital and value theory, while Sraffa's rent theory is improved by Marx's insights into social relationships. Here, then, is the charter for exploring the possibilities for a *rapprochement* between the two schools. After all, at a fundamental level both neo–Ricardians and Marxists are making a common claim, namely, that rent levels are not independent of broader issues of political economy.

6.3   Transport costs, multi–commodity production and differential rent

Differential rent arises as a result of differences in costs of production. In the previous section such differential costs were created by differences in productivity (albeit socially defined). In this section we examine a second source of differential costs, transportation inputs. We do this by employing a von Thünen–like model to examine the way in which differences in transportation costs create extensive and intensive differential rent (see section 6.2).

Throughout we make the following assumptions:

1   A single market exists through which all inputs are bought and output sold.
2   The market is surrounded by concentric bands of land, each of equal area. There are K such zones.
3   Market demand is fixed, thereby establishing the number of land zones/ resource sites devoted to the production of each good (defined as J_n zones for commodity n).
4   There are capacity constraints on each plot of land/resource site, in the sense that no single plot of land/resource site can satisfy the demand for all the products required.
5   There exists a class of landlords levying rent, and a class of capitalists who require land for the production of N commodities.
6   The transportation commodity is supplied from the central market.
7   Full capitalist competition holds, implying equal profit rates.

6.3.1   Extensive differential rent

We will define extensive differential rent here as arising when all producers of the same good use the same technology and labour process to produce the same quantity of output. In this case, for each land-use zone i, if commodity m is produced there, the f.o.b. production price is equal to the cost of production (including transportation costs) plus land rent incremented by the rate of profit:

$$p_i^m = (1 + r) \left[\sum_{n=1}^{N} (a^{nm} p^{n\star} + a^{nm} \tau_{Mi}^i \cdot p_M^i) + a^{mm} p_i^m + Q_i^m R_i^m \right], \qquad (6.8)$$

where $p^{n\star}$ is the price of n at the market; $\tau_{Mi}^i \cdot p_m^i$ is the cost of transporting one unit of good n from the central market to location i; R_i^m is the rent payable if land-use m is employed in zone i; and Q_i^m is the area of land needed to produce one unit of m in zone i. M refers to the location of the central market.

One equation of the form of (6.8) is required for each location. To solve this system, start with a particular land-use pattern that determines the locations where each commodity, m, is produced. For each commodity choose one site as the marginal location, $k(m)$. Under the assumption of identical technologies for identical goods, $k(m)$ will always be the furthest location from the market. The market price under this configuration is equal to the factory gate price plus costs of transportation to the market:

$$p^{m\star} = p_{k(m)}^m + \tau_{k(m) M}^m \cdot p_M^t. \qquad (6.9)$$

Included in equation (6.9) is the rent paid at the marginal location, $R_{k(m)}^m$ (see the makeup of production prices in equation 6.8). The rent paid at all other closer locations producing commodity m is proportional to the savings in transportation costs per acre of output, a_j^m, where:

$$a_j^m = p_m^t (\tau_{k(m),M}^m - \tau_{jM}^m) \cdot y_j^m. \qquad (6.10)$$

where y_j^m is the yield at location j for land use m.

Therefore, total differential rent at j, R_j^m, is equal to:

$$R_j^m = R_{k(m)}^m + 1/(1+r) \cdot a_j^m. \qquad (6.11)$$

In order to calculate rents and prices, however, we need to know the rent paid at the marginal location for each product. One possible solution is to define arbitrarily one of the marginal locations, say, $k(s)$, as the global marginal location where:

$$R_{k(s)}^s = 0. \qquad (6.12)$$

Then the rents at all other locations are calculated as the rent necessary to exclude all other land uses from occupying that location. These rents then allow determination of market prices, and rents at all intramarginal locations.

Finally, we need to specify the price of transportation:

$$p'_M = (1 + r) \left[\sum_{m=1}^{N} a^{mt} p^{m*} + Q'_M R'_M \right], \tag{6.13}$$

where $R'_m = \max_m R^m_M$.

By letting $p'_M = 1$, as a numeraire, equations 6.8 – 6.12 provide a determinate solution for rents and prices for this particular land-use pattern (Sheppard and Barnes, 1986).

We do not know a priori, however, either the land-use pattern or the location of the global marginal plot, $k(s)$, nor can we be sure that each intramarginal location is being used by the activity capable of paying most rent there. To determine which pattern of land uses and rents ensures that each location is occupied by the activity capable of paying the most rent, the following procedure is sufficient:

(a)   Determine the marginal plot of land for each land–use type.
(b)   Choose one of these as the global marginal location, where differential rent is zero.
(c)   For every zone j, determine the land–use type that is capable of paying the most rent, by comparing rental levels from equation 6.11 for all commodities, n. If any location is found to be paying negative rents go to (f).
(d)   Once this has been done for all locations, count the number of land–use zones of each type. If the number of zones of each type does not meet market demand, go to (f).
(e)   This represents a feasible land–use pattern. Calculate the production prices and the rate of profit, r.
(f)   Choose another marginal plot as the global margin; go to (C). If all have been tried, choose a new set of marginal plots for each land–use type, and go to (b). If all possible combinations for (a) and (b) have been exhausted, stop.

This exhaustive procedure will require $\binom{N}{K} . N$ iterations, where N is the number of commodities produced and $\binom{N}{K}$ is the binomial coefficient. A subset of these possibilities will have the appropriate number of land–use zones of each type and non–negative rents everywhere and thus have a rate of profit as defined in (e). These are feasible land–use patterns. Choose the feasible land–use pattern, together with definitions of marginal plots, that has the largest rate of profit. This is the optimal location pattern for capitalists because it maximizes collective profits. This profit–maximizing location pattern is also associated with a determinate pattern of differential rents.

Given that the method of production for a given good is identical at all locations by assumption, this procedure can be simplified. In this case the locations where any good s is produced will be closer to the market than the marginal plot of land for that commodity, $k(s)$, because the only factor that varies with location and affects differential rent is transportation and this increases with distance. It is not clear that feasible land-use patterns will always be those for which the globally marginal plot with zero differential rent is at the furthest location from the market. It is likely, however, that the most profitable land-use pattern will have zero differential rent at the most distant location because this economizes on transportation costs.

Having formally defined and derived extensive differential spatial rent, let us enquire into its broader properties. First, like the neo-Ricardian intensive and extensive rents, it is technically defined. It is based upon the difference not of fertility levels but of transportation requirements. This said, although transportation differentials are the mechanism behind rental payments, the precise level of rents found in any given situation rests on the conflict between workers and capitalists. For only when the wage rate is set — a consequence of social conflict — is the land-use pattern and set of marginal lands established. The determination of rent in the geographical case is thus still rooted in the social relations of capitalism.

A second issue is whether land rent is price determined or price determining. Those rent-paying lands that produce the commodity grown on the global marginal land are strictly non-basic. Because the global marginal land sets the production price of the Nth commodity, and because the global marginal land by definition pays no rent, the rents for the Nth commodity are price determined. In contrast, for the case of the $N-1$ commodities produced on marginal lands $k(m)$ (but not the global marginal land), rent is not a residual payment; rather, rent is a cost of production on a par with wages and capital inputs. This is because the marginal lands associated with each of the $N-1$ commodities pay rent. As such, the production prices for those commodities are necessarily affected by the rent levels charged. In broader terms, treating rent as price determining on those plots implies that the geography of production as expressed in transportation differentials is not peripheral, but is central to the economy.

6.3.2 Intensive differential rent

We assumed above that the same commodity is *produced* and *transported* by the same type of technology. The relaxation of either of these conditions potentially gives rise to the charging of intensive differential rent.

First, a direct counterpart to Sraffa's intensive rent occurs when producers use different types of transportation technologies at different sites to produce the same good. Initially let all producers of the same type of good use the same transport service — the one that is most profitable. Now let there be a rise in demand for that good. Given the existing transportation available, suppose that not enough of that commodity can be shipped in sufficient quantities and at the right times. The result is that some producers must use a more intensive transportation technology that is able to handle larger bulk loads. Presumably, this new transportation service is more expensive, otherwise it would already be in use. Now, those producers who employ the new transportation technique will be competitive with those using the old one only if an intensive rent is levied; it ensures that costs of production are equalized for the two sets of producers. Specifically, intensive rent levels are set so that the cost of transporting 1 unit of any good 1 unit distance is equal for both the old and new transportation technologies. By knowing the sites that employ the new transportation technology one can then readily calculate intensive rents by using equation 6.7. With intensive rents calculated, differential spatial rent can also be calculated using the method discussed in 6.3.1. As a consequence it is possible to portray two types of differential rent when transportation services are included: a differential rent that reflects the differential advantage of distance; and an intensive rent that reflects the difference in transportation technology.

A second case of intensive differential rent is where a difference in production costs arises because of a difference in the actual technique of production. Such differences presumably give rise to different production prices, and thereby differential rents. The addition of this type of rent clearly extends the model. Differential rents now reflect both differences in relative location, and differences in both transport and on-site production techniques.

The consequences of relaxing the assumption that the techniques of production (including transportation) are the same for identical goods are, first, that the global marginal land can be located anywhere. Its location will depend upon how the spatial variation in production methods is related to the spatial variation in transportation costs. For example, if the producer at the most remote location uses efficient production and transportation methods, then the resulting higher profits may more than offset the disadvantage of location. In general, the most remote producer of any good may not pay the least differential rent for that good, and the most remote producer on the landscape is not necessarily the economy-wide marginal location. Second, it is possible that there is no global marginal location where rents are equal to zero. For example, a global marginal producer because of his/her location and/or technique of production has the highest production costs. But because, say, s/he uses the least-cost transportation system s/he will still pay an intensive rent. This parallels Sraffa's argument that, if all producers use techniques 1 and 2 together, the intensive rent is paid at all locations.

In summary, transportation costs are an important source of both extensive and intensive differential rents. As a consequence, we have implicitly shown here that transport rates and the technology of the transportation sector are vital not only in shaping geographical patterns, but also in setting some of the broader parameters that define the economy itself.

6.4   Differential rent and non-land resource sites

In this brief section we turn explicitly to the case of non-land resources. We do so to show the broad applicability of the analyses carried out in sections 6.2 and 6.3 that focus only on land.

Instead of analysing different plots of land each with its own level of fertility, as we did above, we could have obtained exactly the same results by examining different mine sites each with its own ore bearing, or different forests each with its own yield. In each case, there is a marginal land plot, mine site, or forest area that establishes the price of the resource. That marginal plot, mine or forest, however, will vary with the distribution of income because changes in the rate of profit differentially affect prices of capital inputs, and thereby the extraction costs at each location.

Similarly, the conclusions about differential land rent created by differences in transportation costs are also extendible to the non-land resource sector. For example, let us suppose that scattered around a central market are a number of primary resource sites each at a different distance. Assume that each different site produces the same primary resource, employs the same system of transportation, and uses a different technology of production. For each primary production site, a production equation similar to that of equation (6.8) is defined, and a wage–profit frontier drawn. As before, the marginal site is determined by superimposing wage–profit frontiers on one another, thereby fixing the price of the primary resource for a given wage or profit rate. With the marginal land fixed, rents on resource sites are readily calculated. Two corollary points follow. First, there is no 'natural' marginal resource site. The marginal site may be the furthest one from the market, but, given that each resource site has a different cost structure that varies differentially with the rate of profit, it does not have to be. Thus, although it is true that the resource site that is furthest away has the highest transportation costs, such costs may be off-set because of non-transportation costs that are lower than at some closer sites. Second, there is no 'natural' spatial order to the pattern of utilization of resource sites. For example, if demand increases for the primary resource, the new resource sites that are called into existence can be either close to the market or very far away for the same reason given above.

In sum, our broader point is that, although we focused on land when analysing natural resources, the framework we constructed also applies to all natural resources. This is not surprising because the key to the analysis was overcoming the problem of assigning price (rents) to non-produced goods. Providing the resource meets the criterion of non-reproducibility, our framework should apply whether that industry is fishing, mining, forestry or agriculture.

6.5   Absolute/monopoly rent

Under DR I or extensive rent there is one plot of land or resource site, the marginal one, that receives no rent. Marx, however, recognized a second general type of rent that applies even to the marginal site. He called this either absolute or monopoly rent depending upon the mechanism of appropriation. In both cases the peculiar feature of this form of rent is that in general it is levied on all sites regardless of their quality. Much controversy exists over Marx's formulation, however. In fact, within the broader category of absolute/ monopoly rent, there are at least three different variants, which, following the literature, are termed absolute rent (AR), monopoly rent 1 (MR1) and mono- poly rent 2 (MR2) (Lauria, 1982, 1985; Leitner and Sheppard, 1989). Common to each of these three subtypes, following Katz (1986), is the active role of the landlord in appropriating rent. This is seen as we examine each of the three subtypes in turn.

6.5.1   Absolute rent

AR is perhaps the most well known of the subtypes. For Marx it arises out of the transformation of labour values into prices (see Chapter 3). Specifically, AR occurs when the market price of a good is equal to its labour value, and above its corresponding transformed price of production. This is best illustrated by a simple numerical example.

Suppose there are two sectors in the economy, industry and agriculture, where the organic composition of capital (C/V ratio) in the former exceeds the latter. Suppose also that rates of exploitation are equal, and that only one generic good is produced in a fixed quantity in each sector. Following Marx, we can describe the production process in both sectors using either labour value or prices. In labour values, the total value of the product produced in either sector i is equal to: $L_i = C_i + V_i + S_i$, where L_i is the labour value of good i; C_i is the constant capital used in the production of good i measured in labour values; V_i is the variable capital used in the production of good i measured in labour values; and S_i is the surplus value obtained from the production of good i measured in labour values. To express the same process in terms of prices, labour values must be transformed using Marx's formula (see section 3.5.2):

$$P_i = (C_i + V_i)(1 + r),$$

where

$$r = \sum_{i=1}^{2} S_i \bigg/ \sum_{i=1}^{2} (C_i + V_i),$$

P_i is the price of production of good i; and r is the money rate of profit.

Labour values and prices are numerically calculated in the table below:

	C	V	S	L	r	P
Industry	10	7.5	7.5	25	50%	26.25
Agriculture	5	7.5	7.5	20	50%	18.75

The central point to note from this numerical example is that the exchange ratio between the industrial and the agricultural good when measured in labour values is different from when it is measured in prices of production. In labour values the exchange ratio between industry and agriculture is 25/20, but in prices it is 26.25/18.75. In other words, after transformation the exchange ratio moves in favour of industry and against agriculture. More generally, with rates of exploitation equal, one can show that any sector that has a higher organic composition of capital always experiences a more favourable exchange ratio in price terms compared with labour values once Marx's transformation procedure is applied.

Marx's point about AR is that landlords prevent exchange ratios moving against the agricultural sector even *after* labour values are transformed, and by so doing garner for themselves additional land revenue, AR. Specifically, landlords are able through their collective social power to charge an AR to the farmer that is equal to the difference between the labour value of the good and its (transformed) price of production. The effect of such a charge is thereby to maintain a market price of the agricultural product that is above its theoretical production price. In the case of the above example, a landlord will charge farmers an AR of 1.25, making the price of the agricultural good equal to 20 – its labour value. From this example it is clear that the burden of AR falls on the group of industrial capitalists. Collectively they have given up 1.25 units of value per unit of output that should have accrued to them as profit.

From this numerical example, we can now detail the two general conditions under which AR arises. The first is that a low capital–labour ratio must exist in the natural resource sector. Thus, in the numerical example, the organic composition of capital in agriculture is 2/3, whereas it is equal to $1\frac{1}{3}$ in industry. The significance of this condition is that it ensures that the labour value of any commodity produced in such a sector is always greater than its transformed price of production (in the example, the labour value of the agricultural good is equal to 20, but its price of production is equal to 18.75. For a general proof see Barnes, 1984). Such a difference is clearly essential to the realization of AR because by definition this form of rent can be appropriated only when there exists a difference between labour values and prices. The second general condition is that there are barriers to capitalist investment (Scott, 1976; Ball, 1980, 1985; Harvey, 1982; Lauria, 1985; Leitner and Sheppard, 1989). The significance of the second condition is in ensuring that the capital–labour ratio remains low. For, providing investment is blocked, production prices remain consistently less than labour values, thereby allowing continual appropriation of AR. In terms of the example, if investment freely occurred in agriculture in such a way that raised the constant capital there from 5 to 10 units, whereas industry remained unchanged, values and prices in both sectors would be equal. With no difference between labour values and prices, AR disappears.

Both these general conditions, however, have been criticized on a number of grounds. We argue here, though, that these usual criticisms are not as damaging as their critics maintain. There is, however, a more recent critique originating in the capital controversy that points to a fundamental logical contradiction within the theory of AR. Such an objection, we suggest, necessarily leads to either an abandonment or a serious reconstruction of AR.

The first criticism of AR is that measurement of the organic composition of capital is very difficult to attain, and that non-farm resource sector activities such as mining appear to have very high organic compositions of capital, not low ones (Ball, 1985). The counter-argument is that, if it is so difficult to figure the organic composition of capital, then it will be hard to *disprove* the claim made by Marx that such a ratio is low in the resource sector (recent work by Webber, 1987b, however, suggests that such concepts as the organic composition of capital can be operationalized, although this work has still not been applied to this debate). In addition, Marx confined his writings to the farm sector, and therefore, even if we could measure the organic composition of capital in, say, the mining sector and found that it was high, this would still not invalidate the bulk of Marx's writings on AR, which are confined to agriculture.

A second objection often made is that it is unclear how landlords represent a barrier to investment (Scott, 1976, p. 129). If the market is competitive, how do landlords prevent capitalists from investing in a sector in which market prices are consistently greater than production prices? Marx's response to this criticism is very clear. *Given* the assumption that landlords own all the land, then tenant farmers have no incentive to make capital improvements on it because the land will never be theirs. In this sense, it is the very monopoly of land/resource ownership that represents the barrier to investment.

Finally, and this is the most common criticism, it is argued that, if landlords can force prices to rise, there is nothing to prevent them from pushing prices above labour values, thereby realizing even greater AR (Emmanuel, 1972; Ball, 1980). As Scott (1976, p. 129) writes: 'If landlords have the power to extract the difference between value and production price, they have the power to extract even more.' The counter-argument is that Marx is working within the labour theory of value. For this reason, to claim that prices could be set at any magnitude (which is what Scott suggests) 'presupposes that agricultural products are *excluded* from the general laws of value of commodities and of capitalist production ... Hence this is absurd' (Marx, 1969, p. 36). In other words, Marx's general theory of prices based upon labour values is still operating even when AR is levied. In fact, what would be incomprehensible would be if Marx had abandoned the labour theory of value. Its role, after all, is precisely one of establishing order within the sphere of exchange, and preventing the kind of price anarchy that

Scott suggests. In this light, a passage in *Theories of Surplus Value* where Marx discusses this very criticism is particularly valuable:

> But, it may be asked: If landed property gives the power to sell the product above its *cost* price, *at* its value, why does it not equally well give the power to sell the product *above* its value, at an arbitrary price? ... Landed property can only affect and paralyze the action of capitals, their competition, in so far as the competition modifies the determination of the *values of the commodities*. (Marx, 1969, pp. 332–3; emphases in original)

Under our interpretation, Marx is suggesting that landlords only 'affect and paralyze' the organic composition of capital. It is this that establishes labour values, and thereby sets the limits of AR. Landed property cannot directly influence prices, but only 'modif[y] the determination of the *values of the commodities*'. Now, of course, one may wish to question the appropriateness of the labour theory of value, but this is a different objection to the one raised.

In contrast, we argue that a more penetrating critique of AR has its origin in the capital and value controversies. As noted in section 3.5, Marx's major error in his transformation procedure was in not expressing constant and variable capital in price terms. Instead, capital is expressed in terms of socially necessary labour values. Under this system of measurement, one can indeed show that labour values will be above prices of production in those sectors where the organic composition of capital is low. But once Marx's transformation procedure is corrected, the organic composition of capital can no longer be used to predict whether labour values are greater or less than prices. Prices vary with wages and profits and not the organic composition of capital. In fact, as Gibson and Esfahani (1983, p. 94) demonstrate, a low organic composition of capital can be associated with negative absolute rents (this is only a logical and not an economic possibility). The reason for such a perverse result, and one that gets to the central problem of Marx's formulation of AR, is that one cannot use such physically based ratios as the organic composition of capital to determine the size of distributional shares. This was precisely the neoclassical fallacy revealed in the capital controversy. Rather, deviations of labour values from prices, and thus the potential size of AR, depend upon the size of the other distributional variables (wages and profits) and on the socially necessary production methods. Thus, even if the organic composition of capital is low, providing wages and profits are high relative to the total net product (surplus) available for distribution, AR will necessarily be low or non-existent. More generally, this conclusion points to an internal logical contradiction in Marx's theory of AR. Marx's first general condition for realizing AR – a low organic composition of capital – is logically inconsistent with his stated objective. The contradiction between premise and conclusion makes Marx's goal an impossible one. For this reason, one must seriously consider whether it is worth pursuing AR any further.

6.5.2 Monopoly rent 1

Monopoly rent 1 arises because of excess profits accruing to a monopolistic activity that occupies a given site (Harvey, 1982; Lauria, 1982, 1985; Leitner and Sheppard, 1989). It is therefore the monopoly held by the firm/sector itself that gives rise to MR1, not the monopolization of land *per se*. Two different cases are recognizable: first, where the firm's monopoly is related to the nature of the primary resource, and, second, where it is not. An example of the first kind, and it is the one Marx employs, is a vineyard producing a unique type of wine. Here it is the primary resource, land, that enables the producer to charge a monopoly price, thereby garnering excess profits. Such profits are then appropriated by the landlord in the form of MR1. An example of the second case is where a monopoly firm locates at a given site. The monopoly profits accruing to such a firm, which have nothing to do with the site *per se*, are none the less captured by the landlord who appropriates the excess profits made there.

For our purposes, we simply note that MR1 is not consistent with the assumption of full capitalist competition on which we have relied thus far. For prices are set no longer by costs but by mark-ups established by market demand, commodity characteristics, industrial structure and so on. In Chapter 14 we will provide an analytical framework that allows discussion of monopolistic competition, with implications for above-average profits that may allow for the extraction of MR1, but we will not pursue MR1 further here.

6.5.3 Monopoly rent 2

Known also as class monopoly rent, MR2 has its origins in David Harvey's (1974) work. Unlike MR1, MR2 is a result of the monopolistic characteristics of land ownership rather than of those of the enterprise occupying a site. Specifically, it occurs because 'there exists a class of owners of "resource units" ... who are willing to release the units under their command only if they receive a positive return above an arbitrary level ... The realization of this rent depends upon the ability of one class-interest group to exercise its power over another class-interest group and thereby assure itself a certain minimum rate of return' (Harvey, 1974, p. 240). In terms of our earlier discussion, MR2 is the purest expression of the social power of landlords to create scarcity. That power, however, as others have recognized, especially in an urban setting, is continually challenged and negotiated (Lauria, 1985; Katz, 1986). The precise level of MR2 levied depends upon a host of contingent factors such as the institutional setting of the land market, the power of countervailing classes, and the nature of the landowners.

A formal presentation of MR2 can be developed by constructing a three-dimensional equivalent of the wage–profit frontier, where the third axis represents MR2 and reflects the power of landlords to appropriate rent (Figure 6.7; for the analytical details, see Barnes, 1984). Such a wage–profit–MR2 frontier gives

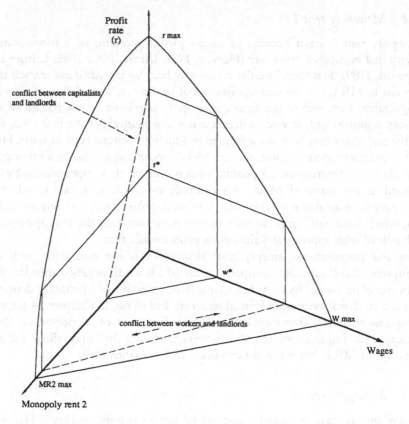

Figure 6.7 Potential conflicts among capitalists, workers and landlords shown on a wage–profit–rent frontier

all possible combinations of income allocations between wages, profits and MR2. In this way MR2 is treated as a distributional variable on a par with profits and wages (this is also seen in embryonic form in Walker, 1975). It should be noted that, because land and labour are both non-produced commodities, there is a linear relationship between MR2 and wages. In contrast, the relationships between wages and profits and between MR2 and profits are non-linear. Whatever the form of the relationship, it can none the less always be shown that as profits or wages or MR2 increase to some finite maximum (represented by, respectively, r_{max}, w_{max} and $MR2_{max}$ on Figure 6.7), the other two distributional variables must decrease to zero (Barnes and Sheppard, 1984).

In terms of interpreting the three axes of Figure 6.7, Katz (1986) argues that Marxists traditionally focus on the active struggle between landlords and capitalists, with workers assumed to be passive. This implies that, with wages fixed, at w^* say, the crucial determinant of MR2 is the conflict between landlords and capitalists over the remaining surplus. With that struggle decided, and MR2 and wages fixed, then the rate of profit is determined. It is clear, as Katz (1986) proposes, that at different places and times the struggle between workers and landlords is primary. In this case, with the rate of profits given at r^*, the important question becomes where along the $W_{max} - MR2_{max}$ frontier wages and MR2 will be fixed (Figure 6.7). Here, landed property and labour compete for income shares after capitalists have taken their 'cut'.

In summary, because of the inconsistency of AR and the special conditions under which MR1 holds, our focus in the remaining part of the book is on MR2. Some may criticize such an emphasis on the grounds that we did not provide any detailed mechanisms for appropriating MR2, citing only the collective power of landlords to garner this form of rent. But our treatment of MR2 is no different from our treatment of wages or profits. For at bottom the shares of the surplus accruing to workers, capitalists or landlords pivot on their respective abilities to wrest income away from other classes and channel it to their own. In short, it depends upon their general social power as a class. We should note that some researchers have begun to examine the specific mechanisms that mediate between the general power of landlords as a class and their appropriation of MR2 on the ground in specific times and places. For example, both Harvey's (1974) class monopoly rent originating out of a geographically captive working class or ethnic minority housing market, and Scott's (1976) notion of scarcity rent, represent attempts to provide such mediations. Clearly, though, more work needs to be done (also see, Edel, 1976; Markusen, 1978; Lauria 1982, 1985).

Summary

The central problem in dealing with natural resources from a political economic perspective is that they have no costs of (re)production. The problem then becomes one of assigning a price on the basis of other criteria. We argued that those other criteria are found by examining Marx's broader view of nature. Specifically, that broader view emphasizes the importance of the social relations of production and distribution in comprehending all aspects of nature, including natural resources. For us this means that rent must be viewed as part of the production system rather than as something that 'nature' creates, and that scarcity must be viewed as the outcome of social relationships rather than as something inherent in 'nature.'

Although accepting this general vision of nature, and the implied role for rent and scarcity, Marxists and neo-Ricardians part company when it comes to analysing exactly how the social relations of production and distribution determine specific levels of scarcity and rent. For the Marxists, scarcity is defined by the class power of the resource owners. In the case of MR2, that power directly corresponds to the ability to levy rents. In the case of differential rent, that power defines the existence of rents but does not directly determine their level. Rather, the level of rent is also dependent upon such factors as the productivity of capital and even the land plot or resource site itself (see Kurz, 1978). Neo-Ricardians, in constrast, define scarcity only in terms of the conflict between workers and capitalists, with landlords passive. Levels of differential rent then depend upon costs of production, but these vary with the rate of profit, not with any inherent productivity of capital. We argued that both the Marxist and neo-Ricardian theories, although very suggestive, are flawed: the former in terms of general logical consistency, the latter in terms of an inadequate recognition of resource owners' power as a class. For that reason a limited synthesis between the two positions was sought. It is here that much more work needs to be done. Such an assessment also applies to MR2.

This said, this chapter has tried to point the way to some of the analytical benefits of examining rent and natural resources by beginning with a political economic perspective. The central advantage, we argue, is the recognition that in analysing natural resources we cannot treat nature as separable from the broader lineaments of society and economy. Based on this, we have shown how the price for any non-reproducible commodity, notably land and natural resources, can be determined as a rent whose magnitude depends upon the relations of production and distribution.

Notes

1 We use the admittedly clumsy term absolute/monopoly rent because within the literature itself these two types of rent are often confounded. We argue in section 6.5, however, that they should not be, and we distinguish between them.

2 The analysis of differential rent for other natural resources, such as mines and forests, follows exactly the procedure outlined for land. In each case, it is the difference in productivity among plots of land, or mine sites, or forested areas that is the impetus for differential rent. We will discuss explicitly non-land natural resources in section 6.4.

3 The lines are parallel because of equal capital–labour ratios on each plot of land. Thus, even though the rate of profit might change, thereby changing prices of produced inputs, each plot of land is proportionately affected in the same way, and thus the wage–profit frontiers remain parallel. The innermost frontier represents the lowest-fertility plot because the 'surplus' or net output on that land plot, given by the position of its wage–profit frontier, is by definition less than any other plot for all combinations of wages and profits.

4 For general reviews of the problems of joint production, see Manara (1980), Schefold (1980), and Steedman (1980). For a discussion of the peculiarities of non-produced commodities, including the recognition of a second type of intensive rent not discussed here, see D'Agata (1983, 1986a,b), Gibson and MacLeod (1983), and Salvadori (1983).

7 The city: incorporating the built environment

Introduction

The purpose of this chapter is to extend the model constructed in Chapter 4 by examining a number of issues that relate to the city. In so doing we demonstrate the flexibility of our more basic model. The chapter is divided into four sections: first, we briefly situate our framework with respect to developments within urban economic geography; second, by drawing upon the analysis of the previous chapters, we present a method for determining intra-urban locations, land rents, profits and prices for a set of producers; third, drawing upon Sraffa's theory of joint production, the model is extended to include durable capital (the urban built environment). We discuss rents, depreciation and both the technical and the economic lifetime of buildings. Finally, the model is linked to retail and service employment using as a framework the Garin–Lowry model. This also enables us to discuss questions of commuting, housing, wages and social conflict.

7.1 The theoretical context

Over the last two decades three alternative approaches to the geography of the urban economy have been developed. Each one, in some way or another, is partly informed by a broader theoretical framework originating in economics.

The approach that is perhaps the most fully worked out is the new urban economics (NUE), whose foundations lie in neoclassical economics. Consisting of an ever-increasing corpus of theory marked by its rigour and mathematical sophistication, the school's modern antecedents are the works of Alonso, Mills and Muth (Richardson, 1977, ch. 2). Initially the new urban economics was criticized for its overly simplistic assumptions (for example, mono-centricity, the absence of durable capital, and a circular city). Subsequent work demonstrated that these assumptions are not essential to the new urban economic model, and can be relaxed. One feature of these models that remains difficult to change, however, is the emphasis on exchange and consumption, thereby implicitly derogating the importance of production and work. Such an emphasis is partly explained by the model of pure exchange that underlies the neoclassical vision as discussed in Chapter 1. As we saw there, this is a vision of individual agents

exchanging their set of exogenously given resources for other sets owned by
different individuals in order to maximize utility. Because the outcome of the
exchange process is not affected by where the resources come from, 'it is [then]
quite unnecessary to consider any process of production in order to grasp the
essential features of [the] analysis' (Pasinetti, 1977, p. 24; also see Barnes, 1983,
ch. 3, for an extended critique of NUE, one that recognizes its various
subtypes).

The second approach arose in the 1970s and is associated with the work of
radical urbanists such as Harvey, Walker, Castells and Smith. Although their
analysis is explicitly set within contemporary capitalism, and draws upon an
avowedly Marxist theoretical framework, Scott (1988a, ch. 1) argues that
such radical urban theorists share with the new urban economists an emphasis on exchange relationships and consumption. This is curious because the
tradition of Marxism is one in which production is central not exchange
(Chapter 1). An examination of Walker's (1981), Harvey's (1982, 1985a, b)
and Smith's (1984) works, however, shows that Scott's claim is exaggerated.
Perhaps a more considered view is that Marxist urbanists are at least
ambiguous about the relationship between production and exchange.
Furthermore, even if we accept Scott's argument, it is clear that for radical
urban geographers the world in which exchange takes place is very different
from the one portrayed by the new urban economics. For example, crucial
to Harvey's theory of class monopoly rent is institutional intervention,
which, in turn, leads to 'unfair exchange' and restricted consumption. For
this reason the end result of the exchange process is not a harmonious Pareto
optimum, as in the NUE models, but one characterized by inequality and
social discord.

The third and most recent approach begins with the work of Scott (1980,
1988a, b) and is based upon the neo-Ricardian school. Scott argues that,
unlike either the NUE or the Marxist approach, the neo-Ricardian view
unequivocally emphasizes the centrality of production. The consequence of
adopting this perspective and applying it to capitalist urbanization is that
metropolitan development is viewed as 'a structured outgrowth of the dynamics of the production system. The system is seen ... as the engine that
drives forward the entire spatial and temporal evolution of the modern
metropolis' (Scott, 1982, p. 185). Specifically, two facets of the urban
economy are emphasized by Scott: first, the technique of production and the
consequent intra-urban production linkages among firms; and, secondly, the
distribution of net income among the social classes engaged in its formation
(Barnes, 1989).

In this chapter we build upon Scott's pioneering work, albeit in an analytical direction (see also, Huriot 1983, 1984, 1985; and Webber 1983, 1984).
None the less, we see his neo-Ricardian stance as only one strand of a broader
political economic view of the city, and it is this broader view that we seek to
elaborate.

7.2 The intra-urban location of the basic sector and rent

The tasks of this section are twofold: first, to define the profit-maximizing intra-urban location pattern for a set of manufacturing plants and producer services within the city (collectively called the production sector[1]); and, second, to define the configuration of land rents within the city that is consistent with profit maximization. The assumptions of the model are as follows:

1   The city is divided into J $(i, j = 1, \ldots, J)$ land parcels, small enough that no parcel can be occupied by more than one producer from the production sector. Non-production sector activities, such as housing and consumer services, also locate in such zones, and they can do so even when a manufacturer or producer service is already located there.
2   N $(m, n = 1, \ldots, N)$ goods (including both manufacturing and producer services) are produced in the city by P production sector establishments $(N < P < J)$. The technology for producing the N goods and producer services is known and is represented by a set of fixed input coefficients a_{ij}^{mn} (where a_{ij}^{mn} is the amount of good m produced at location i required to produce a unit of good n at location j). It is assumed that identical technology is used to produce the same good in different plants. Initially, it is assumed that there is no joint production or fixed capital.
3   There is a single market at the centre of the city through which all finished and intermediate goods are sold. The central market is both a wholesale centre as well as the site of the production of transport services. Transport services are a produced good whose price is determined endogenously.
4   All goods are basic in the sense that every producer is directly or indirectly linked with all other producers through the circulation of commodities.
5   The urban economy is an open one. Imports M $(M = 1, \ldots, N)$ from other cities I $(I = 1, \ldots, JJ)$ are delivered to, and exports to other urban markets are shipped from, the central market.
6   Money wages and profits are equalized throughout the city.

To define the optimal intra-urban location pattern for the P production sector activities requires a simultaneous determination of location patterns, spatial interdependencies among producers, market prices that maximize the rate of profit, and the various land rents charged at the J plots of land. For purposes of exposition this task will be carried out in two stages: first, excluding land rents; and second, with rents included.

7.2.1   Optimal production sector location excluding land rents

First, arbitrarily assign the P production sector establishments to the J land parcels (where $P < J$). For the assignment of sector n on parcel j the market price (i.e. the price the product is offered for sale at the central market) is equal to:

$$p_j^n = \left[\sum_{m=1}^{N} \sum_{i=1}^{J} a_{ij}^{mn} p_i^m + \sum_{M=1}^{N} \sum_{I=1}^{JJ} a_{Ij}^{Mn} p_I^M + \left(\sum_{m=1}^{N} \sum_{i=1}^{J} a_{ij}^{mn} \tau_{Cj}^m + \right. \right.$$

$$\left. \left. \sum_{M=1}^{N} \sum_{I=1}^{JJ} a_{Ij}^{Mn} \tau_{Cj}^M + \tau_{jC}^n \right) p_C^t \right] (1 + r), \tag{7.1}$$

where p_j^n is the production price of good n produced at location j; a_{ij}^{mn} is the technical coefficient of production representing the amount of m locally produced at location i required to produce a unit of n at location j (including wage goods consumed by labour; see Chapter 4); a_{Ij}^{Mn} is the technical coefficient of production representing the amount of M imported from city I required to produce a unit of n at location j; p_I^M is the price at the central market of the imported commodity M from city I; τ_{Cj}^m is the amount of transport services required to ship a unit of m from the central market (C) to location j; τ_{Cj}^M is the amount of transport services required to ship a unit of the imported commodity M from the central market to location j; τ_{jC}^n is the amount of transport services required to ship a unit of good n from location j to the central market; p_C^t is the production price of transport produced at the central market; r is the rate of profit.

Equation (7.1) states that the production price for good n produced at location j is equal to: the cost of local intermediate and wage goods purchased at the market, plus the cost of imported goods purchased at the market, plus the transport costs of shipping both locally produced and imported intermediate inputs from the market to the production site j where n is produced, plus the transport cost of shipping the finished good n back to the market where it is sold. This sum is then incremented by the rate of profit.

Another equation must be added to equation (7.1) specifying the price of transport:

$$p_c^t = \left(\sum_{m=1}^{N} \sum_{i=1}^{J} a_{iC}^{mt} p_i^m + \sum_{M=1}^{N} \sum_{I=1}^{JJ} a_{IC}^{Mt} p_I^M \right) (1 + r), \tag{7.2}$$

It is assumed that, because transport services are produced at the city centre, they do not directly require any transport services in their own production.

There are in total $P + 1$ price equations (P equations of the form given in equation (7.1), and equation (7.2) specifying the price of transport), and $P + N.JJ + 2$ unknowns (P prices of local goods, $N.JJ$ prices of imported goods, the price of transport, and the profit rate). Given that the $N.JJ$ prices of imported goods can be treated as exogenous (they are endogenous only if we look at the complete inter-urban trading system, which we do not here), then, by letting one of the prices equal 1 as a numeraire, the system is closed. Using the arguments presented in Chapter 4, we can show such prices are both positive and possibly unique.

Having solved for prices and the profit rate for one particular assignment of basic sector activities to land zones, we must establish which one of the total $(P).P!/(P_1!P_2! \ldots P_N!)$ assignments is optimal, where P_i is the number of sites occupied by each sector i ($\sum_{i=1}^{J} P_i = P$). We have already discussed in both Chapters 4 and 6 the solution to this problem through the use of wage–profit frontiers, and these procedures are also applicable here. The profit-maximizing location pattern that results depends upon the socially necessary production methods and the distribution of income between wages and profits, i.e. it depends more fundamentally on the class relations between workers and capitalists.

7.2.2   Optimal location of plants with land rents

In the model constructed above the same goods produced in different places will have different market prices, a result of the differential effect of transport costs. Such price differences, however, cannot persist if equal prices are enforced at the central market. As a result, land rents come into existence reflecting the resulting location advantage (section 6.2). To incorporate land rents into the model, it is assumed that there is a single market price for each good that is given by the most expensive, marginal supplier. We thus recognize how the fact that production activities must occupy space implies the inclusion of the non-reproducible commodity of land.

With land rents included, the new price equation for good n produced on land zone j is given by:

$$p_j^n = \left[\sum_{m=1}^{N} \sum_{i=1}^{J} a_{ij}^{mn} p_i^m + \sum_{M=1}^{N} \sum_{I=1}^{JJ} a_{Ij}^{Mn} p_I^M + \left(\sum_{m=1}^{N} \sum_{i=1}^{J} a_{ij}^{mn} \tau_{Cj}^m + \right. \right.$$
$$\left. \left. \sum_{M=1}^{N} \sum_{I=1}^{JJ} a_{Ij}^{Mn} \tau_{Cj}^M + \tau_{jC}^n \right) p_C^t + Q_j^n R_j^n \right] (1 + r), \tag{7.3}$$

where all terms are as before. Q_j^n is the quantity of land required to produce a unit of good n at location j; and R_j^n is the land rent charged per unit of land.

In chapter 6 we demonstrated how rents are calculated from a price equation of the form found in equation (7.3). That procedure, which we will not repeat here, involves identifying both local marginal lands and one global marginal land. By changing both assignments of users to lands and the marginal lands themselves and then comparing wage-profit frontiers it is possible to determine the optimal spatial arrangement of producers within the city, together with the rents paid at each location.

In constructing this model of land rent it is assumed that the same good is produced by the same technology in different plants. As such we are concerned only with extensive differential rent. By relaxing this assumption, or assuming that the transportation sector itself is composed of different techniques of production, it is also possible to calculate the various forms of intensive rent discussed in the last chapter. Furthermore, we can also readily incorporate MR2 into the model following Harvey's (1974) suggestions or those of Huriot (1983). Because we have already provided the framework to add these forms of rent, again we will not pursue them here. Rather, in the next section we turn to an undertheorized type of rental payment, at least within the political economic perspective, that of rents payable on the built environment.

7.3 Rent and the built environment as fixed capital

Following Muth's (1973) pioneering paper, there have been a number of attempts to introduce a durable housing stock into the new urban economic models (Evans, 1975; Arnott, 1980; Brueckner, 1980a, b, 1982; Wheaton, 1982). Typically such models rest on some kind of neoclassical vintage model of growth, whereby, once capital is deposited in the form of housing stock, it cannot be changed into anything else. In the fanciful terminology of capital theory, capital is like 'putty-clay' rather than 'Meccano', 'Lego' or 'ectoplasm'. Neoclassical vintage models, however, were heavily criticized in the capital controversy (Chapter 2). Therefore, we will introduce durable capital stock into the model constructed above by making use of Sraffa's theory of joint production and the various elaborations made to it. (For a lucid introduction, see Mainwaring, 1984, chs 10 and 11, and for a more advanced treatment, see Pasinetti, 1980.)

The urban built environment, whether it is in the form of houses, shops or factories, represents fixed capital, that is, an input with an economic life of more than one production period (see section 4.2). Sraffa treats fixed capital as a special type of joint product, where joint production is defined as a single production process that produces more than one output (see section 3.6). As Sraffa (1960, p. 63) writes, '[t]he interest of Joint Products . . . lie[s] . . . in its being the genus of which Fixed Capital is the leading species'.

In Chapter 6 we briefly discussed some of the features associated with joint production. One of them is the possibility of positively sloped wage–profit frontiers (section 6.2.2). Such an occurrence is the result of negative prices. The issue is a complex one, but it revolves around the fact that under joint production commodities are not producible separately. For that reason, although gross output for any production process will be positive, net output may be negative. For example, to produce 4 units of, say, both commodity 1 and 2 in a joint production process, we require 1 unit of good 1 and 5 units of good 2 as inputs. While good 1's prices are positive, for good 2 they are negative because there is a net output of -1 unit for that commodity. The joint production process, however, can still be profitable providing that the positively priced surplus for good 1 is greater than the deficit for good 2. A positively sloped wage–profit frontier may occur in this example if the wage bundle paid to workers partly consists of the negatively priced good, commodity 2. In this case, even though profits increase, wages can also increase because one of the components of the wage is negatively priced (Mainwaring, 1984, ch.10).

Such results pertain especially to the so-called 'pure' joint products (the classic examples are wool/mutton and corn/straw). Fortunately, however, fixed capital is not a pure joint product, although it shares some characteristics. The first thing we must clarify, then, is the exact nature of fixed capital.

A building, or any item of fixed capital, is a joint product because it is both an input to the production process as well as an output. Specifically, Sraffa treats an item of fixed capital aged t years as a different product from the same fixed capital aged $t + 1$ years. For example, to produce good m, let us assume that we require a building aged t years as an input. In Sraffa's scheme, that same building reappears one year older as an output along with the finished good m. Thus, we have one process producing two goods, the commodity and the building that is one year older. To make this clearer, consider the following relatively simple model where there is only one firm within the city that uses a building with a life of more than one production period.

For ease of exposition assume that the construction industry is located throughout the city. Let the firm producing good m in zone i require a durable building with a physical life of T years. For the moment also assume that the building maintains constant efficiency in the sense that either it requires the same level of maintenance, or employees working within it produce the same level of output, throughout the building's life. Assuming a production period of one year, and letting $p_i^m = 1$, the production equations for *this one firm* over the T time periods that the building is in existence are:

$$\left(\sum_{n=1}^{N} \sum_{j=1}^{J} {}_0 a_{ji}^{nm} \cdot p_j^n + {}_0 a_{ii}^{mm} + {}_0 Q_i^m \cdot {}_0 R_i^m \right) (1 + r) = 1 + {}_1 b^B \cdot {}_1 p^B \qquad (7.4a)$$

$$\left(\sum_{n=1}^{N} \sum_{j=1}^{J} {}_1 a_{ji}^{nm} \cdot p_j^n + {}_1 a_{ii}^{mm} + {}_1 a^B \cdot {}_1 p^B + {}_1 Q_i^m \cdot {}_1 R_i^m \right) (1 + r) = 1 + {}_2 b^B \cdot {}_2 p^B \quad (7.4b)$$

....

$$\left(\sum_{n=1}^{N} \sum_{j=1}^{J} {}_T a_{ji}^{nm} \cdot p_j^n + {}_T a_{ii}^{mm} + {}_T a^B \cdot {}_T p^B + {}_T Q_i^m \cdot {}_T R_i^m \right) (1 + r) = 1 \qquad (7.4c)$$

where ${}_t a_{ji}^{nm}$ is the amount of good n produced in zone j required to produce a unit of good m in zone i in time period t; p_j^n is the price of good n produced in zone j; ${}_t a_{ii}^{mm}$ is the amount of m required in zone i to produce a unit of m in zone i in time period t; ${}_t a^B$ is the amount of the building required to produce a unit of good m in time period t; ${}_t Q_i^m$ is the amount of land required to produce a unit of good m in zone i in time period t; ${}_t R_i^m$ is the rent charged per unit of land used for the production of good m in zone i in time period t; ${}_t b^B$ is the amount of building as an output left in time period t, per unit of commodity m produced; ${}_t p^B$ is the price of the building in time period t; r is the rate of profit.

Equation (7.4) assumes that, for each production period t, there are $N - 1$ commodities being produced that do not require a durable building as an input. Commodity m (the Nth good) does require such a building, and equation (7.4) represents m's production from the beginning time period 0 when the new building is first employed through to time T when the building is either worn out or abandoned. Specifically, in production period 0, N commodities produce N commodities. One of the commodities produced, and used only in the production of good m, is the durable building, B (equation 7.4a). In production period 1, all non-durable commodities are produced again, but because of the durability of B, m is now produced by means of a building aged 1 production period, with $_1 a^B$ units of the building required per unit of output (equation 7.4b). The process of using an increasingly older building to produce good m continues until the end of the economic life of the building, at which point the building is demolished or abandoned (equation 7.4c). (Note that the economic life of a building may be less than its technical life, as discussed below.)

The price of a building of any age is readily calculated. With $p_i^m = 1$, by rearranging equation (7.4c) we derive the price of the oldest building:

$$_T p^B = \frac{\left\{ 1 - \left[\sum_{j=1}^{J} \sum_{n=1}^{N} {_T a_{ji}^{nm}} \cdot p_j^n + {_T a_{ii}^{mm}} + {_T Q_i^m} \cdot {_T R_i^m} \right] (1 + r) \right\}}{_T a^B (1 + r)}$$ (7.5)

Once the price of the building aged T years is known, it is substituted into the T-1st equation to solve for the price of the previous year $(_{T-1} p^B)$, and so on, until prices of the building for each of the T time periods are known.

Once prices of buildings of all different ages are solved we can also define building rent. Building rent, it should be noted, is quite different from ground rent because it represents only the charge on the built structure, and not the levy made on the land itself. Specifically, annual building rent for buildings of age $t + 1$ is equal to the price of the building incremented by the rate of profit at time t minus the price of the same building one year later. Formally:

$$F_{t+1} = {_t b^B} \cdot {_t p^B} (1 + r) - {_{t+1} b^B} \cdot {_{t+1} p^B},$$ (7.6)

where F is the annual building rent.

Providing building maintains constant efficiency, Straffa (1960, p. 65) shows that equation (7.6) is generalizable to:

$$F^* = {_0 b^B} \cdot {_0 p^k} [r(1 + r)^k]/(1 + r)^k - 1.$$ (7.7)

The assumption of constant efficiency, of course, is not a realistic one. As the building ages, more repairs are needed, and it often becomes less suited for the task for which it was originally designed thereby leading to declining output. Because, however, fixed capital is rooted in a wider theory of joint production,

Sraffa is able to define building prices under declining efficiency, as well as the obverse (for the analytical details, see Sraffa, 1960, pp. 65–6; Abraham-Frois and Berrebi, 1979, pp. 86–7). In fact, Sraffa (1960, p. 66) writes that the advantage of his approach is that

It will give the 'correct' answer in every case, no matter how complex, over the life of a durable instrument of production, may be the pattern of falling productivity or increasing maintenance and repairs. It will, besides, make due allowance for any variation in the prices of different materials and services required.

Before presenting the more complex case where all economic activities rent a building, we must discuss the issue of the economic versus the technical life of the building. The outer limit of any building's life is given by its physical durability (represented by T years in the above equations). Buildings, however, may be scrapped before this date if it is economical to do so. As Mainwaring (1984, p. 135) writes: 'there is no reason why a [building] must be operated over its entire physical life-span. If it is profitable to scrap prematurely then competition will ensure that scrapping occurs.' The intuitive reason for early obsolescence is the declining productivity of the building. Although efficient when first produced, the building is increasingly expensive as it ages, thereby making it unprofitable. (In contrast, if all buildings maintain constant or improving productivity there is no premature scrapping. For formal proof, see Abraham-Frois and Berrebi, 1979, pp. 87–90. Alternatively, a building might become obsolete because technical and social change can make it uncompetitive with production in newer buildings – see Chapter 13. An example of this process is the vacating in many cities of older office buildings in favour of newer ones.)

To distinguish more clearly the difference between economic and technical life it is helpful to introduce the notion of truncation. Truncation simply means the premature scrapping of a building in order to secure higher profits. For example, suppose the physical life of a building is three years ($T = 3$). Truncation occurs if the building is scrapped or abandoned after the first or second years. The choice of a truncation date (the year in which to scrap) is really a choice about the technique of production, because each separate truncation date represents a different set of technical coefficients of production, and is associated with a different profit rate for a given wage level. To determine which truncation date is most profitable, draw and compare the different wage–profit frontiers associated with each potential scrapping date. (For the derivation of the w-r frontier under these conditions and numerical examples, see Baldone, 1980.) In our example, we can construct the wage–profit frontiers respectively associated with a building surviving one year, two years, and to the end of its physical life of three years (Figure 7.1.) For any given wage rate, the most profitable truncation is then the one whose w-r frontier lies outside all others.

As can be seen from Figure 7.1, complications may arise when constructing the

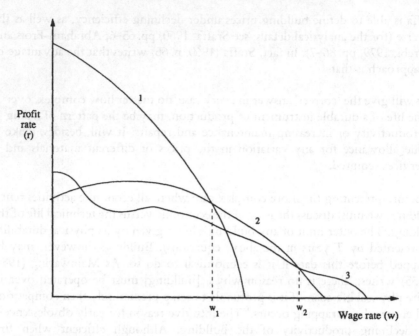

Figure 7.1 Wage–profit frontiers for a building with a physical life of three years

wage–profit frontiers for the different truncation dates because they may be positively sloping. We have already established that under joint production it is possible that some prices are negative, thereby leading to a positively sloped wage–profit frontier for that truncation date. Recalling that fixed capital is a 'species' of joint production, the intuitive explanation of the positively sloped w–r frontier for this case is that, although the net output of good m is always positive, the inefficiency of the aged building causes its net output to be negative, thereby giving the building a negative price. That is, because maintenance costs of the building are so high, or the productivity of the production process is so low, inputs into the building are more than its contribution to the output. In spite of these complications, however, Schefold (1980) demonstrates that: first, prices of all final commodities (all goods except the building) are always positive providing profits are positive; and, second, although the w–r frontier for a given individual truncation may well be positively sloping, the outermost envelope made up from the many individual frontiers is always negatively sloping (see Mainwaring, 1984, ch. 11 for a detailed explanation).

Both Schefold's points are seen in Figure 7.1. The frontier for truncation 1 (scrapping at the end of the first year) is negatively sloping throughout because for the first year the building is an input like any other, i.e. all inputs are treated as circulating capital. Truncations 2 and 3, however, have positive-sloped segments, indicating negative price(s) of the older building(s). The outermost w–r curve made up of the three truncations, however, is negative throughout its entire range. Figure 7.1 also clearly shows that the truncation date depends upon the income distribution. From Figure 7.1, if wages are less than w_1 the building is scrapped after one year, if wages are between w_1 and w_2 it is done after two years, and if wages are greater than w_2 the building is not scrapped until the end of its physical life. In this sense, we once again see how a seemingly pure economic decision, such as scrapping or abandoning a building, is intimately tied to wider social relationships such as class conflict. Finally, although Figure 7.1 does not show it, it is also possible that reswitching of truncations occurs, with a given truncation profitable for two separate wage rates.

The final task in this section is to allow each economic activity to have a building. Because of the cumbersome nature of the mathematical system we will not formally represent this case here (but see Barnes, 1983, pp. 141–4). Baldone (1980, pp. 107–15), however, demonstrates that this complex system yields both an outermost envelope of wage–profit frontiers that is negatively sloped, and a set of positive prices for all final goods. Perhaps the major added complexity that follows from introducing many buildings is the need to take into account the many different ways in which variously aged buildings can be combined together to constitute separate potential sets of techniques or truncation dates. For example, with N buildings in the city, and with each building having a physical life of T years, there will be T^N combinations of the different-aged buildings, and thereby T^N sets of potential truncation dates. Outermost wage–profit frontiers need to be constructed for each of the T^N combinations in order to calculate the profit-maximizing date of scrapping for each building in the city, where each individual scrapping date is influenced by all other scrapping dates. But once scrapping dates are known for each building, the demand for construction services is derived, and thus also the number of construction companies required. Furthermore, it is to be expected that some geographical sectors of the city will have concentrations of buildings with a higher mean age than the average for the city, while other districts will have lower mean ages. The different-aged geographical areas of the city, and thereby the differential construction activity within them, can also be modelled.

In summary, this section has provided an analytical model of the built environment by drawing upon Sraffa's theory of fixed capital. Additional complications stemmed from such an inclusion, but it was demonstrated that the central conclusions and approach of the 'basic model' remain intact. In particular, both production (buildings are a produced good like any other) and income distribution (the prices of buildings are dependent upon the rate of profits) remain pivotal. As before, wage–profit frontiers were used to determine the optimal technique of production, defined in this case as the most profitable truncation date for each building in each zone. Moreover, despite the quirks of joint production, the relevant portion of the wage–profit frontier is always negatively sloping, thereby providing positive prices on final goods. Finally, we are able to calculate prices including rents for buildings of any age. In short, we argue that the model is robust in the face of the complications of the built environment, thereby also indicating the robustness of the more general analytical political economic framework on which it is constructed.

7.4 Residential location

7.4.1 The Garin–Lowry model

At best, the model constructed above deals with only half the urban economy. Questions about the spatial distribution and size of the urban population, the spatial distribution of consumer service activities, the cost, construction and geographical pattern of the housing sector, and the spatial variability of the real wage, have all been ignored. These questions are addressed in this section. This is done by linking the model constructed above to a version of the Garin–Lowry model.

Before discussing the analytical details it is necessary to address some of the criticisms made of the Garin–Lowry model. This is particularly pertinent because major critics include those who advocate as an alternative the political economic perspective adopted by this book.

Apart from technical problems associated with its operationalization (Foot, 1981, pp. 135–6; Webber, 1984, chs 4–7), the Garin–Lowry model is also criticized at a theoretical level (Sayer, 1976; Webber, 1984, pp. 158–64). For example, Webber (1984, p. 160) argues that the 'two central, but related criticisms of the [Garin–]Lowry model ... [are] first, that the model ignores causes; ... [and] second ... that the model and even its dynamic extensions are ahistorical: are not grounded in actual situations'. Both these problems, as Webber (1984, p. 160) makes clear, are the result of the model 'contain[ing] no motor to make things happen'. It will be argued here, however, that such a motor is in effect supplied by linking the Garin–Lowry model to the production-based model of the city constructed earlier in the chapter. By making this link, the dynamics of the production system along with the division of the surplus among classes drive the Garin–Lowry model. The social relations of production and distribution both causally determine the values of the variables modelled by Garin and Lowry, as well as situating the urban economy within a broader system: the production and reproduction of commodities within industrial capitalism. It is for these reasons that there is no incongruity between adopting a political economic perspective and also employing a version of the Garin–Lowry model.

7.4.2 Population

The Garin–Lowry model begins with a set of spatially distributed employment levels for N production sector industries. The Garin–Lowry model assumes that the employment levels and location of the production sector industries are given exogenously, but using the model constructed in section 7.2 and 7.3 we can endogenously derive both variables.

The profit-maximizing algorithm employed in section 7.2 allows us to determine optimal locations for all production sector establishments. Second, we can calculate those physical quantities that must be produced at each production sector industrial site in order to ensure that supply equals demand, defining a socially necessary division of labour within the city (see Chapter 9). By calculating this, we then know the equilibrium physical quantity of commodity m produced at i, x_i^m. The employment in this plant, E_i^m, is then found by dividing the total labour time required by the length of the work week:

$$E_i^m = (x_i^m . l_i^m)/H. \tag{7.8}$$

With levels and the location of production sector employment determined endogenously, we can calculate an expected geographical distribution of workers' residences by using a trip distribution model. Such a model distributes workers employed at zone i among residential zones j on the basis of such criteria as commuting distance, the price of housing and so on. The model we use here is a version of the multinomial equation employed in Chapter 4 to determine trading relationships (equation 4.8). Applied to the residential distribution of workers it says that the proportion of basic workers at employment zone i living in zone j, w_{ji}, is given by:

$$w_{ji} = \frac{e^{-\beta\{kR_j + p_j^h + t_{ji} + (\mathbf{p'b})_j\}}}{\sum\limits_{k=1}^{J} e^{-\beta\{kR_k + p_k^h + t_{ki} + (\mathbf{p'b})_k\}}}, \tag{7.9}$$

where $0 < \beta < \infty$, and is a constant to be estimated. The expression in curly brackets represents the total living costs for workers residing in a zone. This is made up of: the rent on land paid by workers living in zone j (kR_j); the price of housing in zone j (p_j^h); the cost of transport for workers commuting from zone j to zone i (t_{ji}); and the cost of consumption goods in zone j ($(\mathbf{p'b})_j = \sum\limits_{m=1}^{N} \sum\limits_{j=1}^{J} p_j^m . b_j^m$, where b_j^m is the quantity of good m consumed and p_j^m is its price). The value of each of these variables is derived endogenously below.

Equation (7.9) states that the likelihood that employees of zone i will live in zone j increases as the cost of living in j falls relative to the cost at other locations k. The degree to which workers concentrate in the cheapest zone is measured by the index β (section 4.1.1). Specifically, where workers choose to live depends upon land rent, housing costs, commuting costs and costs of consumption. As β approaches infinity, workers are all found in the zone j where these costs are lowest, whereas as β approaches zero workers are uniformly scattered across the city regardless of variations in living costs.

Multiplying employment levels in each zone i by the respective trip distribution equation provides a determination of the number of workers in the production sectors living in each zone j, W_j:

$$W_j = \sum_{m=1}^{N} \sum_{i=1}^{J} E_i^m . w_{ji}. \qquad (7.10)$$

Finally, to determine the total population in each zone j, P_j, multiply the number of workers living there by exogenously given labour participation rates, u_j:

$$P_j = W_j . u_j. \qquad (7.11)$$

In allocating workers to residential zones across the city we necessarily adopted a formal approach. This is congruent with our broader analytical stance. Ostensibly one of the casualties of our abstract method is an inabilty to discuss residential class consciousness and solidarity, which many consider important (Harris, 1984). In fact, in his recent works Scott (1988a, b) makes residential class consciousness pivotal, seeing it as a means of securing the reproduction of labour. We argue that issues of class consciousness and solidarity can be treated within an analytical approach, but we postpone discussion of this until the whole topic of class is examined (Chapters 10–12).

7.4.3   The retail sector

The population estimates derived above for each zone, j, P_j, were based only on workers in the production sector. As is well known from the Garin–Lowry model, such estimates must be further modified to take into account the addition of consumer service sector (retail) employment. To introduce retail services in our model, and their effects on the distribution of population, it is necessary to discuss two issues: first, the retail prices of consumer goods sold at any establishment in zone r; and, second, the pattern of retailing trips made by households living in any zone j.

Taking the first issue, let us assume that there is a retail outlet in each land zone r where there is also located a production sector economic activity. Assuming that retailers purchase their goods from a single wholesale centre at the central business district, prices of consumer goods will vary at different retail sites. Such variations result from the difference in costs among retail outlets. The price of m sold in retail outlet s, equals: the price of m in the central market, plus transport costs of shipping it to zone r, plus land rent and retail labour costs, all incremented by the rate of profit:

$$p_r^{m*} = [p_C^m + \tau_{Cr}^m . P_C + Q_r^m . R_r^m + \left(l_r^m . \sum_{n=1}^{N} p_j^n \, b_j^n / H \right)] (1 + r), \qquad (7.12)$$

where p_r^{m*} is the price of good m sold at the retail outlet at land plot r. All other variables are defined as before, except that the rate of profit is now interpreted as a retail mark-up. It is also assumed that the land rent paid by retail outlets at land plot r equals the land rent paid by any producer sector industry m located in that same parcel.

Inspection of equation (7.12) reveals why retail prices will vary geographically throughout the city. Because of differences in transport costs and land rents, per unit prices of the same commodities will depend upon the location of the retail outlet.

Having established prices of consumer goods throughout the city, we are now able to derive the pattern of consumer shopping trips. This is facilitated by using the same type of multinomial equation employed earlier when examining residential location (equation 7.9). Each family receives a consumption bundle represented by the vector $\mathbf{b}$ ($\mathbf{b} = [b_1^1, \ldots b_K^N]$). The amount of good n represented by the coefficient b_j^n can be purchased from any retail location in the city. For any family:

$$b_j^n = \sum_{r=1}^{J} b_{rj}^n, \tag{7.13}$$

where b_{rj}^n is the amount of good n purchased at the retail outlet at land plot r for consumption at land plot j.

The pattern of purchasing will clearly depend upon prices and the cost of travelling to retail outlets. Given that we know both where production sector workers live within the city and prices at every retail establishment, we can determine the cost to the consumer of purchasing the bundle of commodities that make up their wage from any shop located in zone r. The effective purchase price of consumer good m purchased in zone r by residents of zone j (q_{rj}^{m*}) is:

$$q_{rj}^{m*} = p_r^{m*} + \tau_{rj} \, p_j^l, \tag{7.14}$$

where τ_{rj} is the cost of a shopping trip from j to r. The proportion of the total demand for good m purchased from zone r by residents of zone j is then equal to:

$$h_{jr}^m = \frac{e^{-\beta(q_r^{m*})}}{\sum_{k=1}^{J} e^{-\beta(q_k^{m*})}}. \tag{7.15}$$

The price-responsiveness of the consumers at j with respect to the consumption bundle sold at location r is represented by the value of β. When β approaches infinity, consumers purchase all of their requirements of good m from the cheapest zone. For lower values of β, consumers are more likely to pay more for their good than they would otherwise need. This can be a result of such things as imperfect information, the constraints of daily life, shop loyalty, convenience and so on.

With the degree of consumer price-responsiveness determined by the value of β, equation (7.15) can be used to determine retail revenues for any zone r.

Specifically, the weekly retail revenue for any zone r, V_r, is the product of total customers and the price of the consumption bundle:

$$V_r = \sum_{m=1}^{N} \left(\sum_{j=1}^{J} P_j . h_{jr}^m . b_j^m \right) p_r^{m*}. \tag{7.16}$$

The term in brackets calculates the total purchases of good m in region r, summed over the demand for all residential zones j. This is then multiplied by the price of that good to calculate total revenue for good m, which is then summed up over all consumer goods to give total revenue, V_r. Knowing V_r enables us to calculate employment levels at each establishment, E_r^R:

$$E_r^R = V_r . f_r, \tag{7.17}$$

where f_r is the number of employees hired per dollar of retail sales.

Total employment in each zone i (where $r = i$) is then the sum of retail and basic sector employment:

$$E_i = E_i^R + E_i^B \tag{7.18}$$

Finally, we are able to return to the issue we raised at the beginning of this subsection, namely, the modification of population estimates resulting from the introduction of retail employment. Once retail employment is added to each zone i then, by using the procedure outlined in 7.4.2, we must also distribute retail employees to residential zones and recalculate population estimates. But these new and increased population values will, in turn, require re-estimation of retail trips and the volume of retail trade, thereby changing retail employment levels. In short, the process is a recursive one, where changes in population influence employment levels in retailing, and vice versa, as the multiplier effects of production sector employment work themselves out through the city. Garin (1966) demonstrated that this iterative process eventually stabilizes. As a result, for any given set of employment levels in the production sector, we are always able to calculate the eventual levels of total population and total employment for a city. Garin's iterative solution is provided in the Appendix to this chapter, where the analysis of this section is re-couched in terms of matrix algebra.

7.4.4   The housing sector

In equation (7.9) we suggested that the location of workers living in j depended upon five variables: transportation costs to work and shop, consumption costs, land rent, housing costs, and the responsiveness to living costs. The solutions to the models constructed in sections 7.4.2 and 7.4.3 endogenously determine the first two variables. In this subsection we endogenously derive housing costs and land rents.

Clearly housing is a necessary component of worker consumption and as such is part of the consumption bundle vector **b**. By housing cost we mean the rent applied to the purchase or lease of the physical structure of the residence. These costs depend on the age and depreciation of the building. Although it is possible within our framework to individualize the type of housing for each worker, for analytical simplicity we will assume that all workers live in the same kind of residence, and that homes depreciate at a constant rate. We also assume, as before, that there is a construction company in each zone j that builds new houses to replace old demolished ones, and which obtains inputs from the central market. The price of a new house under such conditions, $_0b^h \cdot _0p_j^h$, is then given by:

$$_0b^h \cdot _0p_j^h = \left(\sum_{m=1}^{N} \sum_{i=1}^{J} a_{ij}^{mh} \cdot p_i^m + Q_j^c R_j^c \right) (1 + r), \tag{7.19}$$

where all variables are as before, and c is the construction company.

With the transport cost component and rent of equation (7.19) different for each residential zone j, it is clear that house prices will vary from one part of the city to another. Because housing is a durable good, this price difference will also be reflected in the annual rental payments for housing. Following the procedure in section 7.3 (equations 7.6 and 7.7), annual housing rent for housing of age t years in residential zone j is equal to:

$$F_{tj}^h = {}_t b^h \cdot {}_t p_j^h (1 + r) - {}_{t+1}b^h \cdot {}_{t+1}p_j^h = {}_0b^h \cdot {}_0p_j^h [r(1 + r)^t]/(1 + r)^t - 1. \tag{7.20}$$

The implication of equation (7.20) is that, because new housing prices are different for each zone, then so too are annual housing rents.

Let us turn now to the second component of the cost of worker housing, land rent. Assume that households in a zone consume the same quantity of land (1 unit), and that residential land rent in zone j is some increasing function of the number of people living in that zone:

$$R_j = f(P_j), \tag{7.21}$$

where f is some positive monotonic function.

Although equation (7.21) implies that the immediate determinant of residential rent is demand (a greater number of households living in zone j implies greater demand and hence higher rents), the broader theoretical context remains production based because population densities depend upon the spatial distribution of production sector economic activities.

One neglected issue in our model is that of land constraints. We implicitly assumed that there is enough room in each land zone j to accommodate all those workers who are assigned there. As Wilson (1974) and others have shown, however, land constraints are readily included within the Garin–Lowry model. Furthermore, there is an implicit constraint operating through rental levels in equation (7.9). Because higher populations imply higher rents, workers will choose less populous zones provided they are to some degree responsive to price differences.

The upshot of this discussion is that, knowing the distribution of workers and their place of employment, we are able to estimate each worker's housing and land rental costs. The one remaining variable that requires discussion with respect to equation (7.9) is the β parameter representing the responsiveness of residential and shopping behaviour to cost differences. We turn to this issue now.

7.4.5 Wages, rationality and class conflict: some reconsiderations

Although the model we constructed above is analytically consistent, we neglected discussing issues of social constraints and conflicts. We consider them here, and in so doing point to modifications of our analysis – modifications that are developed more fully in Chapters 10–12, where we deal with social classes and the inter- and intra-class conflicts that might exist.

In non-spatial models it is usually assumed that wage rates are equal for all workers. This is a reasonable assumption only on the condition that all workers live, work and purchase consumption goods at a dimensionless point. Once we move away from a spaceless economy, such an assumption is problematic. In particular, if workers employed at the same place receive an identical money wage irrespective of where they live, it follows that workers in different zones will receive different real wages because of spatial variations in the cost of the consumption bundle. Hitherto, we have assumed that money wages in a given zone i are set by the marginal worker, where a marginal worker is defined as the person whose combined consumption bundle costs exceed those of any other worker employed in the same zone. Such a solution, given the remarks above, raises at least two issues that need further exploration: first, intramarginal workers are left with a money surplus or 'rent'; and second, different wage rates are paid in different parts of the city because in general the location and living costs of the marginal worker are not the same across manufacturing and producer service companies.

Let us take the first issue. Clearly firms are anxious to eliminate intramarginal workers' surpluses. To do this each firm could pay their workers a different wage rate: one that just allowed purchase of the socially necessary consumption bundle. But apart from the problems of amassing the information to carry out such a policy, it would be impossible to implement on the shop floor.

A more realistic approach from the perspective of employers is to pay a lower wage rate, thereby forcing marginal workers (or any worker whose consumption bundle costs more than the assigned wage rate) to change where they live, shop or work. This points to a problem in our own framework. In the model constructed, we did not impose any income constraints on the workers when making their spatial choices. We assumed that employers always paid a wage that covered their pre-defined consumption bundle, no matter where workers lived and how much they spent on retail consumption goods. Clearly this is unrealistic. It is more plausible to see wages set at some level (determined by worker–capitalist negotiation) which then influences the very spatial choices of workers as they seek places to live and places to shop.

This can be incorporated into the above analysis if we treat the β coefficient as endogenously determined by the constraint of income. The value of β then corresponds in a direct way with wage levels. When β approaches zero, workers are evenly scattered throughout the city, and the marginal worker lives in the zone where the consumption bundle is most expensive, implying higher wage rates. In contrast, if β approaches infinity the marginal worker's consumption bundle presumably will not cost much more than that of the intramarginal worker, which would be the case when wages are low. Our suggestion, then, is that the coefficient β reflects the wage settlement, ensuring that marginal workers select a place to live and a place to shop that allow them to purchase the consumption bundle. By making wage negotiations establish the limits on the marginal worker's spatial choices, it is then class relationships that shape the residential distribution of the city. This further implies that conflict between workers and other classes is over only the marginal worker's consumption bundle; intramarginal workers will still receive a surplus.

Given this interpretation, a number of issues that relate to urban class conflict follow. First, precisely because there are now constraints on the location of workers, workers are likely to be in conflict with both landlords and factions of the capitalist class. As Harvey (1974) makes clear, once there are spatial constraints on the mobility of any class there is the potential for class monopoly rent (MR2). For exactly the same reason, those workers who are more restricted in choice of residences by low wages are also liable to pay higher prices for their retail consumption bundle. When the retailer faces a more captive market, it is possible for him/her to levy a monopoly charge on each good sold because his/her effective spatial monopoly is greater.

Second, endogenously fixing the β coefficient can also lead to potential class conflict between capitalists and landlords, and among capitalists themselves. To keep wages low, capitalists have an interest in pressuring landlords to lower residential rents, and to prevent the charging of MR2 (see Chapter 11 for further discussion). In addition, capitalists operating in the production sector and those that operate in consumer services potentially have mutually opposed interests on two different counts: first, production sector employers are against any retail monopoly prices because of the workers' wages; and, second, there is an implicit conflict over the location of retail sector establishments. In our model the location of shops depends upon demand as reflected in the spatial distribution of population. As such, this location is *not* defined by the outermost wage–profit frontier, as was the case with production sector plants. For this reason, it is very possible that the set of retail locations based on population demands is not equivalent to the set that would maximize profits for *all* capitalists.

Third, if there are different groups of workers receiving different wages at the same site, such as skilled and unskilled workers, then residential segregation on the basis of income differences is a logical outcome of this model. The higher-paid, skilled workers have a broader range of residential sites available to them owing to a less binding income constraint, whereas the unskilled workers would be confined to the zones where living costs are cheapest. This is in stark contrast to Alonso's (1964) suggestion that residential segregation is based on a greater preference by higher-income groups for larger housing space (see Harvey, 1972).

Let us now turn to the second issue, that of differential wage rates. Variation in money wages across the city has been verified empirically by a number of researchers, including Scott (1988a, ch.7). In accounting for such variation, Scott mainly points to differences in commuting costs, but, as we have already seen, there are other sources of variations in wages, including housing and land rent costs, and differential retail prices.

The significance of money wage variation across the city is potential intra-class conflict among workers. This issue is taken up in more detail in Chapter 12, but the existence of differential wages within the city may result in workers with lower wages demanding remuneration from capitalists comparable to that of workers at the higher end of the money wage scale. Higher-paid workers, however, may themselves resist such demands, wishing to keep such differentials in place, thereby precipitating intra-class conflict (for an example, see Clark, 1986).

Summary

Until recently analytical models of the geography of the intra-urban economy were largely synonymous with a neoclassical framework. Certainly radical urbanists provided models as insightful and as conceptually rigorous as the neoclassical ones, but because the forms of discourse of the two schools were so different there was little engagement between them. The recent literature on modelling the city in formal terms from a political economic perspective attempts to move beyond this impasse. The analytical nature of the political economic models challenges neoclassical orthodoxy on its own terms, but it serves more than just a critical function. It also represents a potentially useful way both to clarify existing ideas and to hone new ones. This is what we have tried to do in this chapter. First, following Scott, we provided the analytical undergirding necessary to support a vision of the urban economy rooted in relations of production and distribution that can determine the location, prices and rents paid by basic industry in the city. Second, we showed that this framework can be extended by including fixed capital, thereby allowing us to deal analytically with the cost of the built environment. Certainly, such an extension brings complexities and complications, but at bottom we demonstrated that our framework is sufficiently robust to take into account the naggingly difficult problem of the formal representation of the city's housing, office and industrial building stock. Finally, we linked the Garin–Lowry model with our own to make predictions about where workers will live, shop and travel. Furthermore, the model also allowed us to examine the relationship among wage rates, intra-urban variations in the cost of living and residential choice. Although some may suggest that this is at best misdirected eclecticism, and at worst subversion of our own approach, it remains to be demonstrated in what way it is inconsistent with our broader objectives. In fact, this combination led readily to a discussion of the nature and consequences of urban class conflict.

There remains the question of whether this is empty formalism. We are not claiming that our model or our method are the only ways to represent the geography of an urban economy, but they are potentially useful for both their critical function and their clarifying one. It is easy to think that, because neoclassical models use analytical methods and are flawed, any formal model of whatever ideological stripe is also similarly marred. Such reasoning, apart from embodying an erroneous logical deduction, denies the very breadth of vision for which the political economic approach is celebrated.

Note

1   We use the term 'production sector' rather than 'basic sector', which is the term usually used in the literature, because of potential confusion with Sraffa's concept of basic and non-basic commodities. Recall that for Sraffa the basic sector is the set of industries whose output is directly or indirectly used as an input to all other industries. In contrast, the 'basic sector' as used by researchers in urban economics usually means any manufacturing activity.

Appendix: A matrix version of the Garin–Lowry model

To establish the final distribution of population in any zone j and total employment in any zone i given the location of production sector employment, let us recast the equations used in section 7.4 in matrix form. Let the employment in the production sector for each zone j be represented by a vector $\mathbf{E}_i^B$ consisting of elements E_i^m, where E_i^m is the employment in production sector m located in zone i.

To calculate the distribution of population based only upon production sector workers, multiply $\mathbf{E}_i^B$ by the trip distribution matrix $\mathbf{W}$ (whose elements consist of W_{ji}; equation 7.10), and the diagonal labour participation rate matrix $\mathbf{U}$ (whose diagonal elements consist of u_i). Then equation 7.11 becomes:

$$\mathbf{P}' = \mathbf{E}_i^{B'}.\mathbf{W}.\mathbf{U}, \tag{7A.1}$$

where $\mathbf{P}$ is the vector of population consisting of elements P_i, where P_i is the population in zone i; and $'$ represents the transpose.

We can similarly define in matrix form the vector $\mathbf{E}^R$ consisting of the elements E_r^R, where E_r^R is the level of employment in the retail sector of zone r:

$$\mathbf{E}^{R'} = \mathbf{P'}.\mathbf{H}.\mathbf{B}.\mathbf{P\star}.\mathbf{F}, \qquad (7A.2)$$

where $\mathbf{H}$ is the consumer retail choice matrix consisting of elements defined by equation (7.15); $\mathbf{B}$ is a diagonal matrix with diagonal entries b_j^m; $\mathbf{F}$ is the diagonal matrix of retail employment generating coefficients, f_i; and $\mathbf{P\star}$ is the rectangular (NJ by N) matrix:

$$\mathbf{P\star'} = \begin{Bmatrix} p_1^1 p_1^2 \ldots p_1^N\,0\,0\,0\ldots. \\ 0\,0\ldots0 \quad p_2^1 \quad \ldots p_2^N\,0\,0\,0\ldots \\ \cdots\cdots\cdots\cdots\cdots\cdots \\ 0\,0\ldots0\,0 \quad \ldots 0\,\,0\,0\,0\ldots. \, p_j^1 \ldots p_j^N \end{Bmatrix}$$

The effect of introducing retail employment is to change the population in the city. In turn, however, the change in population affects retail employment. Determining total employment $\mathbf{E}$, and total population $\mathbf{P}$, is thus an iterative process where changes in $\mathbf{E}^R$ cause changes in $\mathbf{P}$ and vice versa.

Substituting equation 7A.1 into equation 7A.2 for $\mathbf{P'}$ gives us the total retail employment directly generated by production sector workers:

$$\mathbf{E}^{R'}(1) = \mathbf{E}^{B'}\mathbf{S}.\mathbf{R}, \qquad (7A.3)$$

where $\mathbf{R} = \mathbf{H}.\mathbf{B}.\mathbf{P\star}.\mathbf{F}$, and $\mathbf{S} = \mathbf{W}.\mathbf{U}$.

Here $\mathbf{S}$ converts employment to population totals and residential locations, and $\mathbf{R}$ distributes retailing with respect to population.

The second estimate of the distribution of population is then:

$$\mathbf{P'}(2) = [\mathbf{E}^{B'} + \mathbf{E}^{R'}(1)].\mathbf{S}, \qquad (7A.4)$$

and, from 7A.3:

$$\mathbf{P'}(2) = \mathbf{E}^{B'}\,(\mathbf{I} + \mathbf{S}.\mathbf{R}.).\mathbf{S}, \qquad (7A.5)$$

where $\mathbf{I}$ is the identity matrix.

The retail employment generated by this population distribution is:

$$\mathbf{E}^{R'}(2) = \mathbf{P'}(2).\mathbf{R} \qquad (7A.6)$$

$$\mathbf{E}^{R'}(2) = \mathbf{E}^{B'}(\mathbf{I} + \mathbf{S}.\mathbf{R}.).\mathbf{S}.\mathbf{R}. \qquad (7A.7)$$

Garin (1966) was the first to show that this iterative process is a convergent one. Garin further showed that the outcome of this convergent process is represented by the inverse matrix $[\mathbf{I} - \mathbf{S.R}]^{-1}$. Total population and total employment are then respectively given by:

$$\mathbf{E'} = \mathbf{E}^{\mathbf{B'}} (\mathbf{I} - \mathbf{S.R})^{-1}.\mathbf{S.R} \tag{7A.8}$$

$$\mathbf{P'} = \mathbf{E'} (\mathbf{I} + \mathbf{S.R}).\mathbf{S}. \tag{7A.9}$$

8 *The labour value circuit*

Introduction

We now turn from the circulation of exchange value and related questions of the price of land and buildings, to an examination of the circulation of labour value in space. We saw that the circulation of exchange value depends on four factors: the socially necessary production methods employed in each region; the socially necessary level of the real wage; the allocation of commodity production among regions; and the spatial structure of the inter-regional economy as expressed in the difficulty of transportation between regions. Given these, a pattern of commodity trade develops under full competition that reflects both the relationships between locations and the production prices charged by competing suppliers of inputs. This implies that a relationship exists between production prices and the circulation of exchange value. The question we address in this chapter is the circulation of labour value during the process of commodity production and exchange.

Section 8.1 explores the relationship between the trade of commodities among firms in different regions and the labour value of those commodities. In section 8.1.1 we investigate how labour values are determined in a fully competitive space economy. In section 8.1.2 we show that, even when the rate of profit is equalized everywhere, the fact that production occurs in space leads to unequal exploitation rates in the Marxian sense in different regions. Section 8.1.3 explores the implications of this finding for the relative importance of class versus regional political alliances. Finally, section 8.1.4 examines the balance of trade in labour value terms, as a prelude to the discussion of unequal exchange in section 8.2. Unequal exchange is explored for both a two-region economy (8.2.1) and a multi-regional economy (8.2.2), and this is followed by a critical re-examination of empirical studies of unequal exchange in the light of the theoretical argument of the chapter (8.2.3).

8.1 The geography of labour value and exploitation

Let us suppose that the space economy achieves some form of equilibrium in its exchange value sphere. In other words, suppose there exists a geographical distribution of production facilities, trading patterns and prices of production that is in trading and pricing equilibrium. In this section we evaluate the circulation of labour value and geographical variations in the rate of exploitation that exist in this idealized vision of a competitive space economy.

8.1.1   Labour values

The transportation sector is once again crucial to this re-examination of Marx's theory in a geographically extensive economy. There have been considerable debates in Marxist economic theory over the definition of productive labour in capitalism. We argue that workers in the transportation sector are productive in the sense that they produce a commodity without which capitalist commodity production is impossible. To repeat our argument of Chapter 4, the transportation commodity is the production of accessibility – something that is essential to any realistic, spatially extensive economy. For this reason, the value of transportation must be included in calculating the value of any commodity.

With this in mind, we define the labour value of commodity n produced in region j as the labour value of the capital goods required, plus the labour value of the transportation necessary to produce those goods, plus the amount of labour required in production, where all requirements represent the amounts socially necessary for production. This is calculated as follows, where the labour value of commodity n produced in region j is λ_j^n.

$$\lambda_j^n = \sum_{i=1}^{J} \sum_{m=1}^{N} \hat{a}^{\star mn}_{ij} . \lambda_i^m + \sum_{i=1}^{J} \sum_{m=1}^{N} \hat{a}^{\star mn}_{ij} . \tau_{ij}^m . \lambda_i^t + l_j^n. \tag{8.1}$$

Here $\hat{a}^{\star mn}_{ij}$ is the amount of capital good m purchased from region i for the production of a unit of commodity n in region j under trading and pricing equilibrium, given the socially necessary production method prevailing in region j. τ_{ij}^m and l_j^n are, respectively, the transportation and labour requirements as discussed in Chapter 4. Such an equation can be written for each commodity produced in each region.

Consider the similarity of the first two expressions on the right-hand side of equation (8.1) with the right-hand side of equation (4.7). We apply the same reasoning used there to show (see equation 4.4) that:

$$\lambda_j^n = \sum_{i=1}^{J} \sum_{m=1}^{N} \hat{a}^{\star mn}_{ij} . \lambda_i^m + l_j^n.$$

These equations can also be written in matrix form, which allows for easier comparison with production prices. Define $\boldsymbol{\lambda}' = [\lambda_1^1, \lambda_1^2, \ldots, \lambda_j^n, \ldots, \lambda_J^N]$ as the vector of labour values for each commodity produced in each region, $\mathbf{L}' = [l_1^1, l_1^2, \ldots, l_j^n, \ldots, l_J^N]$ as the vector of corresponding labour inputs, and $\mathbf{A}\star$ as the matrix of inter-regional shipments of capital goods only that would occur in trading and pricing equilibrium. Then labour values can be determined from (Sheppard, 1987):

$$\boldsymbol{\lambda}' = \boldsymbol{\lambda}' \mathbf{A}\star + \mathbf{L}'.$$

Note that the entries in the matrix $\mathbf{\hat{A}}\star$ depend in part on production prices (equation 4.8). This matrix equation, in conformity with Chapter 3 (equation 3.2), is:

$$\boldsymbol{\lambda}' = \mathbf{L}'[\mathbf{I} - \mathbf{\hat{A}}\star]^{-1}. \tag{8.2}$$

Once we recognize that the labour value of transportation must be included in our calculations of the labour value of any commodity, including transportation itself, then this implies that the labour value of a commodity can, and generally will, vary between regions. This is true for all kinds of commodities including labour power. Certain insights about the geography of labour value and of exploitation follow from this.

First, consider locations where production facilities are highly accessible to other economic sectors, particularly to 'efficient' enterprises within those sectors (where we define efficient firms as those where the labour value of the socially necessary production method is low). At such locations, which can be defined as occupying a position of high centrality within the space economy, it takes less effort to assemble inputs for production, and inputs are efficiently produced in the first place. The labour value of commodities produced there is therefore lower, *ceteris paribus*.

Second, similar variations exist for the labour value of labour power, defined as the labour value of the products consumed by the worker and his/her family per hour worked. Even if the real wage (the consumption bundle per hour of work) is identical in each region, the labour value of all commodities, including consumption goods, varies between regions. This leads to some interesting insights into the geography of Marxian exploitation and Marxian analysis of class in a spatial context, which are explored in the following sections.

At a broader level, this analysis has theoretical implications for the status of labour values. We saw in the debate over the transformation problem that, in an aspatial economy, labour values and exchange values are determined in parallel from knowledge of socially necessary production methods and the socially necessary wage (Chapter 3). While the calculation of labour values in a space economy closely parallels the determination of labour values in an aspatial economy, there is a crucial difference: the inter-regional input–output coefficients depend in part on the spatial structure of production prices (Chapter 4). Thus, if labour values are useful in analysing the regional geographical distribution of economic activities, then we cannot defend (at this relatively disaggregated level of abstraction) the proposition that labour values are an immediate determinant of prices. A similar conclusion was reached by Roemer (1982a), when he attempted to establish a micro-foundation for exploitation and class under capitalism.

Yet this conclusion does not necessarily disprove the law of value – the assertion that labour values implicitly regulate production prices. As Shaikh (1984) argues, the law of value can operate at a deeper level; determining the socially necessary division of labour in such a way that it forces upon society those production methods that lead to a very close relationship between labour and exchange values. Harvey (1982) has argued in a similar vein that socially necessary labour values are defined as those labour values that are necessary to ensure continued capital accumulation, and that when prices diverge from values this threatens capital accumulation and thus cannot persist for very long. We will return to these issues in Chapter 9, where accumulation and the socially necessary division of labour are examined in detail.

8.1.2 Exploitation

In this section we examine the geography of exploitation, and show that in general the rate of exploitation is unequal in different regions. In order to simplify our argument, suppose that the real wage earned by workers is identical in all regions; i.e. workers and their families consume equal amounts of the same mix of commodities in each region weekly, and that the length of the work week is the same. Now, the labour value of labour power in region j (λ_j^L) is equal to the value of the commodities consumed plus the value of the transportation necessary to deliver them to the workers:

$$\lambda_j^L = \sum_{i=1}^{J} \sum_{m=1}^{N} (b_{ij}^m . \lambda_i^m + b_{ij}^m . \tau_i^m . \lambda_i^L)/H. \tag{8.3}$$

Using the previous definition of the rate of exploitation (section 3.3.3), the rate of exploitation in region j is:

$$e_j = (1 - \lambda_j^L)/\lambda_j^L. \tag{8.4}$$

These results tell us two things. First, the value of labour power varies among regions, even if the real wage is the same everywhere. The labour value of labour power depends on the labour value of the consumption commodities at the locations from which they are shipped, plus the labour value of the transportation hired in those regions in order to ship these products to region j. Recalling that the value of the consumption commodities themselves and of transportation depend on the accessibility of these commodities to the inputs necessary for their production, we can summarize as follows. The value of labour power depends on the accessibility of the labour market to the location of efficient wage goods manufacturers, and on the accessibility of those manufacturers to their suppliers of capital goods. The value of labour power tends to be lower in regions with a higher degree of centrality, in this sense, within the rest of the space economy. A second result then immediately follows, that the rate of exploitation will also vary in different regions, being higher in those regions where the value of labour power is lower.

As we noted in Section 3.3.3, a fundamental theorem of Marxist theory is that exploitation is necessary in order for capitalist production to be profitable. This 'fundamental Marxian theorem' may be summarized as: 'the rate of profit is always less than the rate of exploitation' (Morishima, 1973, p. 63; see also, Okishio, 1961). This result is well established for an aspatial economy, but we wish to understand how applicable it is, or whether it makes any sense at all, in a spatial economy. Let us define the average labour value of labour power in a nation, $E(v)$, as equal to the weighted average of the different labour values of labour power in different regions, using as weights the quantity of labour employed in each region. By definition, the quantity of labour employed in region i, L_i, is the sum of all labour required to produce commodities there ($L_i = \sum_{n=1}^{N} l_i^n . x_i^n$). Then the average labour value is:

$$E(v) = \sum_{i=1}^{J} \lambda_i^L . (L_i \Big/ \sum_{i=1}^{J} L_i).$$

The average rate of exploitation, $E(e)$, is defined with respect to the average value of labour power:

$$E(e) = [1 - E(v)]/E(v).$$

On the basis of these definitions, we can make two deductions about the relationship between exploitation and profit making (Sheppard, 1984):

(a)   When labour is allocated according to the socially necessary division of labour, the rate of profit (r) is less than or equal to the mean rate of exploitation for the economy $E(e)$.

(b)   The average rate of profit for the economy as a whole under full capitalist competition is less than the highest regional rate of exploitation in the space economy.

In principle, therefore, there is no restriction on the rate of exploitation in a single region; indeed in some relatively inaccessible regions it can even be negative. There are, however, restrictions on the distribution of rates of exploitation across the different regions. Clearly, if production is to be profitable, the rate of exploitation must be positive in at least one region, and if the distribution of production is such as to ensure that supply matches demand and smooth capital accumulation occurs (see Chapter 9), then the rate of profit is positive only if the mean rate of exploitation is positive. The possibility that the rate of exploitation in individual regions may be less than the rate of profit, or may even be negative, considerably complicates the kinds of accepted beliefs about exploitation and class struggle that stem from the literature on aspatial economies, an issue to which we now turn.

8.1.3   Implications for class alliances

Two broadly differing intepretations of the relationship between class, exploitation and social action can be identified: the macro interpretation and the micro
interpretation. According to the macro interpretation it is taken as self-evident
that exploitation in the sense defined here is an indicator of the degree to which
workers as a class stand to gain by struggling to increase real wages and reduce
the rate of profit. Under this interpretation, in those regions where the rate of
exploitation is low workers may be exploited, but they are exploited less than
their counterparts in other regions. A corollary is that it is possible for workers in
some regions to negotiate high real wages and a low rate of exploitation, as long
as rates of exploitation in other regions are high enough to ensure a reasonable
average rate of profit for the space economy as a whole. Some authors,
particularly in an international context, suggest that this possibility of a 'labour
aristocracy' implies that workers in those regions with a low level of exploitation
are less inclined to join with their comrades elsewhere in a struggle against
capitalism since they already benefit from relatively low exploitation. Others
draw the further implication that workers in regions with low rates of
exploitation are effectively exploiting workers elsewhere. These implications
will be pursued in detail in Chapter 12.

An important qualification to this result must immediately be noted, however.
Workers in some regions of a capitalist space economy can have low or negative
rates of exploitation only because those workers are members of a larger group
that is collectively exploited. While these workers are themselves technically
speaking exploiters, their presence is not essential to the existence of exploitation
in the space economy. In Roemer's (1982a) terminology, while they may be
exploiters they are not culpable. Indeed their presence reduces both the rate of
exploitation and the money rate of profit for capitalist production. In this sense it
is incorrect to identify them as exploiting workers in other regions; there is no
reason to believe that the rate of exploitation elsewhere would fall if for some
reason real wages were to decline in the region with a low rate of exploitation.

The introduction of the spatial dimension produces results here that are
qualitatively different from those suggested by a theory of the aspatial economy.
While the class of workers is exploited in Marxian terms, those in particular
regions may be exploited significantly less than the average or may even have
negative exploitation rates. If exploitation is a measure of how much a group
would have to gain by withdrawing from the capitalist mode of production
(Roemer, 1982b; see Chapter 10 of this book), and if we draw the further
implication that the likelihood that a group of workers would struggle against
capitalism is directly related to the amount that they have to gain from going it
alone, then we would conclude that workers in regions with less exploitation are
less likely to challenge capitalism. Indeed, they may consciously or unconsciously
ally themselves with the goals of capitalist economic growth and thus with
capitalists in their region, forming a regional class alliance, for the simple reason

that they have 'never had it so good.' This is despite the fact that they are still members of an exploited class, and that they may well be highly vulnerable members, in that capitalists may choose to relocate industry out of that region in order to increase the overall rate of profit. Such regional class alliances are commonplace, even though this dimension to the economic analysis of class struggle has received attention only recently (Urry, 1981; Walker, 1985). The kind of analysis attempted here begins to unravel some of the economic reasons that underlie such alliances.

While one reason for regional class alliances may well be the spatial variation of prices and values and their effect on class identity, the model presented here does not pretend to be a realistic representation of even the general contours of actual space economies. We assumed equal real wages everywhere to simplify the theoretical argument – to show that even with equal wages there are geographical variations in the rate of exploitation. On this basis our conclusion is that the rate of exploitation is lowest in regions where the value of the real wage is highest. According to our analysis this would be more likely to occur in those regions that are less accessible to the geographic and economic core of the economy. Yet in fact it is in the core regions and cities of national economies, and of the international economy, that we expect to find the labour aristocracy – regions where real and money wages are high, and the rate of exploitation correspondingly low. The labour value of the real wage depends not only on the value of the individual commodities consumed but on the size of the real wage. Historical, political and cultural factors have undoubtedly contributed to forcing the real wage up more rapidly in core regions with a long history of factory work and unionization, reducing the rate of exploitation there.

On the other hand, however, this model can help explain why it has been possible for struggles to increase the real wage in the core regions to be so successful. Prices and values are lowest in these regions, *ceteris paribus*, because they are places where both commodity production has agglomerated (reducing transportation costs) and where more productive techniques are employed (reducing socially necessary labour time). An absolute increment in the real wage in such a region, therefore, reduces the mean rate of exploitation and rate of profit by a lesser amount than would the same absolute increment in regions where the value and price of consumption goods are higher. If this consideration is coupled with an ability to expand capitalism into new regions where exploitation rates are very high, then workers in the core regions may achieve increased living standards with little loss of profitability. Their success in obtaining part of the surplus would be offset by the new surplus obtained from these new regions, meaning that their increased living standards put less downward pressure on profit rates. The ability of workers from the core to negotiate increased wages will be further enhanced if their skills are necessary to the production of certain commodities.

8.1.4   Circulation of labour value

As with the circulation of exchange value, the circulation of value can be given a geographical interpretation. Consider equation (8.2), repeated for convenience, recalling that we are concerned only with equilibrium flows of capital (represented by $\mathbf{A}^*$):

$$\lambda' = L'[\mathbf{I} - \mathbf{A}^*]^{-1}. \tag{8.2}$$

As discussed in section 3.2, the elements of the matrix $[\mathbf{I} - \mathbf{A}^*]^{-1}$ are interpreted as the sum of all direct and indirect inputs to production. For this inter-regional case, any entry in this matrix represents the total expected quantity of a capital good reaching some other destination sector at a different location by all direct and indirect routes. A direct route means that the commodity is sold directly from the one location to the other. An indirect route is when that output is first sold to some intermediate firm, is incorporated in its product, which is in turn eventually used by the destination firm. Clearly there are many such possible indirect routes via any number of intervening locations, and all of these are taken into account in the matrix $[\mathbf{I} - \mathbf{A}^*]^{-1}$. Equation (8.2) is then interpreted as stating that labour values represent the eventual spatial distribution of the socially necessary labour involved in the production and circulation of all capital goods in all locations (Sheppard, 1987).

In a manner that is also analogous to the case for exchange values, an accounting mechanism showing the geographical circulation of the various components of the value of a commodity can be constructed. We note first that the value of all commodities in the space economy, L, is the sum of the labour value of constant capital, C, plus that of variable capital V, plus surplus value, S:

$$L = C + V + S.$$

Total constant capital is the sum of the labour value of the capital goods used in all commodities in all regions

$$\left(\sum_{j=1}^{J} \sum_{n=1}^{N} \left(\sum_{i=1}^{J} \sum_{m=1}^{N} a_{ij}^{mn} . \lambda_i^m \right) x_j^n \right).$$

Total variable capital is the labour value of all consumption goods purchased by labour

$$\left(\sum_{j=1}^{J} \sum_{n=1}^{N} \left(\sum_{i=1}^{J} \sum_{m=1}^{N} (b_{ij}^m / H) . \lambda_i^m \right) l_j^n \right).$$

Surplus value equals labour value in each region multiplied by the rate of exploitation in that region

$$\left(\sum_{j=1}^{J}\sum_{n=1}^{N}\left(\sum_{i=1}^{J}\sum_{m=1}^{N}(b_{ij}^m/H).\lambda_i^m\right)l_j^n.e_j\right).$$

Thus:

$$L = \left(\sum_{j=1}^{J}\sum_{n=1}^{N}\left(\sum_{i=1}^{J}\sum_{m=1}^{N}\hat{a}_{ij}^{mn}.\lambda_i^m\right)x_j^n\right) + \left(\sum_{j=1}^{J}\sum_{n=1}^{N}\left(\sum_{i=1}^{J}\sum_{m=1}^{N}(b_{ij}^m/H).\lambda_i^m\right)l_j^n\right)$$
$$+ \left(\sum_{j=1}^{J}\sum_{n=1}^{N}\left(\sum_{i=1}^{J}\sum_{m=1}^{N}(b_{ij}^m/H).\lambda_i^m\right)l_j^n.e_j\right). \tag{8.5}$$

This is expressed in matrix algebra as:

$$\Lambda'x = \Lambda'\mathring{A}\star x + \Lambda'Bx + \Lambda'B.\{E\}x, \tag{8.6}$$

where $\{E\}$ is a diagonal matrix with the rate of exploitation for each region on the main diagonal (and zeros elsewhere), B is the matrix of inter-regional shipments of consumption goods. Note the similarity between this expression and that for exchange values (equation 4.9b). Following the same approach as outlined in section 4.1.5, the flows of labour value between any pair of regions can be identified (Sheppard, 1987). Thus the labour value flowing from region i to region j, Λ_{ij}, is defined as

$$\Lambda_{ij} = \Lambda'\mathring{A}\star_{ij}.x, \tag{8.7}$$

where $\mathring{A}\star_{ij}$ is the sub-matrix, of the matrix $\mathring{A}\star$ of inter-regional capital good flows in pricing and trading equilibrium, that includes all capital good flows from region i to region j.

If Λ_{ij} is compared with Y_{ij}, the flow of exchange value from region i to region j, then it is possible to evaluate claims about the existence of unequal exchange among regions; i.e. whether a balance in inter-regional trade of commodities measured in production prices implies an imbalance in the labour value of goods traded (Emmanuel, 1972).

8.2   Unequal exchange and regional development

The existence of logically consistent methods for calculating both exchange values (production prices) and labour values for a space economy, using data on production methods and wage levels, leads naturally to a reconsideration of debates concerning the nature and existence of unequal exchange, and its impact on regional development. This issue was pioneered by Emmanuel (1972). He argued that, when two regions exchange products at equivalent exchange values, the labour values of these commodities are generally not equivalent. We illustrate this proposition with a simple example. Consider a space economy made up of two regions, each of which produces two products. Suppose region A completely specializes in producing commodity 1 and region B specializes in commodity 2, and that region A trades part of its surplus of 1 to region B in exchange for part of region B's surplus of commodity 2. We saw in section 4.1 that if the space economy is in trading and pricing equilibrium the monetary rate of profit is identical for producers of 1 in region A and producers of 2 in region B. The production price for each commodity in each region is then calculated, defining a price ratio for exchange: $p_A^1/p_B^2 = p_e$. When these commodities are then exchanged dollar for dollar, the rate of exchange is one unit of commodity 1 from A for p_e units of commodity 2 from B. It does not follow, however, that the ratio of labour values $(\lambda_A^1/\lambda_B^2)$ equals the ratio of production prices. As noted in section 3.5, exchange values are proportional to labour values only in very special and unrealistic cases.

Recognizing this, Emmanuel defined unequal exchange as being a situation where inter-regional exchange of commodities of equivalent exchange value leads to an exchange whereby the labour value traded from region A to region B is not equal to the labour value traded from region B to region A. This is defined as a redistribution of labour value from one region to another, or a regional trade imbalance in labour value terms, which Emmanuel saw as the foundation for uneven development. He further argued that there are systematic biases in the way in which the ratio of labour values in such an example would deviate from the ratio of production prices, thus producing systematic biases in the direction of unequal exchange.

In coming to these conclusions, Emmanuel relied heavily on Marx's solution to the transformation problem (section 3.5, equation 3.14). He argued that two factors contributed to unequal exchange: differences in the organic composition of capital and differences in wage levels. The region specializing in the product with the lower organic composition of capital exports more labour value than it receives in exchange, *ceteris paribus* (dubbed unequal exchange in the 'broad' sense). The region paying lower wages also exports more labour value than it receives in exchange, *ceteris paribus* (unequal exchange in the 'narrow' sense). This latter case was of particular interest to Emmanuel, who employed this proposition to argue that Third World nations with low wage rates were suffering from unequal exchange and that this helped explain widening global inequalities in economic development (see also Amin, 1974). There are, however, a number of theoretical problems associated with this proposition. We will examine these problems in the two-region case and the multi-region case, before going on to re-evaluate some empirical research attempting to employ concepts of unequal exchange to explain regional development inequalities.

8.2.1 A two-region economy

Most discussions of unequal exchange assume, with Emmanuel and indeed most economists, that there are just two regions producing two possible commodities.

Detailed critical reviews of Emmanuel's theory in this case identify two major theoretical inconsistencies (see Mainwaring, 1974; Gibson, 1980; Foot and Webber, 1983; Barnes, 1985). First, there is a confusion in Emmanuel's writings between the money wage, the real wage and the value of labour power. This is seen most easily if we analyse the relationship between labour values and prices of production. This relationship is defined as follows (Barnes, 1985). Assume, with Marx, that the price of production equals the value of the constant and variable capital used in production, incremented by the overall value rate of profit (as in equation 3.14):

$$P_j^n = (1 + \rho)[C_j^n + V_j^n],\tag{8.8}$$

where the P_j^n is the production price for sector n in region j, C_j^n is the constant capital used in the production of a unit of n in region j, and V_j^n is the variable capital employed in production. This assumes that Marx's solution to the transformation problem was correct. The value rate of profit, ρ, is defined as the sum of all surplus value divided by the sum of the value of all constant and variable capital (section 3.4; Barnes, 1985, p. 731).

By definition,

$$\lambda_j^n = C_j^n + (1 + e_j^n)V_j^n.\tag{8.9}$$

where e_j^n is the rate of exploitation in sector n in region j. Recall that if q_j^n is the organic composition of sector n in region j, then C_j^n equals $q_j^n.V_j^n$. Using these definitions, the price ratio when regions A and B specialize and trade is:

$$P_A^1/P_B^2 = (1 + \rho)[(1 + q_A^1)\,V_A^1]/(1 + \rho)[(1 + q_B^2)\,V_B^2]$$
$$= [(1 + q_A^1)\,V_A^1]/[(1 + q_B^2)\,V_B^2].$$

Similarly, the ratio of labour values is:

$$\lambda_A^1/\lambda_B^2 = (1 + q_A^1 + e_A^1)\,V_A^1/(1 + q_B^2 + e_B^2)\,V_B^2$$

$$= (P_A^1 + e_A^1\,V_A^1[1 + \rho])/(P_B^2 + e_B^2\,V_B^2[1 + \rho]).\tag{8.10}$$

According to Emmanuel's argument, if P_A^1/P_B^2 is larger than λ_A^1/λ_B^2, then a trade of the two commodities at equivalent exchange values leads to a redistribution of labour value from region B to region A. From an algebraic examination of equation (8.10), it may be shown that this occurs if the rate of exploitation in region B is higher, whereas the organic composition is identical in both regions (unequal exchange in the narrow sense). It also occurs if the organic composition is lower in region B, whereas the rate of exploitation and the variable capital invested are the same in both regions (unequal exchange in the broad sense). We conclude from this discussion that it is strictly incorrect to say that wage differences (whether money or real wages) are the source of unequal exchange in the narrow sense. The proper statement is that inequalities in the value of labour power (which bring about differences in the rate of exploitation) are the source of this type of unequal exchange.

The second problem with Emmanuel's formulation is that it assumes that Marx's original solution to the transformation problem between values and exchange values is strictly correct, whereas we know it is not. Mainwaring (1974) and Gibson (1980) show that, if prices and labour values are correctly calculated, then unequal exchange is indeed possible. No predictions can be made about the direction of unequal exchange, however. A lower rate of exploitation and higher organic composition in a region need not imply that this region gains labour value as a result of trade of goods at equivalent exchange values.

As a result, the meaningfulness of any theory of unequal exchange of labour value can also be questioned. Gibson (1980) suggests reformulating Emmanuel's unequal exchange thesis purely in exchange value terms, as follows. Suppose two regions have identical wage levels, and each has available to it production technologies for producing either or both of two commodities. Suppose that capitalists in each region stand to gain from specialization and trade, in the sense that the money rate of profit equalized across both regions once specialization and trade occur is greater than the money rate of profit achieved in either region if it were to remain in autarky. This is possible only if each region uses different production technologies, and each region possesses a technique for the production of one commodity that is more capital intensive than the technique used by the other region to produce the same commmodity (Barnes, 1985). Under these conditions, Gibson shows that, if wages increase in region A relative to region B, then the monetary terms of trade improve for region A. This means that the production price of the commodity exported by region A rises relative to the price of the commodity imported by region A purchases more of region B's product with the revenue made from each unit exported. This conclusion in fact comes closer to Emmanuel's original argument, that high-wage regions benefit from inequities in specialization and trade, while rejecting the relevance of unequal exchange measured in labour values.

8.2.2 A multi-regional economy

As is most often the case in economics, the above results concerning compensation for labour, specialization, trade and regional inequalities are formulated for only a two-region, two-commodity economy. This makes the arguments particularly simple because only one equality (between regions A and B) must be considered, and only two outcomes (greater or lesser inequality) are possible. There remains the question, however, of the applicability of these ideas in a multi-regional context.

In a multi-regional economy, inequalities of exchange are calculated between any pair of regions by measuring the exchange of commodities between those regions. A comparison of exchange value and labour value flows can be made by comparing the ratio of exchange value flowing between each pair of regions i and j (Y_{ji}/Y_{ij}, as discussed in section 4.1.5) with the ratio of labour value flowing between them ($\Lambda_{ji}/\Lambda_{ij}$ as defined in section 8.1.4). It can be shown (Sheppard, 1987) that these two ratios are generally unequal, and thus that unequal exchange in the sense defined by Emmanuel generally exists. Alternatively, the total exchange between some region i and all other regions can be examined, asking whether a net inflow or outflow of labour value occurs when the monetary terms of trade are in balance (as they will be, for example, in trading and pricing equilibrium). The latter approach is more common.

In the multi-regional context, expressions for the ratio of prices to labour values have been developed by Liossatos (1980) and Parys (1982). In each case, however, an important simplification was introduced. Liossatos argues, with Aglietta (1979), that the monetary value of all socially necessary labour by definition equals the total exchange value produced. This means that a multiplier converting abstract labour hours into monetary units can be calculated. On this basis, the difference between the exchange value of production in a region and the monetary value of the labour employed in that production can be calculated. In this case, when the exchange value of output exceeds the monetary value of labour employed in some regions, the opposite must be true in other regions, because the monetary value of all labour equals the exchange value of total output at a national scale. Liossatos then showed that regions with higher than average levels of capital intensity (conditions describing the traditional core regions of many capitalist economies), will produce a quantity of commodities whose exchange value exceeds the monetary value of the labour expended in their production. This is similar to Gibson's conclusion for a two-region economy.

Parys employs a Sraffian model where wages are represented as a monetary payment rather than as a bundle of goods consumed (see note 1, Chapter 3). This allows Parys to keep capital goods costs separate from labour costs. He concludes that if some region, n, has completely specialized in some commodity, j, the ratio of its production price to its labour value increases if either the money wage in that region or the capital–labour ratio of the technology in use exceeds the average for all regions:

$$p_j^n/\lambda_j^n = r.k_j^n + (1+r)w_j,$$

where k_j^n is the capital–labour ratio measured in production prices, and w_j is the money wage. Regions for which this ratio is relatively high are regions that stand to gain in labour value terms when commodities exchange for equivalent exchange value. It is important to realize, however, that both the capital–labour ratio and the wage rate are measured in exchange value terms and thus depend on production prices. This means that there is no indicator of the direction of unequal exchange that is independent of the price system and allows for *a priori* predictions of which regions gain or lose from unequal exchange.

Finally, Gibson (1980) has generalized his theory about wage rates and the monetary terms of trade to the multi-regional case. He shows that if the money wage in a region increases relative to wage levels in other regions, and if the current geographical patterns of specialization and trade remain profitable, then that region will increase its monetary gains from trade. This is because the production prices of its exports will increase relative to the production prices of its imports.

The conclusions of Liossatos, Parys and Gibson all provide qualified support for Emmanuel's thesis that regions gain from high wage rates. This is true whether the gain from trade is measured as increased labour value or increased exchange value, and is true despite the fact that Emmanuel's original argument had some major logical inconsistencies. On the other hand, all three authors examine an unrealistic situation – where each region is completely specialized in a single commodity and where no transportation costs are incurred as a result of trade. As we shall see in Chapter 12, these conclusions must be qualified once transportation and the internal differentiation of regions are taken into account.

8.2.3 Empirical studies

A number of authors have taken Emmanuel's argument at face value and used it to assert that geographical disparities in economic development are due to unequal exchange, transferring labour value from low-wage to high-wage regions (see Amin, 1974; Lovering, 1978). To date, however, there exist very few studies attempting to measure accurately the degree or direction of unequal exchange. Those studies attempting empirical analysis all examine the spatial redistribution of labour value as a result of the exchange of commodities of equivalent exchange value, arguing that regions gaining an excess of labour value can use this as a source of extra surplus value, which is then converted into above-average rates of economic growth.

Three studies are essentially based on Emmanuel's original theory. Ferrão (1985) examines regional variations in the organic composition of capital, the rate of exploitation and the value rate of profit for different districts of Portugal. These three categories are measured in monetary rather than in labour values, and the definition of organic composition erroneously includes variable capital in addition to constant capital in the numerator. Based on these indices, Ferrão notes that there are two groups of regions each of which makes a relatively constant value rate of profit. Noting that the rate of exploitation in one group of regions is typically higher, for any given level of organic composition, than in the other group, he uses Emmanuel's definition of the narrow form of unequal exchange, as reformulated above, to argue that there is unequal exchange between the two groups of regions. Oddly, however, he argues that in Portugal 'higher rates of retention [of surplus value] often coincide with higher values of [the value rate of profit in a region]' (Ferrão, 1985, p. 228). For a given level of organic composition, the value rate of profit is proportional to the rate of exploitation (see section 3.4). Emmanuel's theory, with all of its shortcomings, suggests then that regions with a higher value rate of profit *lose* surplus value as a result of unequal exchange, contrary to Ferrão's argument.

A somewhat more careful analysis at a similar geographical scale is due to Hadjimichalis (1987). Using data on agricultural wages, together with measures of the intensity of non-labour inputs in farming as surrogates for organic composition (tractors per hectare and the percentage of land that is irrigated), he examines the regional differentiation in these indicators within the agricultural sector in Greece, Italy and Spain. High wages and high intensity of non-labour inputs are characteristic of northern Italy and selected provinces of northern Spain, but of few provinces in Greece. Employing Emmanuel's argument, Hadjimichalis infers that in these provinces the agricultural sector has gained from unequal exchange. Provinces with low wages and low non-labour inputs, hypothesized to lose from unequal exchange within agriculture, are in southern Italy, south central Spain and northern Greece. Hadjimichalis does not address the case of regions where wages are high and non-labour inputs low, or vice versa, i.e. where the direction of unequal exchange postulated by Emmanuel cannot be clearly predicted. He does, however, recognize another source of inequality stemming from the heterogeneous nature of the agricultural sector. He argues that petty commodity producers invest more labour, including unpaid family labour, in commodity production; often in products where prices are more unstable. If each labour hour represents an equal quantity of abstract labour, this implies that more labour is being invested and eventually exchanged in return for less labour.

A third study in the spirit of Emmanuel is by Marelli (1983). Using the input–output table for Italy, he employs the theory outlined in Chapter 3 to calculate labour values for each economic sector. He calculates the difference between labour values and Marx's version of production prices (equation 3.14) for each production sector. To calculate the total unequal exchange flowing into or out of each province in Italy, he multiplies the proportion of economic output by sector for that province by the difference between labour values and Marx's production prices, and adds this up across all sectors. Noting that the correlation between this measure of provincial value gained or lost through unequal exchange and gross provincial product per capita is very high (0.88), he argues that there is a close relationship between unequal exchange and regional development. The provinces that he determines are gaining or losing through unequal exchange in all sectors of commodity production also correspond well with those identified by Hadjimichalis for the agricultural sector.

While Marelli's study provides a much more rigorous definition of unequal exchange than those of Ferrão and Hadjimichalis, and shows that it is possible to measure labour value rigorously, he repeats the logical error of the other authors in adopting Marx's original and problematic definition of production prices. Thus we cannot be certain that his conclusions are correct. An alternative approach to the measurement of unequal exchange is suggested by Webber and

Foot (1984), who examine the case of Canada and the Philippines. They use international market prices obtained by export goods as a standard against which to compare labour values, rather than production prices as defined in Equation 3.14. Using a modification of Marelli's methodology for calculating labour values, they calculate the amount of labour contained in each dollar of exports from the Philippines and compare that to the labour contained in Canadian exports. They conclude that there is approximately five times as much labour exported per dollar from the Philippines as from Canada, indicating unequal exchange to the considerable advantage of Canada. Part of Hadjimichalis's study also examines this possibility, arguing that a region where a particular commodity has a low labour value loses from unequal exchange by comparison with a region where the labour value of that commodity is higher (assuming the market price is the same in both regions).

This approach avoids the errors of Marx's solution to the transformation problem. The use of market prices rather than prices of production means that this method also does not measure unequal exchange as the redistribution of labour value when commodities are traded for equivalent exchange value. In addition, all four studies reported here do not examine the flow of labour value among regions but rather note the change in the stock of labour value within a region as the sum of all in- and outflows. The details of inter-regional interdependencies are then missing, and it is not possible to specify which regions gain from which other regions. This is clearly seen in the study by Webber and Foot. Here, the calculated loss of exchange value from the Philippines is not a direct gain for Canada, because only a small proportion of Philippine exports in fact go to Canada. We conclude, therefore, that to date there has been no satisfactory calculation of unequal exchange, even though the methods for such calculations exist. Nor has there been any attempt to measure empirically the monetary gains from trade postulated by Gibson.

Finally, we should note that many other factors contribute to a geographical redistribution of the means for capital accumulation and thus to regional development inequalities. Capital and labour flows are clearly of great importance (and Hadjimichalis considers these). Currently we do not know whether, or the degree to which, unequal exchange in labour value terms underlies these flows, or how important unequal exchange is compared with these other factors.

Summary

If transportation is accepted as necessary to commodity production, the inclusion of transportation costs means that the labour value of any commodity varies systematically among regions. This is also true for the commodity of labour power, implying that the rate of exploitation also varies with location even if the real wage is the same everywhere. It was shown in this chapter how such variations in labour values may be determined from knowledge of the patterns of inter-regional commodity trade.

The geographical variations in labour values and exploitation rates were related to accessibility within the inter-regional economy. The latter has implications for Marxian class analysis, because rates of exploitation may be low or even negative in selected regions as long as they are on average positive. This raised the possibility of a labour aristocracy, and posed the question of whether workers in some places therefore exploit workers elsewhere. It also provided a material foundation for the common observation that in some places there are strong regional class alliances between workers and capitalists.

A method of calculating the exchange of labour value between regions and identifying the existence of unequal exchange was also presented. We examined the theory of unequal exchange, noting some severe logical flaws in its original formulation by Emmanuel. Despite these errors, however, there did seem to be theoretical merit to the thesis that high-wage regions tend to experience improved terms of trade (measured in either labour or exchange value) relative to low-wage regions. There are few quantitative studies of unequal exchange, and some suffer from not properly quantifying unequal exchange, from a reliance on incorrect theory, and from a lack of detailed analysis. The ability, however, to rigorously measure labour values in space should enable far more careful empirical analysis in the future.

9 The quantity circuit and capital accumulation

Introduction

In the previous five chapters we have discussed some characteristics of a space economy under full capitalist competition; showing how the prices for produced and non-produced commodities vary geographically, and how prices, patterns of commodity exchange and the location of commodity production are co-determined. We have also examined the labour value of commodities and its circulation, and examined conflicts between capitalists, workers and landlords that are inherent to the structure of capitalism. This entire edifice is predicated on the ability of capitalists to sell their commodities at a price that allows them to make a profit. Without successful sales, the money and labour invested in production will not be realized, and commodity exchange will break down. The prospect of a profit-maximizing geography of production is of little interest to capitalists if they cannot realize those profits.

To assume that goods will be sold is to assume that markets will be cleared because all commodities are in demand, meaning that there is a smooth exchange of commodities between producers and that capital accumulation by capitalists is unproblematic. This is equivalent to the existence of the socially necessary division of labour, defined as the division of labour that ensures that supply approximately matches demand (section 3.3.1).

Needless to say, in a complex and highly interdependent spatially extensive economy the coordination of supply and demand is difficult, but essential. One way to examine whether markets clear is to follow the circulation of commodities in their physical form, as use values. In successful markets all units of every commodity produced will flow to some demand point where they are consumed. We will investigate the conditions under which this occurs, and the likelihood that those conditions will be achieved under capitalism. To avoid unnecessary complications, we will ignore joint production, fixed capital, and unequal production periods and circulation time. These can be incorporated into the analysis outlined below without affecting the essential results (see Abraham-Frois and Berrebi, 1979; Webber, 1987b). Population growth and the existence of non-reproducible resources give rise to more serious problems, but these will also be set aside for now.

First, we outline the conditions under which supply matches demand in a space economy, given knowledge about the socially necessary production methods in use and workers' consumption standards. This defines the socially necessary division of labour. We shall show that such a division of labour generally exists and can be determined, and furthermore that when this is achieved then profits on investment will be realized and smooth capital accumulation is possible. The determinants of the socially necessary division of labour will be examined, and implications drawn for the role of labour values (section 9.1). Section 9.2 will examine the influence of workers' savings, and capitalists' expenditures on luxuries, on capital accumulation. In section 9.3, we will examine whether the socially necessary division of labour is readily attainable. We will show that it seems to be very difficult to maintain through the disaggregated economic actions of capitalists and workers. We will conclude by examining the implications of this conclusion for the instability of a capitalist space economy.

9.1   The geography of production under dynamic equilibrium

The space economy is made up of many producers of a large number of different commodities, each of whom depends on a number of other producers to manufacture the inputs required for production. If a capitalist space economy is to function there must exist a set of societal processes through which demand and supply are articulated with one another much of the time, or else the whole system of individualistic production mediated through the market would be too uncertain and fraught with risk to persist for very long. The manifest persistence and success (on its own terms) of capitalism is testament to the existence of such processes. In neoclassical theory, coordination of individual actions is automatically achieved by allowing prices to be completely subject to the demand for commodities relative to their supply. Thus the prices of commodities are precisely those prices under which consumers' demand schedules match capitalists' supply schedules. Yet, as we have argued in Chapters 1 and 2, this solution completely suppresses the process of commodity production by turning it into an unproblematic exchange mechanism, and also ignores the social structure of society. The result is a conception of capitalism that builds into its assumptions the conditions necessary to presume that each producer is generally in, or close to, equilibrium, and that this equilibrium is socially optimal. This is inadequate as a description both of the evidently high degree of instability in capitalist economies as well as of the persistent social, economic and geographical inequalities in those systems.

Even though the neoclassical solution must be rejected, it is still necessary to understand how supply and demand are related to one another. Supply matches demand only if for every enterprise the commodities produced during one production period are bought by someone during the next period of production. In other words, the commodities supplied in production period t must equal those demanded in period $t+1$. Generally, we will treat production periods as successive time intervals in the history of capitalist commodity production, which are labelled t, $t+1$, $t+2$, and so on. We will also assume that a production period is one year in length for all commodities.

Now, the supply of commodities is simply the quantity produced during time period t, which we denote symbolically by a vector of production levels:

$$\mathbf{x}_t' = [x_{1t}^1, x_{1t}^2, \ldots, x_{jt}^n, \ldots, x_{Jt}^N],$$

where x_{it}^m is the quantity of commodity m produced in region i during production period t. The demand for that quantity of commodity m produced in region i will depend on expectations about future production levels. Other actors will place orders for a commodity depending on how much of it they currently require per unit of production, and on how much they expect to require during the next production period. If producers expect to manufacture $x_{j,t+1}^n$ units of commodity n in region j in the next time period, then the demand for commodity m from i will equal the quantity of that product that they must purchase in order to make these $x_{j,t+1}^n$ units: $\mathring{a}_{ij}^{mn} \cdot x_{j,t+1}^n$. The total demand in the next production period for commodity m from region i, expressed as orders placed during time period t, d_{it}^m, would be

$$d_{it}^m = \sum_{j=1}^{J} \sum_{n=1}^{N} \mathring{a}_{ij}^{mn} x_{j,t+1}^n \tag{9.1}$$

If we assume that there is no import of commodities from, or export to, other regions beyond the boundaries of our space economy, and if we neglect the possibility of stockpiling supplies, then the requirement that supply equal demand is the requirement that $x_{it}^m = d_{it}^m$. Substituting this into an equation (9.1):

$$x_{it}^m = \sum_{j=1}^{J} \sum_{n=1}^{N} \mathring{a}_{ij}^{mn} x_{j,t+1}^n. \tag{9.2}$$

Equation (9.1) states our condition for coordinating supply and demand — that the output of time period t should equal the inputs required for period $t+1$. Note we have assumed here that commodity exchange is that which represents trading and pricing equilibrium. This means that a location pattern of production consistent with this equilibrium must also exist (section 4.1.3). It is not necessary that all products are produced in all regions; x_{it}^m is non-zero only if commodity m is in fact produced in region i. Equation (9.2) must hold, however, for all sectors that are located in a region, and for all regions.

The conditions described by equation (9.2) are necessary, but insufficient to define the socially necessary division of labour because we do not know what the production levels will be in the next time period. The sufficient condition can be found by returning to our earlier consideration of the socially necessary division of labour. Recall we argued that the socially necessary division of labour ensures that all capitalists will realize their profits; profits that can then be invested in expanded commodity production. Suppose all capitalists reinvest their profits in this way, attempting to further capital accumulation by hiring of more labour and purchasing of more inputs in order to manufacture more commodities. If the mean rate of profit made on investment in period t is r, then, in order to reinvest this productively in increased commodity production, the production level to be achieved during period $t+1$ should be $(1+r)$ times the level of production achieved in period t; i.e. $(1+r) \cdot x_{it}^m$. If such a condition is met then for all producers in all regions, the average rate of growth of production will equal the average rate of profit, because full capitalist competition implies that the mean rate of profit is the same everywhere.

In other words:

$$(1 + g) \cdot x_{it}^m = x_{i,t+1}^m, \tag{9.3}$$

where g is the rate of growth, equal to r by definition. Now equation (9.3) can be substituted into the right-hand side of (9.2) to give the following equation for every sector engaged in production in each region:

$$x_{it}^m = (1 + g) \sum_{j=1}^{J} \sum_{n=1}^{N} \hat{a}_{ij}^{mn} x_{jt}^n. \tag{9.4}$$

Equation (9.4) describes the production levels to be achieved in each sector and region in order to ensure that supply equals demand (9.2), and that each capitalist is able to plough all of his profits back into increased production (9.3). If we know the trading patterns ($\hat{a}_{ij}^{mn}$), and we arbitrarily set the quantity produced in sector 1 of region 1 to equal one, then we can solve equation (9.4) to determine the relative quantities of each commodity that should be produced in each region in order to maintain continued growth and accumulation. Notice the similarity of equation (9.4) to equation (4.4). Technically speaking, the quantities guaranteeing continued capital accumulation are the dual of the production prices at which these goods will on the average be sold. The solution to equation (9.4) will give a vector of such optimal quantities:

$$\mathbf{x}^\star = [x_1^{\star 1}, x_1^{\star 2}, \ldots, x_j^{\star m}, \ldots, x_j^{\star N}].$$

The socially necessary quantity of labour employed in sector m of region i to ensure coordinated demand and supply, and to guarantee capital accumulation, is the product of the optimal quantity ($x_i^{\star m}$) and the labour requirement:

$$\lambda_i^{\star m} = x_i^{\star m} \cdot l_i^m / H. \tag{9.5}$$

A vector of all such values, $\Lambda^\star = [\lambda_1^{\star 1}, \lambda_1^{\star 2}, \ldots, \lambda_j^{\star m}, \ldots, \lambda_j^{\star N}]$ is the socially necessary division of labour − the number of workers in each sector of each region necessary to ensure continued profit realization.

If the spatial division of labour is equal to this distribution then all the production from one time period will be purchased as inputs for increased commodity production in the next period. We can describe this as a dynamic equilibrium because, if it is undisturbed, commodity production will increase at the same rate (g% per annum) in each sector and region, reproducing the same socially necessary division of labour. Prices and trading patterns will also remain unchanged. We might refer to this as the socially necessary growth path; other economists refer to it as the golden path of accumulation. It is an important concept because it suggests that unlimited and crisis-free capital accumulation is possible in a complex capitalist space economy experiencing full competition, with all sectors and regions growing at the same rate. In the Appendix to this chapter we show that, for any geographical distribution of socially necessary methods of production and consumption levels in each region, there is a unique vector of non-negative relative production levels, $\mathbf{x}^*$, representing the socially necessary spatial distribution of production that will allow this to happen.

We also show in the Appendix that, as for exchange values, the socially necessary distribution of production quantities can be interpreted as the only spatial distribution of use values that is unchanged as a result of commodity trade in trading and pricing equilibrium. The overall quantity of commodities is, however, augmented from one production period to the next by the fraction $[1 + g]/g$, which is equivalent to the increase from C to $C + \Delta C$ in Marx's representation of the circulation process (Chapter 4). As discussed in the Appendix, the size of the socially necessary labour force is the product of the amount of labour employed in a sector and the total quantity of trade that this sector has with other sectors. If we put these two considerations together, we can conclude that the amount of the labour force that is socially necessary in a particular sector and region is higher if: its products are required more by society; this region is more competitive as measured by delivered prices of production at the sites where it is required, implying greater trade flows; and it employs more labour-intensive production methods. The competitiveness of its delivered prices depends in turn on the accessibility of that location to cheap sources of inputs, and to other producers requiring large amounts of the product. Recall, however, that accessibility depends both on the geographical distribution of socially necessary production methods and consumption norms and on transport costs. It is therefore crucial to understand the way in which these methods and consumption norms evolve; we cannot simply take them as external to our analysis.

9.1.1   The rate of capital accumulation

In chapter 4 it was established that any change in wage levels leads to a change in: production prices, the rate of profit, trading patterns and the profit-maximizing location pattern. Thus there is a complex relationship between the distribution of income and the geography of production, but we do know that wages and profits are inversely related. Any change in wage levels will also, therefore, change the rate of capital accumulation, because this equals the profit rate, and alter the division of labour that is socially necessary to match supply and demand.

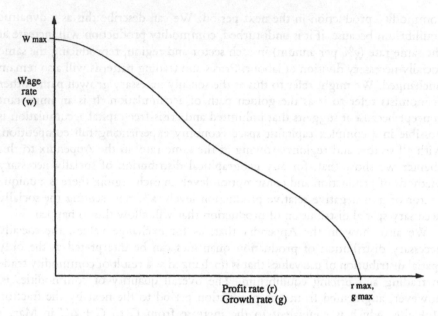

Figure 9.1   A wage-growth frontier

The relationship between wages and the growth rate can be represented by a wage–growth frontier (Figure 9.1). Associated with each point on the wage–growth frontier is a different socially necessary division of labour, **x***. Therefore, one consequence of changes in the distribution of income between workers and capitalists is that alterations would be necessary to the location pattern of production in order to maximize the rate of profit for capitalists. This suggests the hypothesis that changes in the distribution of income, either in individual regions or in the economy as a whole, can bring about changes in the location of commodity production. Thus not only might commodity production leave an individual region because workers there negotiate higher wages, but nationwide changes in, for example, the minimum wage or the influence of unions on wage levels may also lead to inter-regional shifts in the location of industry.

One inference from Figure 9.1 is that wages are a hindrance to capital accumulation and economic growth. It is also clear, however, that growth and capital accumulation are possible for any wage level. This would imply that we cannot simply conclude with Harvey (1982, p. 56) that '[t]he value of labour power ... can be defined as the *socially necessary remuneration of labour power* ... from the standpoint of the continued accumulation of capital.' Capital accumulation at some (low) rate is possible even if wages are very high (but less than W_{max}), because in that case workers consume most of the output and capitalist invest the small amount remaining. Presumably in this case the socially necessary division of labour is strongly biased towards the production of wage goods. This implies that the only upper limit set on wages by the requirement that the economy be capable of accumulating capital is that wages are low enough that the rate of exploitation and the rate of profit are positive. Since a wide range of wage levels satisfy this requirement (anything between zero and W_{max} in Figure 9.1), it is of little use for defining the value of labour power. For Harvey's argument to lead to a deduction about binding limits on the value of labour power, an additional requirement should be added, such as the requirement that the rate of capital accumulation match those levels that prevailed historically in order for capitalists to be willing to reinvest.

A second frequently made argument addresses the existence of a lower limit on wages. It is argued that, since profits take money away from workers, they reduce consumer demand. This raises the spectre of an underconsumption crisis – that wages are too low to allow workers to purchase the commodities produced, leading to an inability to sell commodities and realize profits (Sweezy, 1942; Bleaney, 1976). An examination of Figure 9.1 again shows that no such problem need necessarily occur in a capitalist economy. Very low wages can allow a very high rate of capital accumulation, as long as the appropriate socially necessary division of labour is in place. The profits retained by capitalists also sustain demand in the capital goods industries when they are productively reinvested. Thus, as long as the socially necessary division of labour is strongly biased towards the production of capital goods, low wages allow rapid rates of growth. In arguing this, we do not deny that underconsumption crises may occur, or that they may be resolved by increasing the wages of workers (as many have argued to have been the case between the Great Depression and the late 1960s; now dubbed the Fordist period of capitalism – see Lipietz, 1987; Scott, 1988c). Rather, we argue that if an underconsumption crisis did occur this is because the division of labour placed too much emphasis on consumer goods production given the prevailing wage levels, and that one resolution to such a crisis is to increase wages.

Shaikh has argued that the socially necessary division of labour is defined by labour values, and that this is the way in which labour values pervade the very technical coefficients that others such as Steedman have taken as independent of labour values in their arguments against a law of value (section 3.5.3). Shaikh's argument is based on the conclusion that, when the economy is in the dynamic equilibrium described here, then total surplus value equals total profits, implying that profits are simply a redistribution of surplus value obtained from the exploitation of workers. This equality certainly holds. It does not follow, however, that the labour value of a commodity is dictated by the socially necessary division of labour. The degree of similarity between Λ, measuring socially necessary labour of the first type, and Λ^*, measuring the socially necessary division of labour, is an index of the degree to which the two definitions of social necessity would converge in practice. If it is the case that these two vectors are proportional to one another, then this would be consistent with the law of value as formulated by Shaikh (1981). For this to be the case, then λ_j'' should be proportional to $\lambda^*{}_j''$; i.e. in every sector m and region i:

$$\lambda_j^n / \lambda_j^{\star n} = \left[\sum_{i=1}^{J} \sum_{m=1}^{N} \hat{a}^{\star mn}_{ij} . \lambda_i^m + l_j^n \right] \Big/ [x^{\star n}_j . l_j^n . H^{-1}]. \qquad (9.6)$$

We know of no reason why equation (9.6) should be true, underlining the difficulty of resolving these two definitions of social necessity (section 3.5.3). If there exists a level of wages between zero and $W_{\max}$ for which equation (9.6) is true, then this implies that the two definitions of social necessity are equivalent. This in turn implies that there is a wage level at which the socially necessary division of labour does dictate labour values, in turn suggesting that this is the wage level that is socially necessary (see Harvey above). Even so, the mechanisms driving a capitalist economy towards this particular position on the wage–growth frontier must be specified in order to provide an explanation of how the law of value operates.

9.2 Incorporating workers' savings and capitalists' luxuries

Thus far we have assumed that workers are unable to save money, and that capitalists reinvest all of their profits in purchasing the inputs necessary for increased commodity production. This is known as the Marxian savings hypothesis. In this section we briefly investigate the implications for the space economy of relaxing these restrictions. We should first note, however, that, under the Marxian savings hypothesis, it is necessary that capitalists also make a living wage. Capitalists presumably carry out some productive work in addition to simply owning the means of production, and the hours of direct labour involved in commodity production must include this work. We follow Roemer (1981) in assuming that capitalists are differentiated from other workers not because they possess superior skills but because of their prior ownership of the means of production, implying that they do not 'deserve' a greater real wage than other workers. Thus capitalists also earn the same real wage as workers for the hours of productive labour that they contribute. This basic wage is included in the total wage bill of the space economy. Thus when we say that all profits made by capitalists are saved and reinvested by them, we really mean all *net* profits once this basic wage is paid, and throughout this section capitalists' profits are net profits in this sense.

The profits made by capitalists, whether paid directly to capitalists in the form of real wages that are higher than what is necessary for them to perform their socially necessary productive labour, or whether reserved as profits in some other account, are typically not all reserved for reinvestment. Instead, some proportion of it is spent on a level of consumption that exceeds the real wage of workers. We define this as expenditure on luxuries. Suppose that in all regions capitalists retain a proportion of all profits, s_c, for reinvestment in expanded production while spending the remainder of their profits on luxury goods. The level of consumption socially necessary to reproduce a worker's family in region j has been defined above as the vector $\mathbf{b_j} = [b_j^1, b_j^2, \ldots, b_j^m, \ldots, b_j^N]$. Luxury consumption is then a vector $\mathbf{c_j} = [b_{cj}^1 - b_j^1, b_{cj}^2 - b_j^2, \ldots, b_{cj}^m - b_j^m, \ldots, b_{cj}^N - b_j^N]$, where b_{cj}^n is the quantity of commodity n consumed by a capitalist's family each week in region j. Defining c_j^n as the quantity of commodity n consumed as a luxury by capitalists, then the vector of luxury goods consumed in region j is $\mathbf{c_j} = [c_j^1, c_j^2, \ldots, c_j^m, \ldots, c_j^N]$.

This vector of luxuries includes in part commodities such as caviar that are reserved for luxury use, but also includes quantities of wage goods that exceed levels that are socially necessary for the reproduction of the family. All of these are luxury goods because in a social sense their production and consumption are unnecessary. For that reason they are not included as part of the inputs socially necessary to produce any commodity; in Sraffa's terms they are non-basic in that they are not essential to the process of commodity production.

Capitalists in sector n of region j earn profits equal to a fixed proportion, $r/(1+r)$, of total revenue ($x_{jt}^n p_j^n$). If they reinvest only the fraction, s_c, of total profits, reserving the remainder for expenditure on luxuries, then

$$(1 - s_c).(r/1 + r)x_{jt}^n p_j^n = C_{jt} \sum_{n=1}^N p_j^n.c_j^n, \qquad (9.7)$$

where the right-hand side equals the cost of luxuries, and where C_{jt} is a multiplier to ensure equality between the two. We assume, therefore, that the relative quantities of luxuries consumed are fixed, but that total expenditures on luxuries must match the savings that are not reinvested.

Workers can save money if their money wage exceeds the cost of the socially necessary real wage. We define workers' savings as the proportion, s_w, of the total money wage that is saved in financial institutions and recycled by these institutions as investment capital for capitalists. The total exchange value created in the space economy in production period t is then equal to the sum of: payments to replace inputs, workers' consumption expenditures, capitalists' savings, capitalists' consumption expenditures and workers' savings. The first three of these five components were considered in the accounting identity of section 4.1.5, as represented in equations (4.9a–c), repeated here for convenience:

$$Y = K + W + \Pi. \qquad (4.9a)$$

Or, equivalently:

$$[1 + r].\mathbf{p'Åx_t} = \mathbf{p'Å^\star x_t} + \mathbf{p'Bx_t} + r.\mathbf{p'Åx_t} \qquad (4.9c)$$

The left-hand side of these equations represents the total revenue made on all commodities produced during the t'th production period – the total cost of inputs incremented by the rate of profit. In order to allow for luxury compensation and workers' savings, equation (4.9) must be expanded:

$$Y = K + W + \Pi + (1 - s_c)/s_c \cdot \Pi + [s_w/(1 - s_w)] \cdot W, \qquad (9.8a)$$

or

$$[1 + r] \cdot \mathbf{p'\mathring{A}x_t} = \mathbf{p'\mathring{A}^\star x_t} + \mathbf{p'Bx_t} + s_c \cdot r \cdot \mathbf{p'\mathring{A}x_t} + (1 - s_c) \cdot r \cdot \mathbf{p'\mathring{A}x_t} +$$
$$[s_w/(1 - s_w)]\mathbf{p'Bx_t}. \qquad (9.8b)$$

In these two equations, the second term represents all socially necessary consumption expenditures (by workers and capitalists); the third term represents total capitalists' savings for investment purposes; the fourth term represents total money available for luxury consumption, and the fifth term equals total savings by workers (deduced from the definition that workers' savings equal a fixed proportion, s_w, of total workers' consumption expenditures and savings). If we adopt the Marxian assumption that $s_c = 1$ and $s_w = 0$, then equation (9.8a) becomes equation (4.9a).

Given equation (9.8b), and given the reasonable assumption that capitalists' propensity to save is greater than workers' propensity to save ($s_c > s_w$), the following relationship between the average rate of profit and the socially necessary rate of economic growth can be shown to exist (Appendix to this chapter; see Pasinetti, 1962):

$$\omega \cdot r = g \qquad (9.9)$$

where

$$\omega = s_c - s_w - \kappa s_w [I - s_c Y^\star]/[I - s_w Y^\star], \qquad (9.10)$$

κ is the capital–output ratio; and $Y^\star$ is the net income, approximately equal to $Y - K$. This tells us that in a space economy where both workers and capitalists save a part of their income, in the dynamic equilibrium described in section 9.1 the rate of growth is less than or equal to the rate of profit by a fraction that reflects both rates of savings.

When workers save nothing:

$$s_c \cdot r = g, \qquad (9.11)$$

then the rate of growth is lower than the rate of profit by a fraction equal to capitalists' propensity to save (a classical Keynesian macroeconomic identity). Under these conditions, luxury consumption behaviour of capitalists will reduce the socially necessary rate of growth. By definition, the production of luxuries does not contribute to the economic growth of socially necessary commodity production. Indeed, in an economy where all profits are spent on luxuries then no economic growth or capital accumulation are possible even if the economy utilizes productive technologies and makes a positive profit. This is the case that

Marx referred to as simple reproduction, and occurs when s_c is zero, which from equation (9.11) implies that the rate of growth is also zero.

From equations (9.10) and (9.11), where $\omega < s_c$ because $s_c > s_w$ and $\kappa < 1$, notice that when workers also save a proportion of their wages the growth rate of the economy is, paradoxially, lower than the growth rate when workers do not save. This counterintuitive result stems from the principle, noted by Pasinetti (1962) in this vision of the space economy, that in equilibrium wages are distributed to individuals in proportion to the labour they contribute, whereas profits are distributed in proportion to the amount of capital they own. Thus profits are proportional to savings, and the ratio of profits to savings must be the same for each class:

$$P_c/S_c = P_w/S_w \qquad (9.12)$$

where P_c and P_w represent, respectively, the total money profits accruing to capitalists and workers, and S_c and S_w represent the total money saved by the two classes.

Noting that $S_c = s_c P_c$, because capitalists derive all of their income from profits, and that $S_w = s_w \cdot W$, then equation (9.12) becomes:

$$W/P_w = s_c/s_w. \qquad (9.13)$$

Since the propensity of workers to save is less than the propensity of capitalists to save, W exceeds P_w and the cost of the socially necessary real wage is generally above 50 per cent of the total income made by workers. In this case workers' savings reduce the rate of growth of the economy, because the money saved by workers is a part of the surplus that would otherwise accrue to, and be saved by, capitalists. If these funds accrued to capitalists, then a greater proportion would be saved and would be invested in economic growth. From equation (9.10) it follows that the greater the difference between the capitalists' and the workers' propensity to save, and the smaller the capital–output ratio, the greater the degree to which workers' savings slow the rate of growth of the economy.

9.2.1   The consumption–growth frontier

We have seen that the difference between the rate of profit and the socially necessary growth rate for the space economy depends on the savings activities of workers and capitalists. Yet for any given pattern of savings behaviour there is still an inverse relationship between workers' consumption and the rate of growth. Indeed, one can draw a graph showing the trade-off between the two that is necessary to maintain the socially necessary divison of labour and rate of growth, known as the consumption–growth frontier.

Whether or not workers save money or capitalists consume luxuries, the consumption–growth frontier is identical to the wage–growth (i.e. wage–profit)

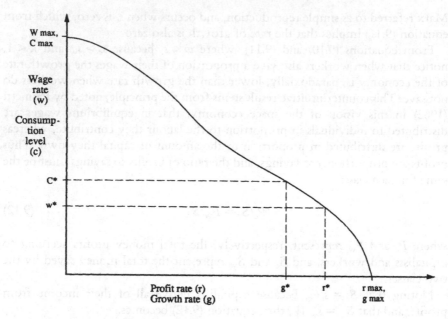

Figure 9.2   A consumption–growth frontier

frontier for the space economy (Figure 9.1). With no savings or luxury consumption, the rate of growth equals the rate of profit and thus workers' consumption equals the real wage. When workers' savings and capitalists' luxury consumption reduce the rate of growth below the rate of profit, however, as shown in equation (9.10), then total consumption will be greater than the wage. This is seen in Figure 9.2 where the frontier drawn is simultaneously the wage–profit and consumption–growth frontier. Here by construction it is seen that, when the rate of capital accumulation ($g^\star$) is less than the profit rate ($r^\star$), then workers' consumption ($c^\star$) exceeds the wages associated with that profit rate ($w^\star$). This difference also suggests new conflicts of interest both within and between the classes of capitalists and workers. Figure 9.3 shows the wage–profit frontiers of two location patterns. At wage rate $w^\star$ location pattern A is profit maximizing. Yet, owing to the difference between $r^\star$ and $g^\star$, it is pattern B that gives rise to the higher growth rate and workers' consumption (Sheppard and Barnes, 1986).

9.3   The instability of dynamic equilibrium

We defined the dynamic equilibrium ensuring that supply from one production period matches demand in the next. It is not enough, however, to establish the existence of this dynamic equilibrium; we must also consider the question of whether there is good reason to expect that such an equilibrium will be achieved in practice. This may be done by asking whether the dynamic equilibrium is stable. If it is locally stable, then when external forces, such as changes in the spatial division of labour or in the socially necessary production methods, push the economy slightly away from equilibrium then the internal processes of capitalism will return it to a socially necessary division of labour and its corresponding rate of growth. This would be evidence that capitalism is robust; that it is capable of responding to external shocks by finding a new path for capital accumulation enabling capitalists to realize profits and allowing regions to grow at the same rate. If the system is globally stable, it suggests that, no matter how great the shock, capitalism will always return to such a dynamic equilibrium. If the system is unstable, however, then a small perturbation away from dynamic equilibrium is followed by the system moving further and further from equilibrium and slipping into a series of crises. This would be evidence ·that capitalism is unable to take care of its own problems, and that instability rather than stability should be expected as the normal state of affairs.

Stability may be examined by exploring the properties of the dynamic equilibrium represented in equation (9.2). It is shown in the Appendix to this chapter that stability depends on the nature of the inter-regional input–output matrix, $\mathbf{A}$, and that our assumption that we have a productive economy implies that the dynamic equilibrium is unstable. If the space economy therefore deviates from the socially necessary spatial distribution of production, then the deviations will only grow greater. Such deviations represent a situation where supply does not match demand – with some sectors in some regions growing more rapidly than average owing to high demand, and others experiencing a realization crisis because they are not growing fast enough to absorb the increased output that

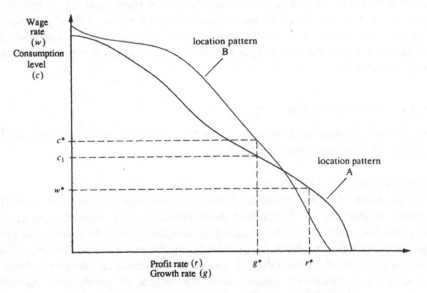

Figure 9.3 Profits and growth for two location patterns

results from the reinvestment of profits. Indeed, the economy can be disaggregated into those parts that do better than average and those that do less well (Sheppard, 1983b; Goodwin, 1976).

Deviations from dynamic equilibrium are bound to happen sooner or later, if only because any space economy is an open system and subject to external influences. It is thus reasonable to conclude that the capitalist space economy as described thus far is inherently unstable. Although a socially necessary spatial distribution of production and labour can be identified that allows the economy to reproduce itself, in practice that distribution is most unlikely to be observed, and would not persist for any length of time. This conclusion supports the view of the capitalist space economy as being ridden with contradictions and highly unstable in nature, in constrast to the harmonious and stable portrayals of neoclassical analysis.

Our opinion, however, is that the degree of instability suggested by this analysis is too great to represent a realistic picture of capitalism. Such a high degree of instability makes it difficult to imagine that capitalism could survive for very long at all, yet it has been the most persistent economic system that humankind has known. Clearly, therefore, our analysis is incomplete; capitalism is simply not as unstable in practice as our analysis here would suggest. One reason is that quantities are not adjusted in response to price signals in the model developed here. Recent theoretical research shows that when quantities depend on prices then, under certain assumptions that capitalists adjust smoothly and not too quickly to realization crises, there is more stability in this framework (Duménil and Lévy, 1987). The remainder of the book is therefore devoted to examining a series of processes that are both sources of, and responses to, the instability and conflict inherent in the system described thus far. While there is little space in this book to do more than sketch out some of these processes, we suggest that a proper understanding of how they influence a capitalist space economy can provide real insight into the disequilibrium dynamics observed in such economies.

9.3.1   Responses to instability

As an introduction to these issues consider the following factors that are both sources of, and responses to, instability.

NON-PRODUCED RESOURCES

The existence of natural resources that cannot be exploited in infinite quantities at the site where they are found root the instabilities of the human economy in its physical backcloth. Over time, capital accumulation requires greater and greater quantities of inputs to produce more and more output. One of the internal limits on this process is the availability of non-produced resources. While the definition of scarcity is intimately linked with the modes of production utilized by humankind, the existence of limits to the exploitation of resources cannot be denied. As resources are exhausted, new production methods must be employed, or new sites found in order to meet demand. Thus the very process of capital accumulation and its ever-increasing demand for non-produced resources leads to shifts in production methods and in the location of resource production; shifts that lead to changes in the inter-regional input–output matrix. Similarly, the search for new resources and substitutes for expensive resources represents ways in which capitalists respond to problems caused by the exhaustion of the resources currently exploited.

INDIVIDUALS AS AGENTS; CLASS ALLIANCES

Thus far we have assumed that a particular production method and real wage dominate in each region and continues to do so *ad infinitum*. This begs the question of the origin of these characteristics, and ignores the fact that economic actors do have some freedom to select technologies and consumption bundles based on their relative costs. Roemer (1981, p. 56) shows that allowing for this kind of behaviour does not affect the validity of the thesis that exploitation is necessary for positive profits, or many of the other properties of the framework developed to date. The thrust of his argument is that, no matter how the real wage and pro-duction technologies are determined (whether dictated or individually chosen), the inter-regional input–output matrix that results will have all of the properties outlined in Chapters 4–9. Thus the existence of a dynamic equilibrium is also per-fectly consistent with allowing for freedom of choice. It follows by the same token, however, that such a matrix is unstable – no matter how it is arrived at.

Even if individuals choose wage goods or production methods that suit them best under the circumstances, it does not follow that this equilibrium is in their individual or collective best interest. The existence of dynamic equilibrium cannot eliminate dissatisfaction or the belief that groups working together can improve their lot, because we saw that a general conflict of economic interest between capitalists, workers and landlords is inherent to capitalism. This forms the foundation for the collective action of classes seeking to change the distri-bution of income, precisely because the circumstances are broadly not of their own choosing (Chapter 10).

Members of a class may work together directly to increase wages, profits or rents, or to change the material conditions of production in such a way as to improve their collective well-being. This kind of collective action leads to conflict between any pair of the three classes (Chapter 11). Furthermore, it is far from clear that class interest coincides with self-interest. Thus, individual members of any class may seek to improve their welfare with the unintended result of worsen-ing the situation for their class as a whole. Such intra-class conflicts may be pursued by allying with other members of the same class, or forming regional inter-class alliances (Chapter 12). These kinds of strategies are likely to be pursued both when there is no obvious crisis, as groups attempt to increase their share, and also in response to crisis as workers or capitalists organize to regain lost profits or deteriorating working conditions.

CAPITALISTS' ACTIONS; TECHNICAL, LOCATIONAL AND ORGANIZATIONAL
CHANGE

Having generally examined possible actions for workers, capitalists and landlords
in Chapters 11 and 12, we will restrict ourselves in the remainder of the book to
detailing the strategies available to capitalists. Similar analyses could be made for
workers or landlords, but we leave this to those more qualified than ourselves. The
following principal strategies exist for capitalists who wish to improve their profits:

Altering production methods   Capitalists possess a variety of ways of reducing
production costs, all of which result in changes in the input–output coefficients of
production. They may adopt new technologies or relocate production. This
might include reducing the direct labour input, by replacing labour with
machinery or by lengthening the effective work week through actions that
shorten time off and increase the pace of work. Other strategies include
relocation or other preventive action to reduce unionization.

   A frequently used strategy is to make major investments at one point in time in
fixed capital that reduces the size of inputs necessary per unit produced for a
substantial period in the future. Such a strategy diminishes the flexibility of future
actions, however, since fixed capital investments must be depreciated before they
can be economically written off, and because fixed capital from bygone eras may
no longer be functional for capital accumulation (Walker, 1981; Harvey, 1982).

Transportation and credit   One way of enhancing profitability in a crisis and of
smoothing out crises is to improve the circulation of capital. This takes the form
of transportation improvements that reduce the time between production and
realization of profits through sale in the market – the elimination of spatial
barriers to the rate of accumulation of capital. It also takes the form of improved
credit institutions to eliminate the temporal barriers between when capital is
available and when and where it is required (Harvey, 1982). Such improvements
can go beyond changing the cost of inputs in response to crises to increasing both
the turnover rate and the speed of accumulation of capital (Chapter 13).

Organizational structure   One traditional response to the difficulties of competition is the centralization of production into major corporations. This both
allows for greater coordination of supply and demand by reducing the degree of
anarchy in the market, and also increases profitability through oligopolies in
production. More recently, it has been argued that circumstances have evolved
where decentralization has become advantageous; circumstances known as
flexible production (Scott, 1988c). The impact of such strategies on profits will be
examined in Chapter 14.

GEOGRAPHICAL EXPANSION BEYOND THE INTER-REGIONAL SYSTEM

An obvious response to crises of over- or undersupply is to seek external markets
for products in excess supply, and external sources for products in excess demand.

In this way it is possible to balance supply and demand through external interaction. This 'spatial fix' is often cited both as a viable strategy and as the explanation for the worldwide expansion of capitalism (Lenin, 1947 [1916]; Smith, 1984; Harvey, 1985c).

THE ROLE OF THE STATE
The state is also historically a participant in attempts to stabilize the capitalist space economy. The internal instability of a capitalist space economy, and the political dimension that is central to class conflicts, both imply that the state is forced to play a central role in stabilizing a capitalist space economy, and cannot be consigned to the role it is given in neoclassical and pluralist theory — as the pluralist arbiter of market externalities.

Summary

We have demonstrated that, for any wage level, there exists a unique socially necessary division of labour allowing for the successful circulation of commodities; all commodities are sold, profits are realized, and all regions experience an equal rate of growth. We dubbed such a growth path, allowing in principle for unlimited capital accumulation, a golden accumulation path. For any geographical configuration of socially necessary production methods and real wage levels, a range of such golden paths exist, one for each level of wages. We also showed that, when capitalists save a proportion of profits for luxury expenditures, this reduces the rate of growth relative to the rate of profit. When workers are paid a high enough wage to save money, which is then available as investment funds for capitalists, this reduces the rate of economic growth. Workers' propensity to save is generally less than capitalists' propensity to save, implying that if the money accruing to workers were given to capitalists then they would save a greater proportion of it for investment in growth.

Finally, the golden path of accumulation is unstable: any small deviation from it leads to a concatenation of problems of realizing profits and selling products that drive the space economy into crisis. It is so unstable, in fact, that there must be mechanisms that stabilize real economies to some degree; mechanisms do not exist in the model developed to date. These include: the exhaustion of non-produced resources; inter- and intra-class conflict and alliances; the actions of individual capitalists seeking to reduce costs of production; the actions of workers seeking to increase wages and improve working conditions; the actions of landlords seeking to improve the rate of return on land; geographical expansion into new areas; and the role of the state.

Appendix: Characteristics of the dynamic equilibrium

We develop four characteristics of the dynamic equilibrium in this appendix. In part 1 we show that there is a unique set of relative production levels characterizing dynamic equilibrium for a given wage level, and socially necessary production methods and consumption levels. In part 2 we provide an alternative interpretation of this distribution. In part 3 we examine the impact of workers' savings and capitalists' luxuries. In part 4 we show that the dynamic process described by this disequilibrium is generally unstable.

Part 1

Consider equation (9.4), written in matrix form:

$$\mathbf{x}^\star = (1 + g)\mathbf{\mathring{A}} \cdot \mathbf{x}^\star, \qquad (9A.1)$$

where $\mathbf{x}^\star$ is the vector of production quantities guaranteeing that supply matches demand, and $\mathbf{\mathring{A}}$ is the matrix of commodity flows to be found in trading and pricing equilibrium. Equation (9A.1) is a linear eigenvalue equation since the elements of $\mathbf{\mathring{A}}$ are not functionally dependent on $\mathbf{x}^\star$. By the Perron–Frobenius theorems, there is a unique positive eigenvalue $(1 + g)^{-1}$ and a unique positive eigenvector $\mathbf{x}^\star$ that solve equation (9A.1), because $\mathbf{\mathring{A}}$ is non-negative and the economy is productive (see Pasinetti, 1977).

Part 2

Applying the reasoning of Part 2 of the Appendix to Chapter 4, equation (9A.1) may also be written as (Sheppard, 1987):

$$\mathbf{x}^\star = (g/[1 + g])[\mathbf{I} - \mathbf{\mathring{A}}]^{-1}\mathbf{x}^\star \qquad (9A.2)$$

By analogy to that Appendix, equation (9A.2) can be interpreted as stating that the spatial distribution of use values that results once all commodities have been directly and indirectly traded between all sectors and regions ($\mathbf{x}^\star$ on the left-hand side) is the same as the distribution of use values that exists prior to this circulation process ($\mathbf{x}^\star$ on the right-hand side). However, the total quantity of use values has increased by the quantity of $([1 + g]/g)$. Furthermore, since the matrix $[\mathbf{I} - \mathbf{\mathring{A}}]^{-1}$ is a measure of the direct and indirect trade flows between regions and sectors, i.e. of the accessibility of each sector and region with one another as a result of commodity flows, the relative production levels in sector n of region j in dynamic equilibrium are proportional to the accessibility of that sector to production elsewhere, weighted by the level of production in each sector and region.

Part 3

Consider the accounting identity of equation (9.8b):

$$[1 + r] \cdot \mathbf{p}' \mathbf{\mathring{A}x}_t = \mathbf{p}' \mathbf{\mathring{A}^\star x}_t + \mathbf{p}' \mathbf{Bx}_t + s_c \cdot r \cdot \mathbf{p}' \mathbf{\mathring{A}x}_t + (1 - s_c) \cdot r \cdot \mathbf{p}' \mathbf{\mathring{A}x}_t +$$
$$[s_w/(1 - s_w)] \mathbf{p}' \mathbf{Bx}_t. \tag{9.8b}$$

In essence this states that:

$$Y = K + C_n + S_c + C_L + S_w \tag{9A.3}$$

where Y equals total income produced, K equals the cost of capital good inputs; C_n is the money reserved for socially necessary consumption by capitalists and workers; S_c is capitalists' savings; C_L is capitalists' spending on luxuries; and S_w is workers' savings. By definition, workers' profits (P_w) equal total savings by workers; capitalists' profits (P_c) equal the sum of capitalists' savings and capitalists' expenditures on luxuries; total savings (S) are the sum of workers' and capitalists' savings, and total wages (W) are the sum of their consumption expenditure and their wages:

$$P_w = S_w$$
$$P_c = S_c + C_L$$
$$S = S_c + S_w$$
$$W = \delta \cdot C_n + S_w$$

where δ is the proportion of the socially necessary wage that is paid to workers (a residual $(1 - \delta)$ is paid to capitalists for their productive work). The aggregate rate of profit must be:

$$P/Y = (P_c + P_w)/Y \tag{9A.4}$$

Now, in dynamic equilibrium, savings must equal investment, I:

$$I = s_c P_c + s_w W. \tag{9A.5}$$

Adding and subtracting $s_w P_c$ to the right-hand side of (9A.5), and applying the definition of P_c to the identity (9A.3):

$$I = (s_c - s_w)P_c + s_w Y^\star,$$

where $Y^\star = Y - K - (1 - \delta)C_n$. Denoting $(s_c - s_w)$ by α, it then follows that:

$$P_c/Y = (I - s_w Y^\star)/\alpha Y. \tag{9A.6}$$

In the long-run equilibrium, we would expect workers' profits to equal the amount of capital owned by workers multiplied by the rate of interest, i, paid to all creditors. We would further expect that the proportion of all capital owned by workers would equal the proportion of total savings made by workers:

$$P_w = i \cdot K_w$$
$$K_w/K = S_w/S,$$

where K_w is the value of capital owned indirectly by workers and K is the total value of all capital. It then follows that

$$P_w/Y = i \cdot (K/Y)(S_w/S)$$
$$= i \cdot (K/Y)[s_w(Y^\star - P_c)]/I. \qquad (9A.7)$$

Substituting the definition of P_c/I from equation (9A.5) into (9A.7), assuming that in dynamic equilibrium the rate of interest equals the rate of profit ($P/Y = i$), and then substituting equations (9A.6) and (9A.7) into (9A.4):

$$P/Y\{1 - (K/Y)[s_c s_w(Y^\star/I) - s_w]/\alpha\} = [I - s_w Y^\star]/\alpha Y. \qquad (9A.8)$$

Now the overall rate of profit expressed as a proportion of net surplus, $P/Y^\star$, or as a proportion of the exchange value of the means of production, P/K, can readily be shown to be related to the rate of growth expressed in terms of the same denominator as follows (Pasinetti, 1962):

$$s_c(P/Y^\star) = I/Y^\star \qquad (9A.9)$$

$$s_c(P/K) = I/K \qquad (9A.10)$$

We have, however, been expressing the rate of growth and profit with respect to total income, and in that case the above simplification does not hold. Instead, denoting the capital–output ratio, K/Y, as κ:

$$(P/Y)\{\alpha + \kappa[s_w(I - s_c Y^\star)]\}/I = (I - s_w Y^\star)/Y$$

or

$$\omega(P/Y) = I/Y,$$

where

$$\omega = s_c - s_w - \kappa s_w[I - s_c Y^\star]/[I - s_w Y^\star]. \qquad (9A.11)$$

Equation (9A.11) is the result reported in equation (9.10) Note that $\omega < s_c$, as long as $s_c > s_w$ and $0 < \omega < 1$.

Part 4

Equation (9.2) may be rewritten in matrix form as:

$$\mathbf{x_{t+1}} = \mathbf{A}^{-1} \cdot \mathbf{x_t}. \tag{9A.12}$$

The standard method of testing a simple linear difference equation such as (9A.12) for stability is to check whether the real parts of the eigenvalues of matrix $(\mathbf{A}^{-1})$ are less than one in absolute value. If so, then the dynamic system is stable. In this case, however, it is easy to show that all of the eigenvalues are *greater* than one in absolute value (Morishima, 1973). This conclusion can be deduced by simply recalling that all of the eigenvalues of $\mathbf{A}$ are less than one (because the economy is productive, meaning that the sum of each row of $\mathbf{A}$ is less than or equal to one; Appendix to Chapter 4 and equation 4.3), and that the eigenvalues of the inverse of a matrix are equal to the reciprocals of the eigenvalues of the original matrix. When all the eigenvalues exceed one, the system is in fact highly unstable.

Disequilibrium: Class contradiction and struggle

'An unstable equilibrium is of limited interest, since it is the one place the system will never be found.' (Goodwin, 1987, p.4)

10 Class and space

Introduction

In several of the previous chapters we alluded to the importance of class action in determining locational outcomes. As yet, though, we have neither provided a definition of class and the ancillary concepts associated with it, nor systematically elucidated the relationship between class action and geographical location. We remedy such omissions in the next three chapters. In this introductory chapter we review some recent literature on class and space, and attempt to link work on social class originating in the emerging paradigm of analytical Marxism (Roemer, 1982a, 1986; Elster, 1985; Przeworski, 1985a) with work carried out by human geographers on the geography of class formation and struggle. The next two chapters then draw upon these ideas to examine rigorously the effects of inter- and intra-class conflict on the geography of economic activity.

Two caveats are necessary, though. First, those who are familiar with a historical or cultural treatment of class will likely find our analytical perspective foreign. We do not claim, however, that our analytical approach to class is either the only one, or necessarily the best. But it is consistent with our general approach to other aspects of political economy, and we also believe that it sheds new light on old questions. Second, our concern is only with Marxist class categories. The justification is that our analytical framework is rooted in a Marxist perspective, and to employ a non-Marxist definition of class would be inconsistent. Admittedly in previous chapters we have also employed a neo-Ricardian analysis, which some researchers argue is predicated upon Weberian class concepts (Shaikh, 1981). Our position, however, is that neo-Ricardianism is an open system inviting theoretical closure; we close it here using Marxist class categories (Barnes, 1989).

10.1 Class-in-itself and class-for-itself

Despite the pivotal importance of class, Marx unfortunately provided neither a sustained theoretical discussion of the concept, nor a rigorous definition of it (as is well known, Marx's direct discussion of class in *Capital* abruptly breaks off after two pages). Marx's writings, however, are suffused with classes and class struggle. Our view is that, like the *homo economicus* concept in neoclassical economics, class is a central entry point for Marx in discussing two pivotal questions: first, the nature of society and the individual's relationship to it; and, second, the motivation for, and the consequences of, human action.

We argue that within the traditional Marxist scheme the nature of society and its relationship to the individual is defined by 'class-in-itself', while the motivation and consequences of individual action are defined by 'class-for-itself'. Both of these concepts were coined by Marx in his *The poverty of philosophy*, but are particularly associated with the work of Karl Kautsky (Przeworski, 1985a, ch. 2).[1] Class-in-itself represents the 'objective' interests that distinguish one class from another. As a consequence, regardless of 'subjective' evaluations about their own economic conditions, individuals are 'objectively' members of a particular class, represented by a set of collective material interests. It is these 'objective' material interests that simultaneously define the individual's place within society (a person's material interests determine the class to which they belong), and the nature of society itself (the constituents of society are the set of 'objective' classes). Class-for-itself, in contrast, is the collective *recognition* of such interests, and the consequent acting upon them. The motivations for collective action are the so-called immediate and fundamental interests associated with each class. Marx argued that such interests may well not be recognized, and even if they are recognized they may not be realized because of conflict both within and among classes. Once, however, the interests of one specific class, the workers, are realized the consequence is widespread social change, culminating in the transformation of the capitalist mode of production.

We shall use the notions of class-in-itself and class-for-itself to organize the rest of the chapter. In section 10.2 we focus on class-in-itself, and break that discussion down into three sub-themes: definitions of class, associated class structures, and collective interests. In section 10.3 we focus on class-for-itself, and examine the processes by which classes recognize their interests and act upon them. To do this we divide the section into two parts: one exploring class formation and consciousness, and the other examining class struggle. In both cases we suggest that the key issue is the relationship between collective and individual action.

10.2   Class-in-itself

10.2.1   Definitions of class

There remains considerable debate on the meaning and definition of class. We shall review here the three most popular Marxist definitions of class, respectively based upon exploitation, domination and property ownership.

CLASS AS EXPLOITATION

The first view is that classes are defined by the exploitation of surplus value. Patricians are one class, slaves another; feudal barons one class, vassals another; and capitalists one class, and workers another. Although the form of exploitation varies in each case, the point of cleavage is always between those who for whatever reason have the ability to expropriate surplus labour, and those whose surplus labour is expropriated. As Lenin (1947, p.492) wrote: 'Classes are groups of people one of which can appropriate the labour of another ...' (see also Harvey, 1982, p.24). Resnick and Wolff (1987) recently added some interesting wrinkles to this traditional account. For them the capitalist class is either an appropriator *or a distributor* of surplus labour. Those capitalists that exploit labour at the point of production take on what Resnick and Wolff call a 'fundamental' class position, while those capitalists that simply transfer surplus value that already exists (for example, bankers) occupy a 'subsumed' class position. Resnick and Wolff then turn their analysis to examining the workers who give up surplus labour. Mirroring the scheme they established for capitalists, workers are called 'productive' when they are part of the fundamental class process, and 'unproductive' when they take on a subsumed class position. Although their elaborations are ingenious, Resnick and Wolff's view of class none the less remains embedded in an exploitation-based definition because the pivot of their scheme remains the expropriation of surplus labour.

The exploitation-based view of class is criticized on at least two grounds (Elster, 1985, pp.323–4). First, it is either too coarse- or too fine-grained. On the one hand, it is too coarse-grained because it does not allow one to distinguish between say, landlord and capitalist, or slave and worker – the former are generic exploiters, the latter generically exploited. On the other hand, it is too fine-grained if, to avoid the first problem, classes are defined by the degree of exploitation. For this would produce an infinite number of gradations, and hence class categories. A second criticism is that it is unclear whether class defined in terms of exploitation can provide an understanding of action. Under an exploitation-based definition, class action results from knowing that one is exploited. But there are two problems: first, given the complex definition of exploitation, and its measurement in labour values, 'no one in a society knows exactly where the dividing line between exploiters and exploited should be drawn' (Elster, 1985, p.323); and, second, throughout Marx's writings there is the insistence that ideology obfuscates the exploitive nature of capitalism. If so, how can exploitation motivate class action when no one perceives themselves as exploited (Roemer, 1988, p.85)?

CLASS AS DOMINATION

A second definition is that classes are defined by power relationships. By this is often meant domination, particularly at the point of production. The early Erik Olin Wright (1978, 1980) is perhaps the best-known recent advocate of this view. In reflecting on his earlier work he writes that it 'rested almost exlusively on relations of *domination* rather than exploitation' (Wright, 1984, p.384). By domination Wright means the relationship of authority at the workplace, and, in particular, the degree to which workers possess autonomy over their own work. With domination the cornerstone of the class definition, Wright then sought to identify 'contradictory class locations' – classes that were betwixt and between the usual bi-polar Marxist class classifications of proletariat and bourgeoisie. Such classes were contradictory in the sense that they take on characteristics of both opposing classes. A prime example is the managerial class. Although dominated by those who control money capital (the capitalists), managers at the same time have a relatively high degree of autonomy in carrying out their work compared with manual workers (who they then dominate at the workplace). In the sense that managers both dominate and are dominated they occupy a contradictory class location. More broadly, Wright (1984, p.385) suggests that the 'tendency of substituting domination for exploitation at the core of the concept of class is found in most other neo-Marxist conceptualizations of class structure' (for example, see Braverman's 1974, work, and the extensions to it suggested by Edwards, 1979).

The principal criticism of the domination view of class is that it applies only to non-market societies (Elster, 1985, pp.327–8). In a market society such as capitalism, a worker is a worker not because s/he is subject to domination after selling her/his labour power; rather s/he is a worker because s/he cannot do anything else but sell her/his labour power, and thereby be dominated. In short, it is the initial distribution of property and wealth that defines workers and capitalists, not that they are dominated at the workplace. In addition, Wright himself in his later work criticizes a domination-based view of classes on two grounds: first, domination does not imply any material antagonism. As Wright (1984, p.385) says, the fact that parents dominate their children does not imply an antagonistic relationship, let alone a materially antagonistic one. This is important because, as we suggested above, Marxist classes are typically defined as possessing 'objective' material interests that are the basis of collective action. But, if Wright's argument is correct, a domination-based view may not be related to material interests at all. A second problem is that a domination-based view of class 'tend[s] to slide into what can be termed the "multiple oppressions" approach to understanding society' (Wright, 1984, p.385). Class is just one of many oppressions, and thereby loses it distinctiveness as an entry point in understanding social life.

CLASSES AS PROPERTY OWNERSHIP

The third definition, and perhaps the most well known, is that classes are defined by the ownership, or lack thereof, of economic resources (assets). Capitalists are capitalists because they own the means of production, while workers are workers because all that they own is their own labour power. Elster (1985, p.322) argues, however, that such a definition is deficient in at least three respects. First, again it is either too coarse-grained or too fine-grained to be useful. It is too coarse-grained because the type of property owned is not specified and therefore one cannot distinguish between say, a landlord and a capitalist. It is too fine-grained because, if to meet the former objection property ownership is defined in terms of particular assets, classes are too numerous and fragmented. Second, such a definition is useful only when workers possess no assets, and capitalists possess them all. Once workers are allowed ownership of assets (through savings, pension schemes, stockholdings and the like), either one ends up with many classes, each one possessing a different level of assets, or one is drawn to the endless debate over the question of how much capital is needed to be a capitalist. The final criticism is that the property-based definition does not cope well with corporate property. The managers of such property, although having control over the asset, do not own it. Thus, managers are counted as workers despite the fact that they are in many respects like the bourgeoisie. The consequence is that in North America 95 per cent of the population are then members of the proletariat (Wright, 1980). But if corporate managers are counted as bourgeoisie then the property-based definition of class is violated.

ANALYTICAL MARXISM AND A REVIVIFED PROPERTY-BASED DEFINITION OF CLASS

Recently, the property-based definition of class has taken on renewed vigour with the analytical Marxist work of John Roemer (1982a), who has developed an approach that circumvents Elster's objections discussed above. In Roemer's account classes emerge from an economic process defined by two fundamental features: first, rational economic actors pursuing their best interests; and, second, an initial inequality in resource ownership. Given these two assumptions, and using the mathematics of orthodox economics, Roemer deduces theoretically the existence of exploitation and a class system. In this sense, exploitation and class are endogenously derived by Roemer, rather than simply assumed. More generally, the importance of Roemer's work is in showing that a necessary entry point for discussing class and exploitation is the fundamental postulate of inequalities in resource ownership. This contradicts some accounts that reverse this order of logic. Unfortunately, under the rubric of a property-based definition, Roemer presents two different definitions of exploitation and associated categories of class. Although recognizing the distinction between the two, Roemer does not assign priority. Roemer's two different systems of class and exploitation are termed, following Van Parijs (1986–7), the Roemer–Elster view and the Roemer–Wright view. We review each in turn.

The Roemer–Elster view begins with rational agents who command a given set of resources. Such agents are intent on maximizing utility. Assuming a capitalist system, and for the moment one without a capital market, agents choose among three options when facing the labour market: working for themselves, hiring labour, and/or working for others. Because everyone must choose at least one option (there is no unemployment), and because of Roemer's (1982a, pp.69–77) demonstration of the irrationality of choosing all three options simultaneously, the three labour market options give rise to five labour market options, which then for Roemer become class categories: the pure capitalist who only hires labour, the small capitalist who hires labour and works for her/himself; the petit bourgeois who only works for her/himself; the mixed proletarian who works for her/himself as well as working for another employer; and finally, the pure worker, who only works for an employer. Classes for Roemer, then, emerge as options within the labour market, with the 'allocation' of individuals to these options depending upon the distribution of property within society. With a given pattern of property ownership among individuals, agents in the economy rationally choose among the five options, thereby determining the class to which they belong. This, of course, is not a choice that individuals necessarily like making. But they do so because of the constraints they are under (their share of society's assets) along with their desire to optimize.

Given this scheme, Roemer formally demonstrates two fundamental principles. First, there is a homology between class and wealth. Termed the class–wealth correspondence principle (CWCP), it establishes that the more wealth (property) an individual owns the 'higher' s/he is in the class hierarchy. Second, there is a homology between class and exploitation, called the class–exploitation correspondence principle (CECP). Defining exploitation as the difference between the socially necessary labour time an individual works and the socially necessary labour time embodied in the real wage consumed, Roemer demonstrates that individuals in those classes that sell their labour (pure workers and mixed proletarian) are necessarily exploited, and individuals in those classes that buy labour (pure and small capitalists) are necessarily exploiters. (The petit bourgeois neither exploit nor are exploited.) As such, there is a symmetry between class status and exploitation status.

The broader importance of these two correspondence principles is in indicating the pivotal role of differential ownership of wealth. The property an individual owns determines both their class position (through CWCP), and whether they are an exploiter or one of the exploited (through CECP). In addition, Roemer demonstrates that his theory is a general one and holds: when labour hires capital rather than the other way around (that is, in the absence of a labour market); when consumers choose among goods (rather than having a set subsistence bundle, which is the usual assumption); and in non-capitalist modes of production, such as feudalism.

In contrast, the Roemer–Wright view of class is based upon game theory and so-called withdrawal rules. Specifically, the economy is treated as a game among players, where the object is to maximize utility. To achieve that end, players may form coalitions with others in establishing the best strategy, where strategies include withdrawing from the game altogether. To establish the optimal strategy a comparison is made between, first, the actual distribution of property (assets) and its effect on class income, and, second, a hypothetical situation in which some members of society have 'withdrawn' from the economy with their per capita share of property. If those who have withdrawn are better off in the hypothetical situation, they are defined as exploited and have an interest in withdrawing. In contrast, those who benefit by remaining are exploiters and have an interest in staying.

Within the framework of this game-theoretical approach, Roemer and, more systematically, Wright (1984, 1985) compare classes and exploitation across modes of production. Associated with each mode is a central inalienable (non-transferable) or alienable (transferable) asset. The asset is central in that it is the axial resource around which a given economy and society pivot. For example, central to feudalism is the inalienable asset of labour; in capitalism it is alienable private property; in statism it is the alienable asset of organization; and in socialism it is inalienable skills. These respective central assets then define the nature of the so-called 'withdrawal rule' associated with each particular mode of production: in feudalism that rule is one where agents can withdraw their labour power; in capitalism individuals can withdraw their share of society's private property; in statism people can withdraw their per capita share of the ability to organize; and in socialism agents can withdraw their share of skills.

With the withdrawal rule defined for a given mode of production, the associated system of exploitation and class is established by comparing the actual distribution of assets within that mode with the counterfactual one in which that asset is distributed equally among all agents. Those who are better off are defined as exploiters, and those who are worse off are exploited. Classes are then defined as coalitions that are either better or worse off after applying the withdrawal rule. For example, in capitalism workers are the coalition that is better off if they withdraw from the game with their per capita share of property (they are necessarily exploited), whereas capitalists are the coalition that is better off by remaining within the game (they are necessarily exploiters). The petty bourgeois, in contrast, are a coalition that is indifferent between the actual and counterfactual situations because they are neither exploiters nor exploited.

As Wright (1984, p.390) notes, the feature common to both the Roemer–Elster and the Roemer–Wright approach to class is:

> that the material basis of exploitation is inequalities in the distributions of productive assets, or what is usually referred to as property relations. On the one hand, inequalities of assets are sufficient to account for transfers of labor surplus [Roemer–Elster]; on the other hand, different forms of asset inequality specify different systems of exploitation [Roemer–Wright]. Classes are then defined as positions within the social relations of production derived from these relations of exploitation.

With social classes so defined, we can now show how Roemer's general property-based definition of class avoids the problems raised by Elster, and noted above. First, classes are neither too coarse- nor too fine-grained because they are defined not by property as such, but by the extent to which property ownership affects exploitation. In Roemer's work, a capitalist is defined not as one who owns capital, but as one who is able to exploit others *because* of the ownership of certain assets. Second, Roemer establishes a precise threshold value of property ownership to determine who is a capitalist, because class divisions are based not on the amount of assets owned but on the effect of those assets on exploitation. Finally, there is room within Roemer's two accounts to deal with the 'contradictory' location of the middle class. The petit bourgeois in both schemes occupy an intermediate position between workers and capitalists.

By way of a summary, we should note two other general features of Roemer's approach. First, it is frequently claimed that Marxism necessarily adopts a structuralist methodological position, one where individuals drop out of the analysis. In Roemer's account, however, his 'derivation reverses the sequence found in Althusserian sociology: here, individuals carve out class positions; there, individuals are carved out for class positions' (Carling, 1986, p.48). In a very real sense under Roemer's analysis individuals choose their class positions, and as a result the distinctiveness of analytical Marxism is precisely in its 'reinstatement of the subject' (Carling, 1986, p.28). (Also see Przeworski, 1985b, who argues that the methodology of Marxism should be that of methodological individualism.) Second, as argued in Chapter 1, Roemer uses the very tools and concepts of neoclassical economics, particularly the *homo economicus* assumption, to prove his case against that scheme. Individual optimization (*homo economicus*) is a key assumption of his work. In one way, this strategy represents a brilliant internal critique of neoclassicism (Ruccio, 1988). Roemer demonstrates that on its own terms the conclusions of neoclassical economics about the social efficacy of the free market do not hold. Roemer, however, clearly wants to go beyond critique and provide a positive alternative. Whether such an alternative is sustainable on the basis of the rationality postulate, though, is controversial (Anderson and Thompson, 1988; Lebowitz, 1988).

WHICH DEFINITION IS BEST?
Having presented three different definitions of class, the obvious question is which one is best. Roemer (1988) argues, not surprisingly, that the most useful one is his version based upon property ownership. For him an exploitation-based definition is simply wrong because of internal logical problems associated with using surplus value as a measure (Roemer, 1988, pp.127–31), while a definition based on domination is of only marginal importance (Roemer, 1988, p.87). Wolff and Resnick (1987, p.143) on the other hand, argue that a definition of class based upon exploitation/surplus value is the only true one. In contrast to these writers, we argue that all three definitions are potentially useful, and that the mistake is thinking that any one of them is definitive. For one must recall the purpose of the class concept, which is to explain both the nature of social life and the actions of individuals embedded within it. These vary dramatically in different contexts. For this reason, to argue that there is a single definition of class denies the variety of class formation and struggle that as geographers we seek to explain.

For example, although in some cases it is very difficult to penetrate the surface appearances of the market to establish whether one is exploited, in other cases it is quite obvious. In that situation one can readily imagine an exploitation-based definition of class being useful in understanding collective action on the part of a group of workers. Roemer's reply that exploitation itself is dependent upon a prior unequal distribution of wealth may be true analytically, but most workers are not privy to Roemer's (1982a) *General Theory* before engaging in class formation and struggle. Again, a domination-based definition may have its difficulties but, especially since the work of Braverman (1974), it is clear that the antagonism at the point of production between workers and managers/owners brought about by rigid measures of labour control may well precipitate the formation of classes and class action (Anderson and Thompson, 1988). Finally, an unequal distribution of wealth is historically an issue around which classes crystallize, and which is an impetus for action and conflict. As Roemer (1988, p.87) writes:

> Certainly the most revolutionary struggles [result from] ... the working class supporting a call for a massive redistribution of private property or an end to the institution of private property. In our times, this call is typical of the great socialist revolutions.

CLASS DEFINITIONS AND SPACE
Concerned with differences over space, geographers are particularly sensitive to recognizing the different contexts in which class action is played out, and hence the advisability of a catholic approach to class definitions. In this light, one role of the geographer is to establish which of the different definitions of class discussed above is most appropriate to each geographical context. Another role, and one with which we are concerned here, is establishing the theoretical difference that the addition of space makes to aspatial theories and categories. In the analytical work presented in previous chapters, we argued that setting aspatial economic theory within a spatial context often disturbs some of that theory's central conclusions. We will argue that this is also the case with the concept of class. Each of the three definitions of class discussed above requires theoretical modification once set in a spatial context.

The first definition, that classes are defined by exploitation, is disrupted by the conclusion that we reached about the geography of exploitation in Chapter 8. There we showed that in inaccessible regions it is possible that workers are negatively exploited; that is, workers earn more than the value of their labour. Negative exploitation occurs because workers in more accessible regions are transferring, through commodity trade, part of their surplus labour to workers in inaccessible regions. If we strictly employ the first definition of class based upon exploitation the effect is to make negatively exploited workers capitalists, because this group is now appropriating surplus value. Such a conclusion, of course, is problematic because in every other way the negatively exploited workers are identical to positively exploited workers.

The same difficulties beset the definition based upon property ownership. In the Roemer–Elster view there is a correspondence between class and exploitation, a result of an inequitable distribution of assets. Within a spatial economy where workers are negatively exploited, the logical possibility exists of an asymmetry among property ownership, class and exploitation. In particular, there is a disjuncture between the class–wealth and class–exploitation correspondence principles. Under the class–wealth principle, negatively and positively exploited workers have the same (minimal) amount of assets and therefore they belong to the same class (both 'choose' to be workers when facing the labour market). But under the class–exploitation principle, negatively exploited workers should be in a different class from positively exploited workers. As such, in a spatial economy the correspondence among class, exploitation and wealth no longer holds. A similar problem occurs in the Roemer–Wright view of class and exploitation. With the existence of negatively exploited workers in a spatial economy, it is no longer clear that all workers would be better off by withdrawing with their per capita share of resources. Specifically, negatively exploited workers may suffer more from opting out of the game than remaining in. If this is so, the definition of class in the Roemer–Wright view is again problematic. Negatively exploited workers who remain in the game in effect become part of the capitalist coalition, yet in terms of their ownership of alienable assets they are identical to other workers who decide to withdraw. This thereby casts doubt on the consistency of a property-based definition of class in a space economy.

The final definition of class we examined was couched in terms of domination. This is not susceptible to the kind of criticism made above with respect to both the exploitation- and property-based definitions. None the less, the domination view begins to unravel when situated within a geographical context. Classes in the domination view are based upon the degree of capitalist control at the workplace. That domination, however, varies tremendously across space. As a consequence, the degree of domination, say, within a semi–conductor plant located in Malaysia, might be very much greater than in an automobile plant in Michigan. If this is so, it is not clear that one can employ the common label of the 'working class' to describe both sets of labourers (for example, see the literature on the concept of a labour aristocracy; Barbalet, 1987).

We are not trying to dispense with class as a category in pointing out the difficulties that beset it within a spatial context. The problems brought by the introduction of geography are quite general, and cut across a wide range of social theories and concepts (Harvey, 1985a, p.xiii). None the less, we do think that geographers should think more carefully about the use of such non-spatial concepts as class, and develop, as some are beginning to do, a theory that is sensitive to, and incorporates the subtleties of, space and place (see Thrift and Williams, 1987; Soja, 1989).

10.2.2 Class structure

The three definitions of class discussed – exploitation, domination and property ownership – are each the basis for a different class structure. Following Keat and Urry (1982), we define a class structure as having three characteristics: first, a

Table 10.1 The relationship between class-in-itself and class-for-itself

	Exploitation	Class definition Domination	Wealth
Class structure	Exploiter/ exploited	Dominant/ subordinate	Owner/non- owner
Objective class interests (a) Immediate	Shorten working day/increase wage	Greater autonomy	Equitable distribution of wealth
(b) Fundamental	Collective ownership	Worker control	Abolition of private property

CLASS-IN-ITSELF

Class formation and consciousness
Class conflict

CLASS-FOR-ITSELF

clear relationship exists among the elements of that structure; second, the relationship among them is systematic and enduring; and, lastly, the structure is abstract. The social structure thus created defines a set of positions or places that are filled by individuals. For the moment we set aside the question of whether individuals are Roemerian – choosing their class places on a rational individualist basis – or Althusserian – class places choose individuals on the basis of capitalism's continued reproduction.

The three class structures that are respectively associated with each of the definitions of class discussed above are shown in Table 10.1. In the exploitation-based definition, the structure is one of opposition between exploiter and exploited expressed in terms of surplus value. In the domination definition of class the structure consists of the relationship between dominators and subordinates, where the issue is the authority (power) at the point of production. The property-based definition of class produces a structure based on the oppositional relationship between owners and non-owners of assets. In each case, the class structure meets the three general criteria discussed above. First, the focus is on the relationship among elements, not the elements themselves. In the case of capitalism, it is the antagonistic *relationship* between workers and capitalists that is important, not the intrinsic nature of workers and capitalists per se. Second, the relationship is systematic and enduring. It is systematic in that, while one class is always a winner, the other is always a loser. Thus, one class always expropriates surplus value while the other gives it up; one class always exerts authority at the point of production while the other is subordinated; and one class always controls property while the other does not. It is also enduring in that such relationships hold across different societies and different epochs. Master/slave, lord/serf and capitalist/worker are all variants on the same antagonistic relationship. Finally, the structure is grasped only when the *abstract* terms in which it is expressed are understood. Specifically, only when one realizes that surplus value, or authority, or property relationships are the key does one pierce the veil of appearances and understand the class structure as it 'really' is.

10.2.3 Class interests

We now come to an issue that many consider is at the heart of the discussion about class-in-itself, that is, class interests. The central feature of class-in-itself is that it is rooted in the forces and social relations of a mode of production, i.e. the base or infrastructure. It is in this sense that class interests are 'objective'. They do not depend upon what people subjectively think, but are defined by the lineaments of an external economic and social structure. Once an individual meets certain criteria (are exploited, dominated, or possess no assets), then by the very definition of that broader structure a person possesses a set of (objective) interests that define their best course of action.

The individual may not be aware of his/her best course of action. In fact, Wright (1978, p.89) explicitly defines class interests as 'in a sense hypotheses: they are hypotheses about the objectives of struggles which would occur if the actors in the struggle had a scientifically correct understanding of their situation.' Wright couches the definition in these terms because a fog of ideology can prevent workers from seeing capitalism's 'true' nature. But once that ideology is demystified workers' 'true' interests are revealed.

Within the broader rubric of class interests Wright recognizes two subtypes: immediate and fundamental. 'Immediate class interests constitute interests within a given situation of social relations' (Wright, 1978, p.89). Table 10.1 outlines the interests of workers under capitalism. In an exploitation view the workers' immediate interest is in reducing the working day and increasing wages. In the domination view, it is to gain greater autonomy at the workplace. In the property-based definition, it is to strive for greater equality in asset ownership (presumably only obtainable in the short term through lobbying of legislators and active political involvement in government using such tools as land reform, inheritance taxes, capital gains taxes, and so on).

In addition to these immediate interests, Wright (1978, p.89) also recognizes 'fundamental interests' defined as those 'interests which call into question the structure of social relations itself'. Specifically, in capitalism the fundamental interest is overthrowing the dominant mode of production, thereby moving to socialism, and eventually communism. Again there is a correspondence between different fundamental interests and each of the three definitions of class (Table

10.1). In the exploitation view, it is to dispense with the role of the capitalist by imposing collective ownership; in the case of domination, it is worker control; and for the property-based definition it is to abolish the institution of private property altogether.

10.3 Class-for-itself

In the previous section we discussed the nature of society and an individual's relationship to it in terms of class-in-itself. Marx suggested, however, that collective action does not necessarily stem from the mere existence of material interests. Therefore, two critical questions are: what are the conditions under which class-in-itself recognizes its class interests and thereby is transformed into class-for-itself? And, having recognized them, what action does that class undertake to fulfill its interests? In this section we address both these questions by, first, exploring the conditions for collective action using the concepts of class formation and consciousness, and, second, examining the consequences of collective action using the idea of class conflict (Table 10.1). More broadly, we argue that in dealing with both issues we need to examine carefully the relationship between collective and individual rationality.

We should note that geographers are active researchers on this broader topic of class-for-itself (whereas little geographical research exists on class-in-itself), examining, in particular, the geographical conditions for 'class capacity'. Coined by Wright (1978, p.98), this term is defined as 'the *social relations within a class* which to a greater or lesser extent unite the agents of that class into a class formation'. Geographers have attempted to define those qualities of space and place that either assist or hinder the process of class consciousness and formation, thereby determining the effectiveness of class struggle. In fact, we will argue below that this work is critical in clarifying the conditions under which either collective or individual rationality is realized.

10.3.1 Class formation and consciousness

Before discussing class formation and consciousness per se, we need first to broach the problem of collective action. Collective action is defined here as the common actions of a group of individuals who are bound together by similar goals and resources. Under this definition class action is a subtype of collective action. In his now-classic *The Logic of Collective Action*, Mancur Olson (1965) argued that, while the actions of rational individuals are always in their best interests, the collective action of a group of rational individuals need not be. As Olson (1965, pp.1–2) wrote: 'it is *not* in fact true that the idea that groups will act in their self-interest follows logically from the premise of rational and self-interested behavior.' This is because of the so-called free-rider problem. As Elster (1985, p.347) wrote: 'Collective action is beset by the difficulty that it often pays to defect. An individual can reap greater rewards if he abstains from the action to get the benefits without cost. This generates a conflict between the interest of the individual class member and that of the class as a whole.' For example, let the collective benefit of a trade union be higher wages. In the case of an open shop a worker can enjoy the benefit of higher wages from union action, and share none of the costs (union dues, loss of pay due to strikes and so on), by not joining. Such workers are free-riders. More broadly, free-riders are a problem for collective action because large groups are an economic public good, and as such benefits cannot be restricted only to those who directly pay for them.

The implication of Olson's work for our purposes is that, because of the free-rider problem, rational individuals may not engage in collective action even when it is ultimately in their best interests. As such, the free-rider problem is a substantial obstacle to class capacity, and thereby class formation and consciousness. This is easily demonstrated by using a simple and familiar example: the model of the prisoner's dilemma (Elster, 1985, p.359). Within the framework of the prisoner's dilemma, collective action yields three types of gains/losses: first, the gains from collective action; second, gains from free-riding; and third, the losses from unilateralism (that is, the cost to an individual when s/he was one of the few to engage in collective action, while the majority abstained). Let there be two parties, an individual I, and everyone else. The returns from undertaking a particular course of action are represented by a two-by-two 'pay-off' matrix (Table 10.2). Each cell represents the outcome of I either engaging or abstaining from collective action vis-à-vis everyone else's decision. There are two entries in each cell, the first indicating the consequences for I, and the second those for everyone else.

From Table 10.2 the pay-offs from cooperation when individual I and everyone else agree to engage in collective action are b for both parties. By comparing this pay-off with the one for both parties abstaining, i.e. a pay-off value of a, the gains from cooperation (collective action) are $b - a$. Similarly, the gains from I free-riding on the collective action of everyone else are equal to $c - b$ (respectively the pay-offs for I abstaining while everyone else engages in collective action). Finally, the costs to I of unilateralism, that is, I attempts collective action but everyone else does not, are $a - e$ (respectively the pay-offs for almost everyone else not engaging in collective action while I engages in collective action).

Suppose in Table 10.2 that $c > b > a > 0 > e$. Then the rational choice for an individual is to abstain from collective action because there is no possibility of incurring a unilateralism loss, while at the same time there is some prospect of a free-rider gain. If all individuals act in a like manner the result is the absence of collective action and therefore class formation (pay-off a), even though the more beneficial decision is collective action (pay-off b). Because collective action clearly does exist, the theoretical task is to examine why the assumptions of the free-rider problem do not apply in those cases. Using the terms employed earlier, we must examine how class capacity operates to overcome the problem of free-riders, thereby enabling class-in-itself to be transformed into class-for-itself. We will argue below that such a task is possible if we take space and time seriously (see Lash and Urry, 1984). This implies that the emerging literature on class formation and consciousness compiled by human geographers is increasingly central to the understanding of collective action.

Within the literature at least six different reasons are given for the absence of free-riding, thereby allowing collective action and class formation (Elster, 1985; Przeworski, 1985b). We will argue that to make sense of each of those six factors we must take into account the geographical and historical context. The addition of geography and history in effect provides the mechanism that turns each of these logical possibilities into the reality of class formation. As Thrift and Williams (1987, p.xiii) write: 'Classes do not wax and wane in a geometrical abstraction but on the ground as concrete situations of conflict and compromise – in a geographical reality.' We argue that understanding that geographical reality is crucial not only for comprehending the practice of collective action, but also for comprehending its existence. Steeped in methodological individualism, analytical Marxists treat collective action as a puzzle or paradox: class action should not exist because of the free-rider problem, but it does. We argue, however, that once individuals are embedded within place and space much of the mystery of collective action evaporates. In fact, the puzzle is why individuals were abstracted from space and place to begin with.

The first two reasons suggested for the existence of collective action are given by respectively Booth (1978) and Roemer (1978). Booth (1978, p.168) argues that people have an 'internal psychic need' for collective solidarity, while Roemer (1978, p.154) avers that collective consciousness 'does not depend on

Table 10.2 A pay-off matrix showing the potential conflict between individual and collective action

		Everyone else	
		Engage	Abstain
I	Engage	b, b	e, a
	Abstain	c, d	a, a

coercion or side payments, but on workers learning to discard the individualist model and adopting collective rationality'. Lacking in each of these claims, however, are the mechanisms that bring about collective solidarity or collective rationality. Presumably such mechanisms rest upon the formation of some kind of class consciousness, which Elster (1985, p.347) defines as 'the ability to overcome the free-rider problem in realizing class interests'. But in making class consciousness primary, one is still not answering the question about mechanisms, but only pushing it back to yet another level. For we still do not know how class consciousness is realized.

To provide a more satisfying explanation it is necessary to examine the geographical and historical context within which class consciousness arises. It is here that the recent work on the constitution of class consciousness across space is pivotal (Foster, 1974; Thrift, 1983, 1985; Walker, 1985; Harvey, 1985a). Such work is concerned with looking at the space- and time-specific institutions, material resources and symbols within the landscape that both help create, and are created by, class consciousness. Couched at all geographical scales – home, neighbourhood, region and nation state – this work helps understand both why people are bound together in collective solidarity, and why individuals might choose collective over individual rationality. More generally, as Doreen Massey (1984a, b) makes clear, 'geography matters!' Only angels dance on the head of a pin; people live in real places – villages, towns, cities and regions. Who you are is in part shaped by where you are. The rich historical economic *geography* of South Wales is vital to understanding the high degree of class solidarity found in mining villages there (Massey, 1984a, ch.5), just as the lack of an industrial geographical past accounts for the absence of class consciousness in the Highlands of Scotland.

Interaction among individuals is the basis for the third and fourth reasons given for the existence of collective action. The third reason is that, once 'externalities' such as feelings of altruism, guilt and injustice influence individual action, the assumption of the selfish free-rider is undermined, thereby creating the conditions for cooperation and class action. More broadly, collective action is realizable because the basis of individual action is no longer the strict material benefits given by the pay-off matrix. The related fourth reason is that collective action occurs when, through interaction, individuals endogenously alter one another's preferences. The fourth reason reduces to the third when the outcome of the interactions is a new set of preference functions that includes altruism, guilt, and so on, as arguments. A different type of endogenous alteration of preferences, though, is the 'demonstration effect'. If individuals are seen receiving a limited pay-off for engaging in collective action, this may induce others to join in, thereby endogenously increasing the pay-off coefficients for everyone. Whether this happens will depend upon what Elster (1985, p.357) calls the 'techology of collective action', that is, 'the functional relationship between the input (total participation) and the output (benefit to the individual collective action)'. Elster discusses three possibilities: a step function (collective action must involve some threshold number of participants); a concave function (initial contributions have little impact, but later contributions yield increasing returns); and, lastly, a convex function (initially small increases in the number of participants yield large benefits, but the increase in such benefits declines with increasing contributions).

Whether collective action is precipitated by externalities or a demonstration effect, the key mechanism is social interaction, which we argue is strongly mediated by spatial propinquity. Externalities such as altruism or guilt, or the demonstration effect, emerge through everyday local interchange with neighbours, friends and fellow workers. But the spatial setting of social intercourse is rarely discussed, let alone systematically investigated, by those who suggest externalities and the demonstration effect as causes of collective action. Once again, there is a need to highlight the geographical. In fact, some work already exists on this issue, although it is not usually linked to the free-rider problem. For example, at an intra-urban level Harris (1984, p.33) discusses the effects of spatial residential segregation, and argues that 'segregation facilitates and hinders social contact. It makes contact between members of different classes more difficult. Conversely, contacts among members of the same class are made relatively easy. Ultimately the significance of segregation for the process of class formation derives from these simple facts.' Again, at an urban level Peet (1987, p.49) suggests that class capacity is higher where 'workers liv[e] in settings which multiply their contacts ... Worker organization is thus easier in large, densely populated cities than in small towns or rural areas'. This finding is then used by him to make sense of the geography of labour unions, and responses to them by capital. Finally, at a national scale, Calhoun (1987) argues that class formation is predicated upon the development of a comprehensive transportation and communication system. Specifically, he claims that the problem with Marx's work is that there is 'no strong account of social organization *per se* ... One result of this is that as classes are deduced from the economic theory, their collective action is presumed to follow simply from rational recognition of common interests' (Calhoun, 1987, pp.56–7). An adequate account, though, must supply the mediating means of organization between rationality and action. For Calhoun this is the large-scale transportation system that developed in the mid-nineteenth century. It was only with the development of these means of interaction and organization that there could be 'co-ordination of activity at a class level' (Calhoun, 1987, p.54). Our argument is that, at all of these geographical scales, place and space are a critical ingredient in class capacity, and more fundamental in inducing collective action.

A fifth factor potentially mitigating the free-rider effect is that workers need not always act rationally. An example is when workers believe that if they engage in collective action others will follow. Such a belief, however, is not rational because if I act in one way it does not logically imply that others necessarily do the same (Elster, 1985, p.41). However, once class consciousness and organization are situated in space and place, we can readily understand why workers believe that if they act in one way others will too. If individuals are spatially proximate, are similar in background, and interact in common institutions, there are clear reasons for expecting that others will behave like them. Once that belief is in place it is collective rather than individual action that is more rational because there is no fear of free-riders. A classic work here is Willis's (1978) *Learning to Labour*, which establishes the material resources and institutions that turn working-class children into working-class labourers, with the implied collective beliefs and actions. Although Willis is not explicit, the processes he describes are intimately tied to a common micro-geography of social interaction. The 'lads' believe the same things and do the same things because they are socialized in the same neighbourhood schools, go to the same neighbourhood clubs and pubs, and have families who work in the same neighbourhood factories. This is not to imply any absolute determinism, but such powerful common experiences are difficult to resist in favour of some abstract individual rationality.

The final reason follows from the incorporation of time into the prisoner's dilemma game (Edel, 1979). In Olson's work, and the extension of it using the prisoner's dilemma model discussed above, it is assumed that the duration of the game is only one period. But, once repeated games are allowed, interaction is possible among players. The result is the development of 'meta strategies' where cooperation is more rational than independence. As Edel (1979, p.753) writes: 'In these longer games, one cannot take other players' actions as given. Other players may respond to one's own moves to collaborate, or retaliate against noncooperative behavior. Thus, the rational calculus may shift toward more frequent collective action.' It has been subsequently shown that, even when games last more than one period, collective action is rational only where there is no finite horizon to interactions, and workers have a low rate of time discount. None the less, like the introduction of space, time clearly disturbs Olson's original conclusions.

Geographers typically do not couch their work on class consciousness and formation in terms of solutions to the free-rider problem. We have tried to show in this section, however, that such work not only fits very comfortably within this broader theoretical analysis, but in many ways makes a significant contribution to clarifying and enriching it, and also calls into question the priority that is given to rational, egotistical individuals.

10.3.2 Class struggle

Given that classes form and pursue their interests (both immediate and fundamental), there is necessarily an antagonistic relationship among them. For, by their very definition, different classes have opposed interests. In the exploitation view the antagonism is between the exploiter of surplus value and the exploited; in the domination view the antagonism is between those who have power at the workplace and those who do not; and, lastly, in the property-based view, the antagonism is between those who own wealth and those who do not.

In many of the accounts of class antagonism it is often assumed, first, that class struggle is pervasive, and, second, that 'might is right' in the sense that the strongest class is always the victor. Using the arguments of analytical Marxists, we will argue here that neither of these two characteristics necessarily holds in the presence of class antagonism across space and place. First, in arguing against the all-pervasiveness of class struggle we examine the issue of class coalitions and its geographical counterpart, defence of place. And, second, in arguing against the view that the most powerful class always wins we discuss the unintended consequences of collective struggle and its geographical counterpart, the devaluation of place.

CLASS COALITIONS
In discussing the relationship among classes we must first establish the nature of the conflict. If it is a zero-sum game (resources are given and constant before the process of distribution begins), it is more likely that there is open conflict than if it is a variable-sum game (total resources change as a result of the very process of distribution). Marx argued that the conflict between landlord and capitalist was of the first kind. There is a fixed amount of surplus value to be distributed, and for this reason the degree of conflict over the division of this finite quantity is high. The conflict between capitalist and worker is more complex, however. This is because the very struggle over distribution changes the amount to be distributed. For example, if workers strike or are locked out in the conflict over distribution, it follows that output and, more fundamentally, surplus value falls, thereby decreasing the quantity to be distributed. In this latter situation, the prospect of class compromise and the formation of class coalitions as a way of avoiding such a loss seems real.

In Przeworski's (1985a, b) recent theoretical work, class compromise becomes a central issue when historical time is introduced into the analysis. Specifically, under certain conditions workers might well agree to curb their open struggle with capitalists if an agreement is struck giving workers increasing wages over the short term. He argues that, although over the very long run workers will be better off under socialism than capitalism, given both the time and costs of such a transition it might be rational for workers to compromise. For example, in moving towards socialism workers presumably engage in prolonged struggle, thereby resulting in lost output owing both to industrial action and to reduced investment on the part of capitalists. In this sense, the game is a variable-sum one where the output to be distributed changes with the distribution process itself. Under the assumptions that the motivation of workers stems from material advantage, and that they have low time discounts (workers value income today far more than income in the future), Przeworski then demonstrates that it is rational for workers to strike a class compromise with capitalists: no overt struggle on the condition of progressively increasing wages. Such a compromise is rational in the sense that workers will be better off materially now under capitalism than under some hitherto unrealized and uncertain state of future socialism.

The issue of class coalitions is also a complex topic. With three classes there are in effect three coalition strategies: the two stronger classes align and neutralize the weaker class; the weaker class aligns with one of the stronger ones and collectively neutralizes the third; and, lastly, the two stronger classes engage in conflict, allowing the third to survive and possibly emerge as the victor.

Although these are the three usually recognized coalitions, there is also a fourth that arises precisely from the inclusion of geography, termed here 'defence of place'. In this case, all three classes align with one another in a particular place against one or more classes located in a different place. The conclusion that workers in certain places are negatively exploited (Chapter 8) shows that such alliances may have a real material foundation. Although not usually discussed by non-geographers, we will argue that this form of coalition is prevalent, and vital to understanding the space economy.

In the last few years writers from a radical perspective have increasingly recognized that allegiances to neighbourhoods, cities and even regions often cut across class boundaries (Urry, 1981; Harris, 1983, 1984; Harvey, 1985b, pp.148–64, 1986; Clark, 1986; Hudson and Sadler, 1986; Cox and Mair, 1988; Leitner, 1990). This occurs, as Harvey (1985b, p.148) writes, because the various classes living within a place wish

> to preserve or enhance achieved modes of production and consumption, dominant technological mixes and patterns of social relations, profit and wage levels, the qualities of labour power and entrepreneurial–managerial skills, social and physical infrastructures, and the cultural qualities of living and working.

In this view, coalitions form among classes in a place to defend what has been achieved, and, if possible, to build upon it. In addition, class compromise is also occurring. Workers forgo their struggle with capitalists on the condition that their community, their place, is preserved (Cox and Mair, 1988).

There is now an enormous amount of work on the various types of coalitions that 'defend place', including both growth coalitions concerned with achieving more, and business/local state/worker coalitions concerned with maintaining what is already there. The evidence suggests that such coalitions are unstable (Harvey, 1985b, pp.148–62), but while in existence their effect is clearly to fracture classes. In defending place, workers are in effect pitting themselves against other workers in other places. The fragmentation of the working class is not a necessary occurrence but, as Cox and Mair (1988) argue, it is a likely one given the 'local dependence' of workers. Furthermore, this co-option of workers is carried out using a range of strategies from outright threats to appeals to local pride (Cox and Mair, 1988).

UNINTENDED CONSEQUENCES OF COLLECTIVE ACTION

In both the geographical and non-geographical literature it is implicitly assumed that, because capitalists are the strongest class, it is their interests that are ultimately realized at the expense of others. This need not be so, however. In fact, if Marx's prediction of socialism is to come true it cannot be so. Elster (1985) argues that there is frequently a conflict between individual and collective capitalist rationality, one that potentially thwarts the realization of capitalist interests, thereby also undermining the idea that 'might is right'.

In the course of pursuing their own goals, it is possible that rational individual agents bring about an outcome that is either better or worse than intended. Following Elster (1985, pp.22–4), suppose an individual desires to increase her/his income and acts to realize that goal. If only one individual undertakes that act, let that individual's income rise from a^0 to a^1. But now let us suppose that all individuals have the same goal and simultaneously undertake the same action. If there are unintended consequences from action they are either beneficial ($a^2 > a^1$) or detrimental ($a^3 < a^0$), where a^i ($i = 2,3$) is the income realized *after* collective action. Elster calls the beneficial consequences 'the invisible hand' (following Adam Smith) and the detrimental consequences 'counterfinality' (following Sartre) (Elster, 1985, p.24).

Elster (1985) argues that there are two reasons for the occurrence of unintended consequences. First, they arise because individuals 'entertain beliefs about each other which are such that, although any one of them may well be true, it is logically impossible that they all be' (Elster, 1985, p.44). For example, if each individual believes that s/he will benefit from an action as long as others do not copy it, and adopts that action under the assumption that *all* others will not follow, then collectively this is a case of mutually inconsistent beliefs. In the vocabulary of philosophy, to act on this basis is to commit the fallacy of composition. That is, the outcome of one person pursuing a goal can be quite different from the outcome if all individuals simultaneously pursue the same goal. Second, unintended consequences arise because people misjudge the 'technical relations' of action. Because of the complexity of relationships involved, an individual might not be able to work out in advance all the repercussions of his/her action and their potential feedback effects on welfare.

Elster argues that the first cause of unintended consequences is the most important when examining Marx's work. As an example, he cites Marx's falling rate of profit thesis. Here individuals pursue what is to them a rational course of action: investing in new, technologically superior capital equipment, thereby lowering costs and increasing individual profits. Yet in making this decision capitalists commit the fallacy of composition. For when all individual capitalists increase the capital-intensity of investment, the organic composition of capital increases, thereby *lowering* the general value rate of profit for everyone. Although this example shows the power of the fallacy of composition in producing detrimental consequences, we will argue in Chapter 12 where we further discuss Elster's ideas that ignorance of technical relations (the second cause of counterfinality) is perhaps the most important contributor to deleterious collective consequences within a *spatial* economy.

In broader terms, Elster argues that the idea of unintended collective consequences is the best way to make sense of Marx's notion of dialectics and contradiction. Indeed for Elster (1985, p.48) it is 'Marx's central contribution to the methodology of social science'. In this view the importance of dialectics is in focusing analytical attention directly on the causes of the instability, and the anarchic nature, of the capitalist free market. Such a conclusion is also pertinent to Marxist geographers who have discussed the idea of spatial dialectics. The debate between Peet (1981) and Smith (1981) on this issue, though, was at best inconclusive, while Soja's (1980) paper on the socio-spatial dialectic was concerned less with dialectics and more with formally establishing the theoretical position of 'space' within Marx's base/superstructure scheme. If, however, spatial dialectics is defined in terms of the unintended geographical consequences of collective action it takes on a renewed vigour.

We will examine the geographical manifestation of the unintended consequences of collective action in some detail in Chapter 12, but Harvey (1982) already provides the beginnings of an analysis (which he couches in terms of the contradiction between individual and class interests). In a passage brilliantly anticipating Elster, Harvey (1982, pp.389–90, 189) writes that:

> Individual capitalists, acting in their own self-interest and striving to maximize their profits under the coercive pressures of competition, tend to expand production and shift locations up to the point where the capacity to produce further surplus value disappears. This is, it seems, a spatial version of Marx's falling rate of profit thesis ... [T]hat the behavior of individual capitalists tends perpetually to de-stabilize the economic system ... is ... I would submit, the fundamental proposition that lies buried within the falling rate of profit thesis.

Specifically, Harvey argues that, in their quest to maximize profits, capitalists continually search out new, lower-cost locations. But, in so doing, other capitalists are also prompted to relocate to lower-cost sites. The immediate (unintended) consequence is the devaluation of existing places as firms scramble to move. Fixed capital and pools of skilled workers are abandoned in such cities as Hamilton, Youngstown and Sheffield as capitalists move on in their search for higher returns. However, this strategy of relocation cannot be successful. Capitalists fail to understand that, because everyone moves, no one gains. In fact, the very act of relocating and leaving behind vast quantities of fixed capital in 'devalued places' unintentionally precipitates the fall in profits that such relocation was designed to avoid. More broadly, 'competition necessarily leads individual capitalists to behave in such a way that they threaten the very basis for their own social reproduction' (Harvey, 1982, p.68). This is counterfinality with a vengeance.

Summary

This chapter had two central tasks. The first was to provide a discussion of, and a role for, social classes within our broader analytical approach. This is important because there is often a tacit belief that an analytical approach is somehow incompatible with any discussion of class. We tried to show that such a view is misfounded. Of course, our presentation of social class is not the contextual one that a historical or social geographer adopts, but this was never our intention. Rather, our purpose was the narrower one of laying out the logical relations among a set of abstractly defined class categories and concepts. In so doing, we are not rejecting the importance of context. In fact, in several places we pointed to the importance of including it. But our task here was not one of empirical investigation, but one of conceptual clarification. Specifically, we wanted to provide a framework for discussing class that is both compatible with our general analytical scheme and also enables us to discuss the issue of instability, which is the focus of this part of the book.

Our second task was to examine the analytical effects of introducing space and place into the usually aspatial presentation of class relations. Our central conclusion was that geography makes a difference. Specifically, a number of aspatial class relationships and concepts require either serious modification or elaboration once geography is included (for example, the various definitions of class). In addition, the introduction of space and place also clarifies certain processes that are traditionally presented as opaque or paradoxical (for example, the conditions under which the free-rider problem breaks down). In fact, incorporating the effects of space and place in a discussion of social class denies the presumptions of methodological individualism and individual rationality that analytical Marxism celebrates. This said, we think the work of Elster and other analytical Marxists is at least suggestive for geographers. Their work on unintended consequences of action, class coalitions, class compromise and even the free-rider problem provides a broader framework in which to set and integrate work already carried out by geographers. But this is not a one-way avenue of ideas. Geographers, with their sensitivity to space and place, have as much to give social theorists of class as they have to gain from them.

Note

1   The idea that class-in-itself 'spontaneously' gives rise to class-for-itself is criticized by
    Przeworski (1985a, ch.2). Our use of these two terms, however, is primarily
    expository and pedagogical. We do not believe that the transformation between
    class-in-itself and class-for-itself is spontaneous, and, in fact, later in the chapter we
    sketch out a potential mechanism connecting the two, one where space and place are
    central.

11 *Location and inter-class conflict*

Introduction

Having introduced the relationship between class and space in a general way, we turn in this chapter to an examination of the links between inter-class conflict and the geography of production and circulation. In undertaking this task, we generally assume that each class pursues only its collective interests in struggling with other classes. As a consequence, counterfinality does not arise because there is no conflict between the individual and the collective. This does not imply that we consider the issue of collective versus individual interests unimportant; rather, we delay discussion of it until the next chapter. This said, inter-class conflict can overlap with intra-class conflict, and in those circumstances we have not tried to separate them here.

In discussing inter-class conflict we will focus on three classes: landlords, capitalists and workers. Furthermore, we will examine inter-class conflict in terms of the three possible pairs of conflicts: workers versus capitalists, workers versus landlords, and landlords versus capitalists. Occasionally, however, we will go beyond the specific class pair to discuss the role of the 'third class' within the conflict. In discussing these three types of conflict we argue that for each pair there are three potential sources of antagonism corresponding to the broad definitions of class discussed in the previous chapter. In particular, conflicts arise between pairs of classes over: income distribution (the monetary equivalent of the surplus value definition of class), power relations (corresponding to the domination definition of class) and property ownership (corresponding to the asset definition of class).

11.1 Workers vs capitalists

11.1.1 *Geographical conflicts over income and consumption levels*

The first type of worker–capitalist conflict we discuss is the struggle over income and real consumption levels. We divide the discussion into two parts: a theoretical discussion of the nature of that conflict; and an examination of the thesis proposed by a number of researchers that this conflict is delimited by certain fixed bounds.

THEORETICAL EXPOSITION
In Chapter 4 we argued that conflicts over income, represented by the wage–profit frontier, are inescapable and enter into the very determination of the

profit-maximizing geography of production (section 4.1.2). To recapitulate briefly, for every possible permutation of locations for sectors producing in different regions a wage–profit frontier is constructed. The location permutation associated with the highest rate of profit for a given level of wages is the profit-maximizing location pattern. Figure 4.3 showed the wage–profit frontiers for three locational permutations: A, B and C. The profit-maximizing locational permutation is the one whose wage–profit frontier lies outside all other frontiers for a given wage rate. Examination of Figure 4.3 shows how the profit-maximizing location pattern critically depends upon the distribution of income between workers and capitalists; at w_1 location pattern B is optimal but at w_2 it is pattern C. More broadly, the geography of production depends upon the conflict between the two classes of workers and capitalists. Thus, in the above example, location pattern C is interpretable as arising when workers are strong as a class (wages are high, profits are low), and location pattern B as arising when capitalists are strong (profits are high, wages are low). Because we need a prior specification of class conflict in order to determine income distribution, and thereby locational patterns, class conflict is essential to understanding the geography of production. In making this claim we are not suggesting that geographical patterns are epiphenomenal – that they are explainable only by the non-spatial variable of class conflict. Rather, as argued in the previous chapter, class and space are themselves tightly interwoven.

A similar argument holds with respect to the relationship between the rate of capital accumulation (g) and workers' consumption levels (c). This is represented by the consumption–growth frontier discussed in section 9.2.1. When capitalists reinvest only part of their profits, implying that the rate of profit (r) exceeds the rate of growth (g), there is a potential geographical conflict between workers and capitalists over consumption levels. From Figure 9.3, with a wage level w^*, the profit-maximizing location pattern is A, which implies a rate of profit of r^*. Assuming that associated with r^* is a rate of growth g^*, then from Figure 9.3 we see that the location pattern maximizing profits does not necessarily maximize consumption levels. Whereas the profit-maximizing location pattern A gives a consumption level c_1, location pattern B provides for the higher consumption level of c^*. As a result, there is a potential conflict between workers and capitalists. Capitalists will seek to obtain location pattern A because it maximizes profits, but workers will seek to obtain location pattern B because it maximizes their consumption levels.

SOCIALLY NECESSARY LIMITS ON CONFLICT?

The wage–profit and consumption–growth frontiers given above suggest that any trade-off from zero profits/growth to zero wages/consumption is possible. This view has been challenged recently. It is argued that there are 'socially necessary' constraints on both wages and profits and consumption and growth levels, by which is meant that certain fixed levels of wages/profit and consumption/growth are necessary for the reproduction of the economy.

Harvey (1982, p. 56), for example, argues that there is a single level of wages that is 'socially necessary . . . from the standpoint of the continued accumulation of capital'. His argument is that, in order to sustain a given rate of accumulation, consumption (wage) levels must be set so as to meet the effective demand necessary for the continuance of that level of growth. As Harvey (1982, p. 55) writes: 'If there is a general rise in the standard of living of labor . . . it is because the accumulation of capital requires the production of new needs.' From Chapter 9, however, it is clear that Harvey's argument about the existence of a socially necessary wage rate is not justified, on two counts. First, Harvey assumes that there is a single, 'natural' level of accumulation (growth) necessary for reproduction. This is not so. Provided that outputs are greater than or equal to necessary inputs in the next production period, the economy is reproduced. As a result, the rate of accumulation (growth) can be high, low or even zero while still reproducing the economy. Secondly, Harvey argues that in order to sustain a given rate of growth only the wages of workers are adjusted to satisfy the constraint of effective demand. Although this is plausible, it is not logically necessary. The surplus that is produced at the end of the production period can be used in three different ways: workers' consumption, capitalists' consumption or reinvestment. It is logically possible, therefore, for capitalist consumption and reinvestment, say, to increase and thereby meet the new levels of effective demand that are required to sustain a higher rate of accumulation, while worker consumption remains constant. More generally, accumulation can occur whether wages are high or low, and although the wage (consumption) rate is an important determinant of the rate of growth it is not the only one.

A second argument for a socially necessary wage comes from Burawoy (1979), and more recently from Przeworski (1985a). Both argue that workers live under a hegemonic system, and are beguiled by use values. As such, workers are primarily interested in their material standard of living, rather than in revolution. For this reason workers are willing, in effect, to cooperate in their own exploitation provided that the capitalist system is able to meet their material aspirations. The resulting class compromise produces a hegemonic wage – a wage rate that workers are willing to live with, and live on, without causing disruption. Although intriguing, this thesis suffers from not taking into account the diverse and changing geographical contexts in which production is currently carried out. In particular, a number of geographers and economists argue that fundamental change is currently underway in both the labour process and the geography of production and is undermining the traditional class compromise. Indeed, real wages have been stagnating or falling since the mid-1970s in most capitalist countries. By changing both the geographical location of investment (divesting from hitherto core industrial areas in favour of peripheral regions where labour is unorganized, inexperienced and relatively unskilled – Massey, 1984a; Amin and Smith, 1986; Clark et al., 1986) and the nature of technology (a move toward more flexible neo-Fordist techniques with a high degree of control over the labour process – Aglietta, 1979; Lipietz, 1987), the old class compromise is crumbling under new regimes of accumulation and new spatial divisions of labour (see O'Connor, 1984; Clark, 1986). Even Przeworski (1985a) recognizes that, on the terms of his own model, class compromise, and the associated socially necessary wage rate, do not occur where one class has a clear upper hand, as seems to be the case now.

The effect of both Harvey's accumulation–driven socially necessary wage rate, or Burawoy's and Przeworski's hegemonic wage rate, is to close the system of price equations by setting wage levels at some 'socially necessary' level. Our own view is that, although the rate of accumulation and the extent of hegemony may be factors influencing the wage rate, they are not the only ones, and it is best to keep the wage rate open to additional influences.

The counterpart to a socially necessary wage rate is a socially necessary profit rate. The argument is that to reproduce the economic system – that is, for capitalists to continue reinvesting – profit levels must be at some socially necessary level. In many ways this argument is more compelling than the one suggesting the existence of a unique socially necessary wage rate. While labour is a necessary input into production, the decision to produce anything at all is with the capitalist. The return on capital is thereby a key behavioural incentive to continue investing, and therefore it seems more plausible to assume an expected level of profit rates below which there is little or no investment (see Przeworski, 1985a, for an elaboration of this thesis). This view is most associated with Keynes' notion of capitalists' 'animal spirits', where expectations about profits determine the rate of investment.

Clark et al. (1986) have recently combined Keynes' idea with the work of some geographers who emphasize the importance of local, place-specific employment relations in determining investment rates. In this view, the socially necessary rate of profit that allows continued investment in a given place is set by both general future beliefs and the expectations about a unique set of social, cultural and economic relations found within that community. Although it is plausible that there exists a minimum rate of profit to enable continued production, there is no reason why that rate of profit should be permanently fixed. As with the case of wages, one can demonstrate that growth and investment can occur at both high and low rates of profit. In fact, if capitalist consumption were very high, high profit rates could be associated with low levels of investment. In other words, at best there is a minimum level of profits that cannot be exceeded, but there is no reason why profit rates cannot be above such a level for continued reproduction. Furthermore, Webber (1988) empirically demonstrates for the Canadian economy that there is considerable variation among sectors in their profit rate, with little evidence of any convergence to some socially necessary level. In fact, the sectoral levels of profit do not even correlate with the amount of investment in that sector.

In sum, although it is unrealistic to assume that wages and profits and consumption and growth can be established, arbitrarily there is also little justification for believing that they are set at some single, unique socially necessary level. Following our remarks in Chapter 1, there is a case for carrying out context-specific studies that identify the particular constraints on wages and profits and on consumption and growth at given places and times.

11.1.2   Power relationships: domination at the workplace

In examining the conflict between workers and capitalists we have focused thus far on the struggle over income shares. Another important aspect of class struggle is over working conditions and the labour process (Braverman, 1974; Edwards, 1979). We now examine all changes in production methods whereby employers assert their domination by redistributing economic surplus away from workers without directly lowering wage rates. These include initiating processes of rationalization, which may or may not include laying off labour; reduction of fringe benefits such as vacation time; reorganizing workers' time more productively through job reclassification, or by employing 'scientific management' techniques; or, lastly, introducing new and technically more efficient machinery. Each of these changes is interpretable in terms of its impact on the input–output matrix ($\mathbf{A}$).

The first three changes to the labour process – rationalization, reduction of fringe benefits and the reorganization of workers' time – are all ways of directly or indirectly increasing the length of the working day, because they reduce the ratio of number of hours that workers are paid to the hours they spend actively engaged in production. The effect is to reduce the wage good input required per unit of output, thereby lowering the cost of production for the capitalist.

Formally, recall from equation 4.1 that:

$$a_j^{mn} = \hat{a}_j^{mn} + b_j^m l_j^n / H, \tag{11.1}$$

where a_j^{mn} is the amount of good m, as both a capital and a wage good, that is required to produce a unit of good n in location j; $\hat{a}_j^{mn}$ is the amount of capital good m required to produce a unit of good n in location j; b_j^m is the amount of wage good m required per week l_j^n is the amount of labour required to produce a unit of good n in location j; and H is the length of the work week. If the length of the work week increases for the various reasons given above, then labour costs fall per unit of n produced, thereby lowering the technical coefficient a_j^{mn}, and also decreasing costs of production.

The fourth type of change to the labour process – introducing a new, technically more efficient production technology – is different from the first three in that it directly causes a decline in the value of the input–output coefficient, a_j^{mn}. In this case, technical change is either capital saving (decline in $\hat{a}_j^{mn}$) or labour saving (decline in l_j^n), or both.

The effect of both types of change is that the capitalist, by increasing domination, lowers input costs, thereby raising profits. As a result, there is a congruency of interests for capitalists in undertaking technical change that simultaneously increases domination of workers at the workplace and increases profits. (We will show in Chapter 12, however, that for a spatial economy this congruency does not always hold because of a form of counterfinality; that is, although domination may increase with technical change, profits can fall.)

In contrast, for workers the conflict over the labour process at the point of production is not usually over monetary remuneration per se, but over the degree of domination to which they are subject. In fact, for workers, domination and conflict over wage levels are only imperfectly related to one another, as the example in Figure 11.1 suggests. Suppose that an existing technique A represents a more desirable labour process for workers compared with technique B, whereas at current wage levels, w_1, technique B is more profitable for capitalists. One can imagine some circumstances in which workers might struggle to retain technique A (e.g. it is less hazardous to their health or offers better working hours) even though for any ruling rate of profits wages might be higher if B were adopted ($w_1 > w_2$ at rate of profit r_1).

More generally, this suggests that there is an asymmetry in the motives between workers and capitalists. Capitalists are primarily interested in increasing domination at the workplace through the methods discussed because of increased profits. In contrast, as our example suggests, one can imagine workers resisting domination even if wage levels are potentially less than they might be after new work practices are introduced, because working conditions are central to their daily lives.

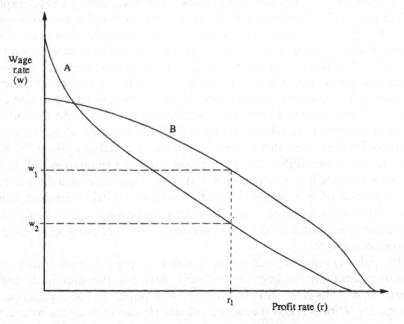

Figure 11.1 Two wage–profit frontiers representing different labour processes

11.1.3 Property relationships: communally owned means of production

We suggested in Chapter 10 that for John Roemer property relationships underpin worker–capitalist conflict. For him, the unequal distribution of property is the very basis of class formation and exploitation. Rather than focusing on this general theory, we examine here the effect of property relations on our production-based model by focusing on the consequences of collective ownership of property. We should note that in so doing our analysis is not restricted only to socialist states; rather, it applies equally as well to worker co-ops, and, to a lesser extent, to publicly owned enterprises within capitalism.

The first point is that if a plant is worker-owned or publicly owned it is very likely that exploitation will still exist as defined in Chapter 3. This is because, from Chapter 9, it is necessary for a portion of its surplus to be reinvested if the plant is to continue operating.

Ignoring location for the moment, assume that a single worker-owned co-op uses the average technology and sells at the production price. Assume also that all profits are paid directly as wages, implying from Chapter 3 that the rate of exploitation of workers in that firm is equal to zero. If profits are equal to zero there are no funds for reinvestment, implying that the growth of the firm is also zero. Worker-owned co-ops, however, often initially require large capital expenditures to replace deteriorating and inefficient capital stock (such firms are usually sold to workers precisely because they are old and inefficient). Thus, under our initial assumption, there appear to be only two strategies available to raise money for reinvestment and growth while maintaining a zero (exploitation) profit rate. The first is to sell the commodity produced at an above-average production price. In a competitive environment, however, this is not viable. (A variant of this strategy is to use above-average technology, thereby garnering windfall profits. But to employ such technology is to presume that the co-op has the funds to purchase it, which is precisely the issue here.) The second strategy is for workers to contribute directly towards an investment fund for new capital equipment. But in formal terms this is similar to exploitation. At the end of the day workers are not receiving in wages the full value of their labour because of a difference between the value of their labour and their labour power. This said, there are at least two differences between this form of exploitation and the form found under capitalism: workers within co-ops are receiving something in return for the portion of their wages that is diverted into capital investment, namely, ownership of capital equipment; and workers can be assured that the fraction of their wage that is subtracted is spent on investment and not luxury (capitalist) consumption goods.

Although exploitation is likely even when property is community owned (thereby potentially leading to a conflict between the individual and the collective), it is less clear what will happen with respect to the antagonism over domination. With workers themselves collectively determining the nature of the labour process, it seems probable that domination at the worksite is minimized.

None the less, one can still imagine a rift developing among workers. For example, one technique may provide for less desirable working conditions but a higher value of output, compared with another. Within a workers' co-op there might be a split between those who favour worse conditions but higher wages, and those who prefer the reverse. Similarly, splits can develop between those who work efficiently and those who do not, especially when the profits from higher output are necessary to finance equipment replacement and expansion.

Clearly these effects of communal ownership are drawn in only a sketchy way. Our main intent was simply to indicate that our analytical framework can cope with such cases, and suggest the likely consequences in terms of class conflict.

11.2 Workers vs landlords

Harvey (1985a, p. 38) writes that:

> The split between the place of work and place of residence means that the struggle of labor to control the social conditions of its own existence splits into two seemingly independent struggles. The first, located in the workplace, is over the wage rate, which provides the purchasing power for consumption goods, and the conditions of work. The second, fought in the place of residence, is against secondary forms of exploitation and appropriation represented by merchant capital, landed property, and the like. This is a fight over the costs and conditions of existence in the living space.

Our concern in this section is with the second of these issues, the fight over living space. Of course, as Harvey reminds us, the two types of struggle are ultimately linked. In particular, because the cost of living space is one element defining the value of labour, any struggle over it necessarily spills over and affects the relationship between capitalists and workers. (In addition, there are other ways in which the worker–landlord nexus impinges on the worker–capitalist relationship; see Harvey, 1985a, pp. 41–56 for an insightful exposition.)

Although recognizing Harvey's pioneering work in this area, Katz (1986) argues that a problem with his work is the derogation of workers' abilities to resist the landlord at the homeplace. There is no struggle between workers and landlords in Harvey's account, only the subordination of the former by the latter. Katz further argues that Harvey is driven to this view because of his capital-logic argument. Specifically, because land is just another form of capital for Harvey ('fictitious capital', Harvey, 1985b, ch. 4), and because under a capital-logic approach the needs of capital always have priority, workers are necessarily conceived as passive in their confrontation with landowners. As Katz (1986, p. 71) writes, the ' "class struggle" approach to space is completely missing from Harvey's analysis of rent ... [R]ent is analyzed as a relation between landed and industrial capitals: the *possibility* that worker struggle (a) exists and (b) leaves its mark on rent and land markets is not considered.' This leads Katz (1986, p. 66) to argue for a 'reconceptualization of both rent and space as the objects and locations of cross-class struggle'. In particular, this 'means paying attention to those moments of practical and analytical critique of the commodity form of shelter and space – both those moments when workers struggle individually or in limited ways to reshape the balance of power and distribution of the social surplus between capital and labor, as well as those moments when the demand for the abolition of the rent relation is itself posed – even if it appears in the fleeting moment of a house squat before the police arrive' (Katz, 1986, p. 75).

Central to Katz's analysis, therefore, is a recognition of the *active* resistance by workers against landowners. Parallelling our definitions of class, there are at least three different forms in which worker–landlord struggle can be manifest: over levels of rent charged, over the condition and control of housing property, and, lastly, over the ownership of private property. We examine each in turn.

11.2.1 *Conflict over rent levels*

As noted in Chapters 6 and 7 there can be a conflict of interests between workers and landowners over rental payments. We will attempt to sketch out here the mechanisms underlying this relationship for both differential rent and monopoly rent 2.

We argued in section 7.4.5 that the money wage rate is established by a marginal worker, defined as the person whose expenditures on land and housing rent, consumption goods and commuting are greatest. We also argued that a ceiling is established on the total cost of the marginal worker's consumption bundle, set through negotiation between capitalists and workers. Two components of the wage bundle represent housing costs: the rent on land, which in our model in section 7.4.4 is related to the number of people living in a zone; and the rent on the housing stock itself, which depends upon the building's initial price of production, its current age and the prevailing interest rate.

Let us first look at the effect of landowners increasing both these two components for the case of the marginal worker. There are three possible outcomes. First, the higher costs of housing are passed on to the employer as higher wage payments; the conflict between workers and landowners is therefore transformed into a conflict between landowners and capitalists. Second, the worker(s) can resist such a charge (strategies of resistance are discussed in the next subsection). Finally, the worker's price-responsiveness coefficient β (equation 7.9) endogenously increases, with the result that s/he moves to a place with cheaper housing costs. The same options are also available to intramarginal workers (i.e. those whose monetary wage is less than the price of the four components identified above), but in addition they have a fourth choice: to pay the extra rent demanded from the 'surplus' that remains once expenditures on the four components of the wage bundle are met.

All but the first of these four options point to a direct antagonism between workers and landowners. Workers either must directly resist higher rents, or are forced to move, or intramarginal workers are made to forgo some of the monetary surplus that they accrue. Such conflict between workers and landlords, however, is further exacerbated by the levying of the second type of rent: monopoly rent 2.

Monopoly rent 2 is the remuneration accruing to land that is levied over and above the differential rent for a given property. In particular, Harvey (1985b, p. 65) argues that monopoly rent is set at a 'positive return above some arbitrary level', and can be realized because 'as a class ... [land]owners have the power always to achieve some minimum rate of return'. More fundamentally, monopoly rent is a consequence of the monopoly that the ownership of space confers. For, unlike most commodities, each unit of land is unique, differentiated on the basis of location. As such, the owner of each plot has the potential to exact a monopoly rent. Whether this monopoly rent is realizable, however, depends upon the existence of a group of consumers who are constrained in their location. Harvey argues that there are indeed a set of consumers who are spatially restricted in terms of housing location – the workers. Specifically, Harvey (1985a, p. 40) argues that the spatial mobility of the working class is limited by three factors when choosing living space: the length of the working day, the wage rate and the cost of transportation. The result of such constraints is spatial entrapment. And, once trapped, landlords charge monopoly rents knowing that workers are unable to move easily to other areas.

MR2 can be incorporated into our model in the way we have already hinted in section 7.4.5. With an upper ceiling on wages (and also the length of the working day, H, as suggested by Harvey) set by class relations, workers are presented with a limited choice of locations. Such circumscribed choices are then the preconditions for the levying of MR2. The options for workers are the same as those discussed above for differential rent, but an interesting twist is that the very levying of the MR2 produces circular and cumulative effects whereby those workers who are subject to such rent are ripe for even higher rental charges in the future. This is because the very levying of MR2 further limits the residential choices available to workers (there is less money available for transport or rent elsewhere), with the result that they are spatially constrained even further. In this view, high rents in inner-city working-class areas are not simply because inner-city land is expensive and workers live on this land because of a low preference for residing in the suburbs. Rather, workers live here because of constraints on choice; constraints that drive up rent levels further through the levying of MR2.

11.2.2 Power and property relationships: community organizations

Our examination of rent relations between workers and landowners above was quite mechanical. We need to add to this account the power relationships between the two classes. Specifically, we require an account of the mechanisms by which one class tries to effect control, and how the other resists, thereby making 'space ... what it is – quite literally, a contested terrain' (Katz, 1986, p. 77).

Here the work of Lauria (1982, 1985) is very useful because it is one of the few attempts to portray on the ground the groups that contest the power of land-owners. In particular, Lauria shows that dual to the landowners' institutions of power within the property market are countervailing neighbourhood organi-zations that oppose and resist such power (see also Stoecker, 1988).

Two broad strategies of resistance can be identified. First, there can be direct opposition to landowners by forming tenants' unions and the like. By forming such countervailing collectivities, renters can resist more forcefully the land-owners' demands for both differential rent and MR2. The second strategy is to undermine the power of landowners by taking over property ownership. This requires the establishment and maintenance of community-controlled institu-tions, such as land trusts, clearing houses, credit unions and development corpor-ations. The effect of these strategies is, on the one hand, to circumvent the normal application of differential rent (the pattern of differential rent given by the market is disturbed because market principles are no longer in force), and, on the other hand, to undermine the preconditions for realizing MR2 (the con-dition of monopoly that underpins MR2 is weakened).

Some might question whether the conflict described here between landlords and community-based groups is really an example of class struggle. For the locally based countervailing organizations need not comprise only workers. Cer-tainly the petit bourgeoisie are often involved in community organizations, in which case what forms is a limited class coalition that 'defends place' (Chapter 10). In other circumstances, however, residential segregation is only class based (Harris, 1984), in which case the landowner–community struggle is also a class struggle. Clearly, we must examine the particulars at hand to resolve this ques-tion. Another potential objection to our analysis of the worker–landowner nexus is that within the working class there are considerable differences between workers who own their house and those who rent. In such cases conflict might be primarily intra- rather than inter-class based. It is to this issue that we now turn.

11.2.3   Property ownership: the housing class debate

In discussing the relationship between workers and landowners we implicitly assumed that workers do not own property. A considerable literature now exists, however, that disputes such a premise. In the United States, Canada and even in Britain the working class are increasingly owning their own homes (Agnew, 1981; Pratt, 1986; Saunders, 1978, 1979). Although there are a number of theoretical implications of such a finding, the one on which we focus here is that, when some workers are landowners, conflict is no longer only between the two separate classes, workers and landowners, but arises within the same class, the workers. Where this generates internal divisions within the working class, the consequence is a potential shift in the balance of class power.

Known as the housing class debate, two different theoretical views have emerged with the recognition of wide-scale working-class homeownership (for a good review see Harris and Pratt, 1987). We argue here, however, that these two views are neither as incompatible nor as different as usually presented. At bottom, both theories are united in making the common claim that home-ownership causes internal fragmentation within the working class, thereby weakening it.

The first view, known as the domestic property approach and associated with the work of Saunders (1978, 1979), suggests that separate (Weberian) housing classes arise with homeownership. This is because workers who own houses benefited from the inflation in housing prices over the last thirty years, and as a result have access to a large pool of capital. Saunders argues that because of that accessibility the material interests of homeowners are very different from those of non-owners, and therefore it is legitimate to speak of the two groups as distinct classes. Saunders' position has subsequently been criticized by Marxists who argue that, because class is defined only within production and not consumption relations, one cannot claim that homeownership is a basis for class formation. We will argue, however, that through modification Saunders' argument can be made compatible with the traditional Marxist view of class.

In previous chapters, we have suggested that factionalization can occur among the working class – for example, the development of a labour aristocracy based on differences in exploitation or domination. If this position is accepted, there seems no reason why we cannot extend the argument to include the creation of internal divisions within the working class because of differential property ownership. In other words, maintaining our definition of the working class as emerging in production relations (Chapter 10), we can still recognize the occurrence of working-class factionalization where, in this case, housing classes are based on differences of asset ownership, rather than exploitation or domi-nation. In this view, homeownership is just one of the many potential causes of internal division within the working class. In making this argument, we are not suggesting that factionalization must occur (a point made by both Agnew, 1981, and Pratt, 1986), but it might. Again we have to look at the particulars of the context.

The second view is the Marxist approach known as the incorporation thesis. The argument here is that homeownership is allowed under capitalism because of the political and ideological benefits to capitalism that ensue. In particular, homeownership is yet another means by which capitalism prevents workers from seeing their immediate or fundamental interests by, for example, inculcat-ing the possessive individualism of capitalism, legitimating private property and instilling financial conservatism and stability at the workplace because of mortgage debt. As Harvey (1985a, p. 42) writes: 'a worker mortgaged up to the hilt is, for the most part, a pillar of social stability.' From our perspective, the central feature of the incorporation thesis is its claim of working–class factionalization: while homeowners are incorporated presumably there is a second group of non-homeowning working class who are not, and who are therefore not ideologically committed to capitalism in the same way as the first group. Again the empirical evidence indicates that homeownership sometimes leads to the incorporation of workers, but not always. However, as long as some workers are incorporated, factionalization can occur, thereby weakening the working class as a collective.

The broader implication of both the domestic property and incorporation theses is that, once some workers own their own homes, the conflict between workers and landowners is displaced by a conflict within the working class itself. In turn, this spills over to affect the class relations between workers and capitalists. In putting forward this thesis we are not subscribing to the functionalist argument that workers are necessarily split over the homeowning question because it functionally benefits capitalism. Rather, we support Pratt's (1986) contention, which is in line with Elster's general argument, that a potential unintended consequence of individual workers pursuing homeownership is a loss of class power, thereby assisting the reproduction of capitalism.

11.3 Capitalists vs landowners

In *The Limits to Capital* Harvey (1982, p. 358) asks 'why the revolutionary force of capitalism, which is so frequently destructive of other social barriers that lie in its path, has left landed property intact . . .?' He suggests three reasons: first, it separates labour from land as a means of production; second, it helps legitimate the institution of private property; and, third, rent provides a mechanism for allocating capital in the 'correct' spatial pattern. While these are clear benefits, there are also antagonisms between landlords and capitalists, thereby giving landed property a contradictory role within capitalism (a contradiction that Harvey readily acknowledges). As before, we recognize three types of conflict between capitalists and landowners: over rental levels, over the conditions of tenure (power relations) and over the ownership of land itself.

11.3.1 Conflicts over rent

The three different types of rent we identified in Chapter 6 – extensive and intensive differential rent, and monopoly rent 2 – are each associated with a different degree of conflict between landlords and capitalists. Such conflict varies from passive acceptance of rent on the part of landlords to the active seeking of it. In this section we examine the different types of rent and respectively link each with a different level of conflict, and then review the thesis that there is a socially necessary level of rents that limits the struggle between landowners and capitalists.

FROM PASSIVE TO AGGRESSIVE RENT-SEEKING LANDOWNERS

The first case is of landlords passively accepting rent, and is found in the form of extensive differential rent (section 6.2). Given the institution of private land-ownership, extensive differential rents accrue to landlords through a competitive bidding process among capitalists; landlords play no role in the process other than passively receiving payment of a rent that they themselves had no part in setting. Of course, the levying of this type of rent is predicated upon a general monopoly power that landlords possess, but this does not imply active struggle on the landlords' part.

In the case of intensive differential rent, in contrast, landowners are more aggressive in accruing rent. In particular, their more active role manifests itself as the power to prevent rents from slipping below previously achieved levels. To summarize the argument from sections 6.2.1 and 6.2.2, in both Ball's account of intensive rent and the reconstructed version of it using a Sraffian framework (section 6.2.3), a ratchet effect is in operation. Intensive rents are constant or increasing, but never declining. This occurs because, as a class, landlords possess the power to prevent competitive market forces from operating. This power does not bring them into open conflict with capitalists; rather, landlords set clear upper limits to rental levels. None the less, landlords are clearly more than just passive agents.

The final source of rent, MR2, is one where conflict between landlords and capitalists is most overt. Here MR2 is a reflection of the landowners' naked power as a class. Rather than passively awaiting extensive differential rent, or partially intervening in the market by setting limits on intensive rent based upon pre-viously established levels, landowners aggressively seek to increase rent. This is seen in the profit/MR2 trade-off in Figure 6.7. Clearly, MR2 is realized through different mechanisms, but its defining feature is the active role of the landowners.

A SOCIALLY NECESSARY RENT LEVEL?

A related issue, and one complementing the earlier discussion of capitalist–worker relations, is the possibility of a socially necessary rental level. The idea here is that rent levels are set in such a way that, first, they are high enough to allow the continued existence of landlords as a class (and the benefits that follow therein for capitalists), but, second, they are not so high that they endanger continued capitalist reinvestment. Reinvestment can be endangered either because land-owners appropriate too much surplus directly from capitalists in the form of rent, or indirectly because their high rents produce high wage levels, thereby restricting the exploitation of workers. In either case, if rent levels are excessive the reproduction of the system as a whole is jeopardized.

Harvey (1982, pp. 363–4) seems to suggest this view when he writes that '[t]he landlord is perpetually caught between the evident foolishness of taking too little and the penalties that accrue from taking too much'. Between these two poles, however, there exists a 'terrain of compromise' (Harvey, 1982, p. 363), which is the socially necessary rental level.

From the discussion above it is clear that socially necessary rents are fully applicable only for the case of monopoly rent. Only for this type of rent do landlords possess full control over the rents they receive. In the cases of extensive and, to a lesser extent, intensive rents, competitive pressures within the capitalist class set rental levels, and not the power of landlords as a class. As a result, the applicability of a socially necessary rental level is more limited than Harvey perhaps suggests. Furthermore, as with the case of socially necessary limits on wages and profits, it can be seen from Figure 6.7 that, in principle, a wide range of MR2/profit combinations are possible, implying that restrictions on rental levels need to be established on a case–by–case basis.

11.3.2 Power relationships: conflict over land tenure

In addition to the conflict over rent, there is also a struggle between landlords and capitalists over land tenure, that is, the conditions under which land is leased to the capitalist by the landlowner. To examine this type of conflict, we look at two cases: extensive and intensive rent.

The extensive rent example follows from a simple model developed by Sheppard (1987). Assume a von Thünen landscape, where commodities are bought and sold in a central city for *given* 'world' prices. Assume further that for each commodity production costs are uniform, and that there is no inter-dependent production. Under such conditions we can describe two different land–use patterns – one associated with profit maximization, and the other associated with rent maximization. The important point for our purposes is that in general these two land–use patterns need not coincide. For this reason the issue of who has the power to control land use is critical, and is also one that is likely to generate conflict between the two classes.

In formal terms, assume two crops, A and B. At distance s from the market:

$$p^*_A = [1 + r_A(s)] \cdot [p_A + t_A(s)] \tag{11.2}$$

$$p^*_B = [1 + r_B(s)] \cdot [p_B + t_B(s)] \tag{11.3}$$

where p^*_n is the exogenous 'world price' for good n; p_n is the production cost of good n; $t_n(s)$ is the transport cost of good n over distance s; and $r_n(s)$ is the rate of profit made on good n at distance s from the market.

Under a profit-maximizing land-use pattern, crop A will be grown at distance s only if:

$$1 + r_A(s) > 1 + r_B(s). \tag{11.4}$$

which implies that:

$$p^*_A/[p_A + t_A(s)] > p^*_B/[p_B + t_B(s)]. \tag{11.5}$$

In other words, for specialization to occur in good A at distance s, profits for A must be higher than those for B (equation 11.4). But, for this to occur, the ratio of 'world' prices to local costs for A must be greater than the corresponding ratio for B (equation 11.5). With world prices given exogenously, and transportation rates the same for goods A and B, it is therefore only local costs and transportation charges that determine spatial specialization.

Let us now compare the profit-maximizing land-use pattern with the rent-maximizing one. Add to equations (11.2) and (11.3) a rent term, $Q_n R_n$, where Q_n is the quantity of land used to produce crop n, and R_n is the rent per acre for crop n ($n = $ A,B). By rearranging the two augmented equations, crop A would be grown at distance s under the rent-maximizing pattern only if:

$$\{p^*_A - [1 + r_A(s)] \cdot [p_A + t_A(s)]\} y_A > \{p^*B - [1 + r_B(s)] \cdot [p_B + t_B(s)]\} y_B, \quad (11.6)$$

where y_n is the reciprocal of Q_n, and represents the yield per acre.

Equation (11.6) states that landlords will specialize in crop A at distance s if and only if the production of A yields the greatest total rent at that point, and *not* the greatest rate of profit. Furthermore, the profit-maximizing and rent-maximizing rules for specialization coincide only if $y_m = y_n$ for all m,n, because only then are rent differences solely a result of local price differences. When this not the case, it is quite feasible that the sequence of land-use zones deduced from the profit-maximizing rule are very different from those deduced from the rent-maximizing rule.

Clearly this simple model is restrictive in that prices in the model are given exogenously, as are trading patterns. None the less, Sheppard (1987) argues that the necessary conditions for specialization given by the model are quite general. If this is so, our result is an important one in pointing to the central role of land tenure in the relationship between capitalists and landlords where extensive differential rent is charged. For, when landlords have the power to control land use, a specialization pattern given by equation (11.6) holds. In contrast, when capitalists have that control, a specialization pattern given by equation (11.4) will exist. In reality, of course, both rules in all likelihood apply, each in a different geographical area. Whatever the pattern, however, the likely consequence is a continual conflict between landlords and capitalist farmers over land use.

Paralleling the case of extensive rent, there is also the potential for conflict over conditions of land tenure between landlords and capitalists when intensive rents are charged. In particular, it is a struggle over the technique of production used on a given plot of land. Such a case has already been presented in section 6.2.3, and we simply summarize those results here.

Recall from section 6.2 that, for intensive rent to be levied, different lands must employ different techniques of production. Whether we follow Ball's or Sraffa's account, more productive techniques with higher per unit costs are necessarily introduced on some lands to meet demand and satisfy land-supply constraints. One interesting case to which this gives rise is that under certain circumstances capitalists can choose among different production techniques that satisfy both the demand and land constraints but yield different levels of profits and rent. In this case there is a potential for conflict between landlords and capitalists because landlords will want capitalists to use the technique that maximizes rents, whereas capitalists will wish to pursue the technique that maximizes profits. In particular, such a result occurs when there are multiple equilibrium solutions to profits and rents, which arise when there is a positive relationship between wages and profits. Consider Figure 6.5 in which there are two techniques 1 and 1,2, and where the $w - r$ frontier for the latter is positively sloped. At wage rate w^* there are two solutions: r^*, where intensive rent is levied, or r^{**}, where there is no rent. The solution adopted depends crucially on the nature of land tenure, and the respective power that capitalists and landlords possess. If landlords are able through their tenure agreement with capitalists to restrict the kind of technique used (in this case 1,2), then landlords garner intensive rents. If capitalists, however, have control over the conditions of production used on the land, then technique 1 is employed.

The broader implication of both cases is that, although there is no direct conflict between landlords and capitalists over the realization of extensive and intensive differential rent itself, there is the potential for a high degree of conflict over the conditions of land tenure. In fact, direct inter-class conflict over rent might be low, while that over land tenure might be high. The landscape of production, therefore, will crucially depend upon both direct and indirect forms of class antagonism between landlords and capitalists.

11.3.3 Conflict over land ownership: capitalists, landowners and fictitious capital

Clearly capitalists and landlords are not divided over the legitimacy of the institution of private property per se. After all, private property ownership is in many ways the *sine qua non* of the existence of capitalism, and thereby also of these two classes. Rather, the cleavage point between the two, as we have just argued, is over the levying of rent and the control of land use.

One resolution to these conflicting interests is for capitalists to wrest owner-ship of land from the traditional landowning class, thereby both gaining control over land use and appropriating rent that is otherwise a subtraction from surplus value. Such a resolution, however, is clearly not adopted by all capitalists. For example, it is well known that most office functions are carried out in rented space, whereas heavy industrial production is usually undertaken on sites owned by the firm. To examine further the circumstances under which capitalists either buy land or rent it, we discuss Harvey's (1982, ch. 11; 1985b, ch. 4) idea of treating land as 'fictitious capital'.

Harvey argues that, because of the contradictory class interests of capitalists and landlords already identified, land must be transformed into a form that is easily integrated into the general circulation of capital, thereby reconciling the conflicting interests of the two classes. Such a resolution is effected by making land 'a pure financial asset, a form of fictitious capital', where '"[f]ictitious capital" amounts to a property right over some future revenue' (Harvey, 1985b, pp. 97, 95). Land, therefore, is viewed as nothing more than its ground rent capitalized at the rate of interest, just as the value of stock is nothing more than its future dividends similarly capitalized. By treating land in this way, Harvey argues that the interests of landowners become congruent with those of any other capitalist; that is, ensuring that the circulation of capital proceeds unimpeded. For our purposes the important implication of treating land as a form of fictitious capital is 'that the power of any distinct class of landowners . . . [is] broken, [and] that ownership of land becomes from all standpoints (including psychological) simply a matter of choosing what kinds of assets to include in a general portfolio of investments' (Harvey, 1985b, p. 97).

Although we are not yet willing to abandon the distinctiveness of a separate landowning class (even if land is treated as a financial asset, some people choose to own only that asset), Harvey's point is well taken. In fact, by treating land as a form of fictitious capital, we can now address the question initially raised of the circumstances under which capitalists purchase the financial asset land, and the circumstances under which they rent it. We do this by combining Harvey's treatment of land as fictitious capital with the work of Scott (1988a) on production complexes.

A hallmark of any financial asset, including land as fictitious capital, is risk. Interest rates and the return on assets easily change, thereby making their prices unstable. In his recent work, Scott recognizes two major types of production complex, where each is subject to different types of risk and responds to it differently. Vertically disintegrated production units operate in unstable product markets, and to deal with such risk engage in a high degree of non-standardized interlinkages with other small production units. Through subcontracting and external economies of scale, risk is externalized. In contrast, vertically integrated production units operate in stable product markets, and their interlinkages are standardized among relatively large production units. Risk is internalized and controlled by taking advantage of internal economies of scale. For both types of production complex, land is an input like any other. But, in its form as a financial asset, land is also different in that associated with it is a risk. Given the nature of the two complexes, it is logical that vertically disintegrated units, such as office activities, will attempt to externalize the risk by renting, whereas integrated units, such as heavy industrial producers, will internalize it by purchasing the asset land.

To summarize, we have argued that one resolution to the tension between landowners and capitalists over rent and land-use control is for capitalists to purchase the land that they use. But any casual inspection of the geography of economic activity will show that not all capitalists adopt such a strategy. To suggest why some capitalists resolve the tension by purchase of land and others do not, we linked together Harvey's work on fictitious capital and Scott's work on production complexes. Clearly, this argument needs to be fleshed out further, but it provides a beginning point for analysis.

Summary

Our task in this chapter was to examine inter-class relationships in our production-based model. We did this by considering the relationship between each pair of classes in terms of three potential sources of conflict: income (as a surrogate for surplus value), domination and property ownership. That we can analyse class relationships in terms of these three sources of conflict suggests the robustness of our class definitions discussed in Chapter 10. It further reinforces the point also made there that a class, and its struggle with other classes, is defined differently in different circumstances.

Of the three pairs of conflict that we examined, perhaps the worker–capitalist one is the most clear cut. For there is little doubt that the relationship between the two is antagonistic. A central point in our discussion is that worker–capitalist conflict enters into the analysis right from the beginning – be it defining wage or consumption levels, struggling over labour practices at the workplace, or, following Roemer, defining the very basis on which class and exploitation arise. Although some have argued that there are predefined limits to the conflict between labour and capital, we prefer to leave such limits open, only closing them when there is a sufficient warrant to do so in the case at hand.

The relationship between workers and landowners is far less straightforward. On the one hand there are clear spillover effects on to the capitalist–worker relationship. If landowners hike residential rents, capitalists as well as workers may suffer and resist such moves. On the other hand, there are also spillover effects in the other direction where workers, in organizing community strategies of resistance, are calling into question the private property relation that underpins both landowners and capitalists. Finally, workers may well endure internal conflict as some members of that class purchase their own homes from landowners (thereby becoming landowners themselves), while others do not.

A similarly complex relationship holds between landowners and capitalists. In one sense they are antagonistic in that the landowner's appropriation of rent and control over land use gives her/him a different set of material interests compared to the capitalist. But in other ways, the two classes also have common interests in that both are legitimated by private property and, if Harvey's treatment of land as a form of fictitious capital is also accepted, both are interested in the unimpeded circulation of capital.

In short, although our conceptual scheme from Chapter 10 suggested an orderly and definite set of interests and relationships among classes, we have seen in this chapter that even at a theoretical level this cannot be sustained. As we saw, it was very difficult both to exclude intra-class relationships in the discussion of inter-class ones, and to avoid introducing the 'third' class in discussing the other pair of classes. If this level of complexity is unavoidable at a formal, abstract level, it speaks to the even greater turmoil of conflicting and intersecting class interests on the ground.

12 *Location and intra-class conflict*

Introduction

In our discussion of inter-class conflict in Chapter 11 it was generally assumed that individuals are motivated by, and act upon, their class interests. In other words, there are no free-riders; individuals pursue collective rather than individual rationality. Such an assumption, as we argued in Chapter 10, need not hold. Often a conflict arises between individual and collective maximization; a conflict that at best makes class formation problematic, and at worst undermines the prospect of class formation altogether.

We argue in this chapter that conflicts can occur within classes, and that they are the result of one of two reasons: either a direct discord, where an individual intentionally acts in such a way that his/her actions adversely affect another individual in their class; or an indirect form of conflict that emerges from unintended consequences – a process, following Elster (1985), that was labelled counterfinality in Chapter 10. In this chapter we will examine capitalists, workers and landowners in turn and suggest that each one is subject to intra-class conflict because of either a direct or an indirect discord between the individual and the collective. Because we emphasize indirect unintended consequences as a source of conflict, we begin by discussing further the idea of counterfinality.

12.1 Unintended consequences and intra-class conflict

Elster (1985, p. 44) argues that the most important reason for the occurrence of unintended consequences in general, and counterfinality in particular, is

> when several individuals simultaneously entertain beliefs about each other which are such that, although any one of them may well be true, it is logically impossible that they all be. An important special case arises when a particular description that may be true of *any* agent, for purely logical reasons cannot be true of *all*. If the individuals having these mutually invalidating beliefs about each other all act as if they were true, their actions will come to grief through the mechanism of unintended consequences ... In condensed jargon, counterfinality is the embodiment of the fallacy of composition.

In order to clarify Elster's point it is useful to state the general form in which mutually invalidating beliefs are found for any given individual. That form is: 'I will perform a given act under the assumption that others will not simultaneously undertake the same act.' If everyone holds such a view it is mutually invalidating because, although one person can believe it, not everyone can without contradiction. It is precisely because of such a contradiction that unintended consequences then ensue. Where Elster is silent, however, is in providing reasons for individuals holding mutually invalidating beliefs. He just assumes that they do. Contrary to Elster, we suggest that it is often implausible for an individual to hold mutually invalidating beliefs. For to do so implies that the individual is unable to recognize that others are equally capable of perceiving opportunities for gain and acting in similar ways. For example, why would a capitalist believe that other capitalists do not see the individual benefits of, say, employing a new capital-intensive technique of production and like her/him also move to introduce it? Given Elster's broader project, the ironic conclusion is that anyone who believes that others will *not* act in a like manner is behaving in an irrational way.

Elster (1985, p. 23) also offers another explanation of unintended consequences, which does not invoke mutually invalidating beliefs of the type discussed above. In this second view (which for a peculiar reason he calls 'trivial') unintended consequences occur because individuals misjudge what Elster (1985, pp. 23–4) terms the 'technical relations' involved in action. Contradiction arises not because people misjudge the intentions of others, but because of the complexity of the situation in which they find themselves. In particular, unintended consequences occur because individuals cannot work through all the ramifications that follow from their action and everyone else's. To clarify this point let us again use the example of individual capitalists each introducing a more capital-intensive technology, which, following Marx, produces the unintended consequence of a general decline in the rate of profit. If we interpret the counterfinality here in terms of 'technical relations', the falling rate of profit occurs because capitalists cannot work out all the consequences of everyone undertaking the same action of introducing more capital-intensive technology; it is *not* because each capitalist thought that others would not introduce a similar change. We will argue in this chapter that ignorance of technical relations is the more general case of unintended consequences. (This is not to reject Elster's idea of mutually invalidating beliefs, but we need to know on a case-by-case basis why individuals within the same class think that other class members do not believe the same things as they themselves.)

As suggested in Chapter 10, the significant implication of linking unintended consequences and ignorance of technical relationships is that it highlights the critical role of geographical relationships. It is precisely in a highly interdependent *spatial* economy that individuals find it extremely difficult to work through all the consequences of their own and others' actions. As a result, although an individual undertakes an action that appears beneficial, the end result can be deleterious for everybody because of unintended consequences.

The importance of unintended consequences and counterfinality for the task of this chapter is that they potentially give rise to an indirect form of intra-class conflict. In following their own ends, individuals unintentionally create debilitating results for all others, thereby causing a rupture within the class itself (intra-class conflict). We will now examine the specific mechanisms that engender such conflict for each class, beginning with the capitalists.

12.2   Intra–class conflict among capitalists

The small amount of literature that exists on direct conflict among capitalists suggest that the principal source of intra-class antagonism is competition for survival. As Bowman (1982, pp. 574–5) writes: 'insofar as capital accumulation requires a redistribution of capital in favor of more efficient producers, the continued reproduction of capitalism is decidedly opposed to the interests of some firms.' This is a direct form of intra-class conflict because every capitalist has a vested interest in ridding the market of other capitalists representing her/his competition. The forms in which that direct conflict is manifest are the well-known strategies of takeovers, mergers, price wars, and so on.

Although Bowman (1982) argues that direct conflict is an important source of intra-capitalist antagonism, he recognizes that indirect conflict represented by counterfinality is often more critical. This view is shared by Harvey, who incisively illustrates for a space economy that the most important effects of direct competition among capitalists is the indirect one of calling into question the viability of the whole capitalist system. Assuming perfect competition on a bounded geographical plane, Harvey (1982, p. 389) writes:

> If one capitalist expands output and shifts location to maximize the prospects of realizing values ... then other capitalists are forced to follow suit in order to defend their competitive position. The aggregate long-run effect on a closed plain is that the search for individual excess profits from relocation forces the average profit rate closer and closer to zero. This is an extra-ordinary result. It means that competition for relative locational advantage on a closed plain ... tends to produce a landscape of production that is antithetical to further accumulation.

The remaining part of this section takes up Harvey's general argument by focusing on indirect intra-class conflict in a spatial economy, and the logic of counterfinality associated with it. In particular, we examine the intra-class conflict that originates from three types of individual capitalist decision making: the location decision, the decision to introduce technical change, and the decision to specialize and trade. In each case, the general mechanism creating intra-class conflict is the same. In undertaking any of these three different actions the capitalist necessarily changes the inter-regional input–output matrix **A**. But because of the complex spatial interdependencies of production the nature of such changes is not readily predictable by the capitalist. In other words, and casting the argument in terms of section 12.1 , when undertaking any one of these three types of decision individuals are ignorant of their wider effects on technical relations. As a result, an outcome can occur that was not intended by anyone, and is potentially detrimental to all. Such a conclusion, we should emphasize, is the direct consequence of introducing spatial relationships. In this sense, by working through the unintended consequences of capitalist decision making, we are demonstrating in theoretical terms how geography matters.

12.2.1 The relocation decision

In Chapter 4 we determined the optimal location pattern for all capitalists using wage–profit frontiers. The best location pattern, it will be recalled, is the one whose associated w-r frontier lies outside all others. Within this framework we will now show that a capitalist's decision to relocate potentially creates intra-class conflict.

Given the inter-regional input–output matrix A discussed in Chapter 4, we determine the best location pattern for capitalists by first defining a vector S of binary numbers, with dimension NJ by 1, with entries s_j^n; s_j^n is equal to one if good j is produced in region n, and zero otherwise. Different vectors S then define different location patterns for production. The profit-maximizing location pattern is found by choosing, from the set of all possible 2^{NJ} vectors S, the one that maximizes profits (equivalent to the vector S associated with the largest eigenvalue of the matrix; for the analytical details, see Sheppard and Barnes, 1986).

To discuss intra-class conflict within this framework, assume that a profit-maximizing geography of production exists and is defined by matrix A_1, which is associated with a vector S_1, and a rate of profit r_1. Now, assume that for whatever reason producers in a particular place see an opportunity to pursue their own individual interest by relocating to another region in such a way that undercuts their competitors' prices, thereby creating excess profits in the short run. This change in location implies a change in the vector describing the geography of production from S_1 to S_2, leading eventually to a new trading equilibrium described by input–output matrix A_2 with equalized rate of profit r_2. The central question for our purposes is, does the change from S_1 to S_2, which appears to be profitable for the individual capitalist, result in r_2 always being greater than r_1?

This question can be answered in the affirmative if $A_2 < A_1$. In that case, by the Perron and Frobenius theorems of non-negative matrices, $r_2 > r_1$. But, by definition, a relocation of production means that some inputs to the regions with new production sectors will be larger in A_2 than in A_1. Note that the condition that $A_2 < A_1$ is sufficient but not necessary. Thus it is still possible for r_2 to exceed r_1. But this cannot be guaranteed, and the possibility that individual relocation decisions may be detrimental to the average collective rate of profit must be acknowledged. We conclude that self-interest and collective profit maximization need not coincide for capitalists because of ignorance of technical relations; that is, capitalists are not able to predict the effect of relocating on overall profits because of the complex spatial interdependencies involved. Such counterfinality, in turn, potentially creates a division within the capitalist class.

12.2.2   The decision to introduce technical change

A central debate in Marxist economic theory is over whether technical change implies a falling rate of profit (Rigby, 1990). Marx argued that there is a tendency for profit rates to fall because technical change is labour saving, which thereby reduces the organic composition of capital and thus the value rate of profit. In addition to its economic significance, Marx's theory of the falling rate of profit is important methodologically; recall from Chapter 10 that Elster and Harvey take it as the exemplar of counterfinality. Over the last thirty years, however, Marx's conclusion about a falling rate of profit has been criticized, thereby also undermining Elster's argument that it is an example of unintended consequences. In particular, both Okishio (1961) and Roemer (1981) show for an *aspatial* economy that, if capitalists adopt cost-reducing production methods and if the real wage is constant, then once price changes occur in response to the introduction of the changed technique the new equalized rate of profit is always higher than before. Such a proof, for Roemer (1981, p. 98) at least, 'settles, in a fundamental way, the Marxian conjecture of a falling rate of profit due to competitive innovations by price-taking capitalists. It is essentially the end of the classical story.' We will argue here that such a conclusion is premature. For a spatially extensive economy we will show that the decision to undertake technical change can cause the overall rate of profit to decline, thereby engendering counterfinality and intra-class conflict.

Roemer's (1981, pp. 97–8) argument that profits increase with technical change is based on the assumption that the cost of production at old prices falls in those sectors introducing technical change, while remaining constant in all other sectors. This is sufficient to ensure that the equalized rate of profit associated with the new prices after technical change is greater. This occurs because all input–output coefficients remain unchanged in every sector other than the one introducing technical change (see Sheppard, 1990). This condition need not hold in a spatial economy where production prices are determined by the procedure outlined in Chapter 4. This is because the changes in production prices that ensue from technical change affect commodity flows, and changes in commodity flows represent changes in transportation inputs in all sectors and regions. As the relative prices of different suppliers of some inputs alter, commodity producers will shift their purchasing behaviour to buy more from those regions where the price has fallen in relative terms, and less from those regions where it has increased. This implies that in most sectors some inter-regional trade coefficients (a_{ij}^{mn}) and transportation input coefficients (a_{ij}^{tn}) decline whereas others will increase. With some coefficients increasing, it is possible that the total cost of production based on the altered coefficients, but calculated for the old prices, will increase for some sectors in some regions. If this occurs, Roemer's condition for technical change to increase profits does not apply, thereby making it difficult to predict profit rates in a changing space economy.

To clarify our argument, consider the following simple numerical example. Assume that there are two regions each using different technologies to produce commodities 1 and 2. The technology for this case is as follows:

	Inputs		
	C_1	C_2	C_t
R_1C_1	3	2	1
R_1C_2	2	3	2
R_2C_1	4	3	1
R_2C_2	2	4	3

where R_iC_i is commodity i produced in region i; and t is the transportation commodity (produced in each region).

Suppose only sector 1 in region 1 introduces a new technology, combining just 2 units of commodity 1 with 2 units of commodity 2. As a result, the factory gate price of commodity 1 falls relative to that of commodity 1 in region 2. Producers in region 2, therefore, purchase more of good 1 from region 1 than before because of an increased price advantage. The consequence is that transportation requirements in region 2 presumably increase, say, to the amounts given below:

	Inputs		
	C_1	C_2	C_t
R_2C_1	4	3	2
R_2C_2	2	4	4

If this happens, at the old prices (before technical change) the new mix of inputs is necessarily more expensive than the one holding before technical change (even though buying such inputs represents the most profitable response that this sector can make to the new prices). More generally, our example shows that Roemer's sufficient condition for an increase of profits after technical change – that no production method is more expensive at the old prices – need not hold when the demand for transportation is endogenous (for a fuller analytical discussion of these issues, see Sheppard, 1989).

In sum, we have shown that counterfinality is possible when capitalists introduce a new technology of production. Given the complexity of price changes resulting from technical change in a spatial economy, capitalists are unable to know the long-run effect of technical change on the collective rate of profit. Their only guide is the immediate effect on profits, which from their perspective clearly increase. But, as we showed, the longer-run outcome can be a decline in average profit rates, thereby generating a conflict between individual and collective interests.

12.2.3 The decision to specialize and trade

The decision to specialize at a particular place is the dual of the decision to relocate. The relocation of the production of commodity m from region i to region j is a decision to specialize more in m at j and more in other commodities at i. For this reason the same potential intra–class conflict among capitalists discussed in section 12.2.1 is also pertinent for trade and specialization decisions.

Consider initially the case of two regions, 1 and 2, each producing two commodities in the absence of transportation costs. In the case of autarky, where no commodity trade exists between regions and each region sets its own profit rate:

$$\mathbf{p_1}' = (1 + r_1)\mathbf{p_1}'\mathbf{A_1} \tag{12.1}$$

$$\mathbf{p_2}' = (+ r_2)\mathbf{p_2}'\mathbf{A_2} \tag{12.2}$$

where $\mathbf{p_j}$ is the price vector in region j ($\mathbf{p_j} = [p_j^1, p_j^2]$); $\mathbf{A_j}$ is the 2×2 input–output matrix for region j; and r_j is the rate of profit in region j.

In this two–commodity case each region produces both goods and trades them internally. Suppose now that trade occurs between the regions. The necessary and sufficient condition for inter-regional trade to exist is that each region achieves higher profits after trade than before it. Formally, to establish the existence of trade we seek conditions such that the rate of profit after trade, $r_j^\star$, exceeds both r_1 and r_2.

In our two–region, two–commodity example there are only two possible complete specialization patterns. Either region 1 produces good 1 and region 2 produces good 2, or the converse. To establish which, if either, of these two specializations will actually occur in a non-spatial economy, compare the autarkic price ratios of the two commodities produced in each region. In so doing, and following Steedman (1979), we are in effect applying the principle of comparative advantage. Specifically, if $p_1^1/p_1^2 < p_2^1/p_2^2$, then it is most profitable if region 1 specializes in good 1 and region 2 in good 2, and vice versa if the inequality is reversed. When the autarkic price ratios are equal, the pattern of trade is underdetermined because every trading possibility, including that of no trade, is equally profitable. Note that by comparative advantage we mean the idea that a region specializes and trades in the good that it produces *relatively* cheaply compared with the other region.

Once comparative advantage is known, then by setting the inter-regional exchange ratio for the two commodities at a value somewhere between the two autarkic price ratios $[p_1^1/p_1^2 \lessgtr p^{1\star}/p^{2\star} \lessgtr p_2^1/p_2^2$, where $p^{i\star}$ is the with-trade price of good i ($i = 1,2$)], we can demonstrate that trade always raises the profit rate in both regions. Specifically, assume that $p_j^1 = 1$ and rearrange equations (12.1) and (12.2) for, respectively, goods 1 and 2 to solve for the autarkic profit rate in region j:

$$(1 + r_j) = (a_j^{11} + a_j^{21} \cdot p_j^2)^{-1} \tag{12.3}$$

$$(1 + r_j) = (a_j^{12}/p_j^2 + a_j^{22})^{-1} \tag{12.4}$$

Assuming that region 1 specializes in good 1 and region 2 in good 2, and if $p^{1\star} = 1$, the equalized with-trade profit rate is:

$$(1 + r^\star) = (a_1^{11} + a_1^{21} \cdot p^{2\star})^{-1} \tag{12.5}$$

$$(1 + r^\star) = (a_2^{12}/p^{2\star} + a_2^{22})^{-1} \tag{12.6}$$

If the exogenously determined inter-regional price ratio falls between the two autarkic price ratios, it is clear by comparing equations (12.3) and (12.5) that the rate of profit in region 1 rises because $p^{2\star} < p_1^2$, and that it also rises in region 2 because, by comparing equations (12.4) with (12.6), $p^{2\star} > p_2^2$. That is, trade raises the rate of profit because the price of the imported commodity is less than in autarky, whereas the price of the exported commodity is higher.

We will now argue that, in a spatial economy, this simple price-related comparative advantage principle used by Steedman (1979) does not always operate. For this reason there is the potential for counterfinality and intra-class conflict.

When examining a spatial economy, we need to add the production of transportation services to equations (12.1) and (12.2). Let us assume that both regions possess their own transportation services (which are not traded), and that specialization and trade occur. In this case, the production equations for either of the two potential trading specializations, either region A specializing in good 1 and region B in good 2, or vice versa, are given by:

$$\mathbf{p}^{\star\prime} = \mathbf{p}^{\star\prime}(\mathbf{I} + \mathbf{R}^\star)\mathbf{A}^{\star\prime}, \tag{12.7}$$

where $\star$ denotes after trade, and:

$$\mathbf{R}^\star = \begin{Bmatrix} r_i^\star & 0 & 0 & 0 \\ 0 & r_j^\star & 0 & 0 \\ 0 & 0 & r_1^\star & 0 \\ 0 & 0 & 0 & r_2^\star \end{Bmatrix} \quad \mathbf{A}^\star = \begin{Bmatrix} a_{ii}^{11} & a_{ji}^{21} & a_{1i}^{t1} & a_{2i}^{t1} \\ a_{ij}^{12} & a_{jj}^{22} & a_{1j}^{t2} & a_{2j}^{t2} \\ a_{i1}^{1t} & a_{j1}^{2t} & a_{11}^{tt} & a_{21}^{tt} \\ a_{i2}^{1t} & a_{j2}^{2t} & a_{12}^{tt} & a_{22}^{tt} \end{Bmatrix}$$

where p^m_i is the production price of good m in region i; $r_i\star$ is the rate of profit in region i; and a^{mn}_{ij} is the amount of good m produced in region i required to produce a unit of good n in region j.

Assume that regions 1 and 2 specialize respectively in commodities 1 and 2 and after trade each region employs only its own transportation sector. The effect of including transportation costs in the calculation of with-trade profit rates is then found by rearranging the first two equations that comprise equation (12.7) and setting $p^1_1 = 1$:

$$(1 + r_1\star) = (a^{11}_{11} + a^{21}_{21}p^{2\star} + a^{t1}_{11}p^t_1)^{-1} \tag{12.8}$$

$$(1 + r_2\star) = (a^{12}_{12}/p^{2\star} + a^{22}_{22} + a^{t2}_{22}p^t_2/p^{2\star})^{-1} \tag{12.9}$$

By comparing equations (12.8) and (12.9) with equations (12.5) and (12.6) it is clear that profit rates are lower in the with-trade case because transportation costs are now levied. More generally, the effect of adding transportation costs, where previously there were none, is to lessen the gains to trade, where gains to trade are equal to the increase in profits resulting from trading. Clearly, the extreme case here is where transportation costs are so great that they prevent any gains to trade, thereby militating against trade itself. Now, presumably capitalists can work out in advance the increase in production prices that stems from higher transportation costs in the event of trade. As a result, they should be able to decide rationally if trade is worth pursuing. Their calculations will be based on the pre-trade price of transportation. Once trade commences, however, the transportation price will change because profit rates and other prices alter. As we argued in Chapter 2, there is a complex relationship between the rate of profit and prices. If the rate of profit, say, increases after trade, the relative price of transportation can decline, stay constant or increase. If it increases, the price of transportation may be so large that the gains to trade (calculated at the old transportation price) are eroded, thereby making it a sub-optimal specialization pattern for capitalists. As a consequence, a potential unintended consequence of deciding to maximize profits by specializing and trading in a spatial economy is again the counterfinality result of a lowered average profit rate.

Once we introduce many regions that are able to specialize in more than one commodity the potential for counterfinality increases further. The reason is the same as given in section 12.2.2 – that capitalists' decisions to specialize and trade are based on old, pre-trade prices. Once regions specialize, relative prices change differentially, and regions will respond by altering their pattern of purchases, potentially leading to an increase in their transportation costs. With some transportation coefficients increasing, the total cost of production calculated at the old prices (before trade) may increase for some sectors in some regions, thereby making the post-trade locational pattern less profitable than the pre-trade one. More generally, in a multi-regional, multi-commodity case, individual capitalists are simply unable to predict in advance the prices of all their inputs, including their changed requirements for transportation. It is ignorance of technical relations, therefore, that creates the potential conditions for counterfinality.

In summary, we have argued in this section that the decision to relocate, to employ a different technique of production or to specialize and trade are all variants of the same process. Each ostensibly offers the individual capitalist a higher rate of profit. But, once geography is included, counterfinality can occur in the form of a general decline in average profits, thereby pointing to a clear conflict between individual and collective maximization. Such a result, combined with our conclusions from Chapter 9 that, once the economy is out of equilibrium it is very difficult to return to it, suggests that there is nothing in the general process of intra-class conflict that is stabilizing; in fact, quite the reverse. Thus, if the initial state was one of disequilibrium there might be no mechanisms for restoring stability; indeed, disequilibrium might well be exacerbated.

12.3 Intra-class conflict among workers

We have already alluded above to potential conflicts within the working class (section 10.2). In this section we provide a more systematic tratement. In particular, we examine the issue of direct conflict among workers in terms of the segmentation of the working class, and indirect conflict in terms of capitalists' responses to gains by workers. In discussing both issues we draw upon Offe and Wiesenthal's (1980) broader analysis of the logic of workers' collective action, and we begin by summarizing their argument.

Offe and Wiesenthal (1980) argue that there are two fundamental reasons that workers continually struggle to realize collective action. First, unlike (money) capital, which is homogeneous and liquescent, living labour because of its 'insuperable individuality' is from the beginning differentiated and indivisible (Offe and Wiesenthal, 1980, p. 74). Admittedly many of these differences are socially constructed (skill, education, ethnicity, gender), but, provided that they are perceived as real, they generate real consequences. The most important consequence is that workers are driven to a very different mode of collective organization vis-à-vis capitalists. To be successful, capitalists need only organize their 'dead labour' (capital) by merging and integrating it into a larger block, whereas workers can never merge or integrate but at best only associate. In particular, workers must overcome their indivisibilities and heterogeneous interests by engaging in a form of political practice that Offe and Wiesenthal label 'dialogical communication'; that is, workers must continually converse with one another, and work through and resolve their distinct interests and separateness. In contrast, just one person among capitalists can speak for everyone because the units of resource they own are all alike (capitalist communication is 'monological'). In short , the very nature of the resource controlled by workers and capitalists (respectively, living and dead labour) implies that workers face far greater obstacles in organization as a class than do capitalists.

The second reason that workers find it difficult to organize collectively is that they lack ownership and control of capital. As a result, they are always reacting to capitalists' actions and rarely initiating their own. In this sense, workers are organized by capitalists before they ever engage in their own organization. This is not to deny the importance of worker resistance at the point of production (Burawoy, 1979; Edwards, 1979), thereby making the workplace a terrain of struggle. But, following Harvey (1982, pp. 111–19), we suggest that in general this resistance has its limits. As Harvey writes, 'struggles on the shop floor ... are part of the perpetual guerilla warfare between capital and labor ... From this standpoint, such struggles must indeed by viewed as frictional and transient, which is not to say that they are politically or ideologically unimportant' (Harvey, 1982, p. 117).

In the remaining part of this section we will argue that the two points Offe and Wiesenthal raise are central in understanding the two forms in which intra-class conflict among workers is manifest – respectively, segmented labour and capitalists' responses to workers.

12.3.1   Direct conflict: working-class segmentation

The first of Offe and Wiesenthal's (1980) points is that workers from the beginning are defined by different interests. We will explore three different types of interests here – skill, gender and ethnicity, and geographical location. Implicit in our discussion is that in each type dialogical communication may not be sufficient to overcome the difference in interests, with the result that direct intra-class conflict can occur.

Consider first skill differences. Although some argue that Marx thought that all labour is alike, this claim is not really sustainable.[1] Marx certainly wrote about abstract labour time as equivalent to an unskilled labourer working at average intensity, but this is simply the metric to which different concrete labour skills are reduced. It does not imply that all workers are identical. In fact, in his writings Marx explicitly recognized different levels of skill among workers, for example between the relatively few who possess specialized knowledge, and the bulk who do not. As Marx wrote, the application of 'natural sciences to the process of material production ... leaves [most] ... worker[s] with the knowledge of only a few manipulations. ... It remains true, of course, that a small class of *higher workers* is formed, but this is minute in comparison with the mass of workers who have been deprived of all knowledge' (Marx quoted in Elster, 1985, pp. 264–5, fn 3; emphasis added).

Although the term 'labour aristocracy' is often used to describe this 'minute' faction, Barbalet (1987) argues that this term had a very specific meaning for Engels and Lenin, who developed it; a better concept is Marx's own notion of the 'collective worker'. Whereas labour aristocracy refers only to a specific type of worker in a particular form of capitalist production (a subcontractor in nineteenth-century England), the 'collective worker' applies to all the ways in which historically changing divisions of labour (machinofacture to post-Fordism) within the production system create internal splintering, and thereby potential intra-class conflict within the working class. Furthermore, Barbalet argues that the concept is not restricted just to differences in skill but also includes divisions over the associated intellectuality and authority of the labour position.

To use a recent example, we have witnessed over the last few years a massive enlargement of that initial 'minute' fraction of workers possessing specialized knowledge – the rise of the 'professionals'. In fact, some argue that the professionals are now so large a group that they represent a separate class – 'the new class'. Wright, as we saw in Chapter 10, tries to deal with this occupational group by speaking of them as occupying a contradictory class location. But, if we accept the argument put forward by Barbalet, there is an alternative interpretation. The rise of the professionals in the last two decades represents a massive fracturing within the working class along the lines of skill, intellectuality and authority because of changed production relations within capitalism. Furthermore, because of such pronounced differences among these three axes, professionals as a group have very different interests from, say, blue-collar workers (Urry, 1986; Thrift, 1987).

A second source of segmentation is gender and ethnicity. In general both these issues are problematic for a fundamental Marxist framework. This is because of the base/superstructure model of economy and society that such fundamental Marxists hold. In this model, both ethnicity and gender are placed in the superstructure because by definition they do not affect the forces or relations of production that constitute the base. As a result, if ethnicity or gender are discussed at all they are reduced to more fundamental economic relationships that lie within the base. For example, although Marx recognized an apparent ethnic antagonism between the English and Irish working class (see Elster, 1985, pp. 22–3), he argued that this ethnic division is not a real one (i.e. it is not explainable in terms of ethnicity itself), but rather is a reflection of fundamental *economic* relationships. Thus, Marx wrote that the ethnic 'antagonism [between the English and Irish workers] is artificially kept alive and intensified by ... all the means at the disposal of the ruling class. *This antagonism is the secret of the impotence of the English working class*' (Marx quoted in Elster, 1985, pp. 21–2). What Marx was doing here was reducing ethnic differences between the English and Irish working class to a functionalist argument about the economic well-being of capitalism; ethnically based antagonisms are simply a ploy by capitalists to weaken the working class. With his deep suspicion of functionalist explanation, Elster argues that Marx's reduction of ethnicity (or gender differences) to the economy cannot be sustained (for the blow-by-blow arguments see Elster, 1985, pp. 392–4); rather, ethnicity and gender are significant in and of themselves.

The view that ethnicity and gender are to a degree autonomous of the economy is, however, increasingly recognized by some Marxists (Edwards, 1979, ch. 10; Himmelweit, 1984). Often in such work ethnicity and gender are treated as 'entrance barriers' to certain occupations. Proponents of this view, however, have generally failed to explain why such barriers exist in the first place. (Why is there prejudice along lines of gender and ethnicity and not along some other lines?) While it is clearly unsatisfactory that ethnicity and gender go unexplained, for present purposes this more recent work does support our broader point that ethnicity and gender represent distinct interests and potentially lead to intra-class conflict.

The final source of internal fracturing within the working class is geography itself. In Chapter 8 we established that workers in different locations are likely to be exploited at different rates. This result, it will be recalled, arose solely because of the inclusion of geography. Following Urry (1981) we can imagine that such different exploitation rates might well be the basis for allegiances between workers and capitalists in particular places. Workers in a location where exploitation rates are low or even negative may very well ally themselves with capitalists to defend their place (see Chapter 10). Although these 'structured coherences' might be unstable, as Harvey (1985b, ch. 4) contends, they point to quite different interests among workers in different places.

12.3.2　Indirect conflict: responses by capitalists

The second reason given by Offe and Wiesenthal for workers' experiencing difficulties in achieving collective action is their lack of control, and thus their necessarily reactive (as opposed to proactive) posture towards change. As already suggested, this does not imply that workers are passive. After all, by organizing collectively workers have achieved some remarkable gains since the time of the first industrial revolution. For example, the period from the Second World War until the mid-1970s was one in which at least a large portion of workers benefited in terms of wages and working conditions from a Fordist social contract between the state, big business and big labour organizations (Scott, 1988c). However, precisely because of the gains made, some commentators have recently argued that this contract is now crumbling in the face of a vigorous attack on organized labour. In particular, to increase profits and to undermine workers' control on the shopfloor, capitalists are restructuring the nature of economic activity through such strategies as introducing: new spatial divisions of labour (firms are situating routine, unskilled activities in geographically peripheral sites both at home and abroad to tap into cheap, greeenfield and non-unionized labour markets – Froebel et al., 1980; Massey, 1984a); new technologies (Fordist mass production techniques are being replaced by neo-Fordist ones – Storper and Walker, 1989); new labour practices (a move towards flexible labour – temporary, part-time and sub-contracted labour – Morris et al., 1988); and new forms of industrial organization (large vertically integrated production units are being replaced by smaller, vertically disintegrated units, where subcontracting and thereby control of labour is much stronger – Holmes, 1986; Scott, 1988a). More generally, the progress that workers made in the past in resisting capitalists and making gains, such as under Fordism, took decades to achieve. Therefore if capitalists, who possess the power to instigate change, introduce entirely new kinds of production processes and geographical relationships to control labour, as they are doing now under a post-Fordist regime, then workers are at a disadvantage until such time that they are able to introduce new strategies of resistance and gain.

Let us reconceive this relationship between workers and the response of capitalists in terms of both our broader model and the idea of unintended consequences. In effect, in our discussion above we are suggesting that for the twenty years after the Second World War the wage component of the regional input–output $\mathbf{A}$ matrix was increasing, partly because the coefficients b_{ij}^{mn} were enlarging. The dampening effect that this had on profits, however, was offset by *in situ* technolological change (reducing the capital good inputs $\hat{a}_{ij}^{mn}$). In fact, Harvey (1987) argues that during this time, in return for increasing annual wages, workers did not vigorously contest changes in the labour process. In addition, the high rates of profit possible in peripheral locations where wages remained low, when averaged with the lower profits associated with high-wage locations, still enabled capitalists to achieve a reasonable mean profit rate. By the mid-1970s, however, productivity gains were stagnating in the industrial nations, i.e. the $\hat{a}_{ij}^{mn}$'s were no longer falling rapidly enough to compensate for increasing real wages, with the result that profits fell. The response by capitalists was to initiate wide-scale changes in the inter-regional input–output matrix $\mathbf{A}$ that left workers worse off. In particular, *the unintended consequence* of resisting capitalists and increasing real wages in some locations was the emergence of an entirely new set of geographical and production relations that eroded hitherto realized gains made by workers (O'Connor, 1984; Lipietz, 1987).

In this interpretation, the rise of neo-Fordism in the 1980s is the embodiment of counterfinality. Workers engage in collective action in the attempt to gain higher wages and better working conditions, but the end result is one under neo-Fordism where earlier gains are lost. Here, the cause of unintended consequences cannot be Elster's mutually invalidating beliefs, because workers are already engaging in collective action (free-riders are a non-issue). Rather, workers come to grief because of an inability to predict future consequences, which translates in our terms into an ignorance of technical relationships. We should note, however, that these unintended consequences for workers do not occur autonomously (as they do for capitalists' decisions; section 12.2), but are the result of another class exercising its power to change 'the game'. In addition, the result of this indirect conflict might well be increased intra-class conflict, making the task of resisting capitalist strategies more difficult in the future. This, at least, is the upshot of Clark's (1986) work on the intra-class conflict among workers in the mid-West auto industry. The unintended consequence of workers' gains over the previous three decades in the automobile industry was the break-up of Fordist relationships in the late 1970s – a break-up that precipitated intra-class conflict as different branch locals of the same union competed with one another to avoid the ensuing lay-offs.

In summary, we have argued in this section that powerful direct and indirect forces are operating on workers that produce intra-class conflict. The direct conflict stems from the 'insuperable individuality' of workers; that is, workers from the beginning are different from one another. In contrast, indirect conflict ironically can originate from the very success of the working class in resisting the demands of capital.

12.4   Intra-class conflict among landowners

Very little work exists in geography on the class formation of landowners and the internal tensions that beset such a group. The analysis that we present here is therefore preliminary, although it rests on what we consider a solid theory of unintended consequences. The discussion of intra-class conflict among land-owners is broken down in accordance with the two main types of rent identified in Chapter 6: differential rent and monopoly rent 2.

12.4.1   Conflict over differential rent

Landowners can pursue three main strategies in improving their position as individuals, and hence there are three potential sources of intra-class conflict: raising differential rents, making land-use improvements, and changing land use by altering tenure conditions. We examine each in turn.

The first strategy of raising differential rents is unlikely to produce direct intra-class conflict. If one landowner attempts to raise rents, and no one else does, the user of that land is likely to move to a cheaper plot, with the result that only the single landowner is deleteriously affected. (In contrast, if all landowners collectively raise differential rents, the result is monopoly rent 2, which we discuss below, and not differential rent.) However, there is one indirect source of landowner counterfinality stemming from inter-class conflict. If some land-owners raise differential rents charged on housing leased by workers, capitalists might be forced to provide higher wages for their employees. If in order to pay for the higher housing costs the wage increase is very large, capitalists might well undertake some action, be it economic or political, to undermine the power of landowners. The historical example here is of the English corn laws, which initially enabled landowners to exact large differential rents, but which were eventually defeated at a political level by the capitalist class.

The second strategy is making improvements to the land, for example by introducing drainage or irrigation schemes. Such improvements would increase yields, thereby increasing rents per acre. If we assume that demand for agri-cultural products is fixed, then increasing the supply of goods on one land implies that less goods are required from other lands. In other words, fewer acres of land are needed to produce the same quantity of goods. The result is that the marginal and near-marginal lands will cease production. When differential rents are then recalculated for the fewer number of plots in production, it is very possible that a number of landowners, including those who introduced the land improvement, are worse off than before the change. More generally, the complexity of interdependent production is such that it is almost impossible for any landowner to work through the consequences of making land improvements. In this case, the unintended consequences of improving the quality of land are that some members of the class go out of business, while others (potentially including those initiating the change) experience declining rents.

The third strategy is that of enforcing change to a new land use that increases rents. Recall from Chapter 6 that the procedure employed to allocate producers to land plots involved making use of wage–profit frontiers. The land-use pattern that emerged was one that maximized profits, in the sense that it was associated with the outermost wage–profit frontier. From Chapter 11, however, we saw that there is potential conflict between landowners and capitalists over the issue of land use. In particular, we showed in section 11.3.2 that the land-use pattern associated with profit-maximization is generally not the same as that associated with rent-maximization. Thus when profit-maximization has prevailed there is scope for individual landowners to increase rent by changing the type of activity employed on their land. This is very similar to Neil Smith's (1979) notion of a rent gap, a gap between the rent levied on current activities in a given location and the higher rent expected from alternative land-use patterns. (Using the inner city as an example, potential rents can be greater if tenement housing is replaced by upmarket apartments.)

In terms of our broader model, when individual landowners change land uses to increase rents, this has the effect of altering the relationships within the regional input–output **A** matrix, thereby changing the wage–profit frontier. For example, transportation requirements would be different because activities are now in different locations. Although it is possible for the reconfigured wage–profit frontier to produce higher differential rents for everyone, that need not necessarily occur. Some landowners may benefit from higher rents, but others might equally experience declines. It is also possible that the whole class is worse off. In short, there is again simply no way of predicting the outcome before the change is initiated. Interestingly, Smith (1987) himself comes to a similar conclusion. After initially arguing that the rent gap is both a necessary and sufficient condition for an increase in rent (thereby benefiting the landowning class as a whole), he now recognizes that it is only a necessary condition (see also Ley, 1987; Badcock, 1989).

12.4.2 Conflict over monopoly rent 2

A central characteristic of monopoly rent 2 is that it is levied on the marginal land where differential rent is zero. As a means of establishing potential intra-class conflict over MR2, we can therefore examine what happens to this marginal land under the three kinds of changes discussed in the preceding section: an increase in rent, a change in the condition of land use and a technological improvement to the land.

The outcome of any one of these three changes is an altered input–output **A** matrix. Changing rent levels or changing the conditions of land use imply relocation, and therefore an altered pattern of specialization, while land improvements directly change the capital and labour coefficients of production. The question is, do these three types of change always imply an improvement in MR2? If not, there is potential conflict between the individual and the collective.

To address this question we can use the same reasoning employed when examining the mirror issue for capitalists (section 12.2). Recall that capitalists choose among certain strategies (to relocate, to introduce technical change or to trade) that change the **A** matrix. We concluded there that the complex changes in the **A** matrix make it impossible to predict the eventual effect of such strategies on the collective rate of profit. Such a conclusion also holds here with respect to the marginal land and the level of MR2 levied. With profits and wages given, either a different land-use specialization or directly changed capital and labour coefficients might increase MR2 on the marginal land, but there is no guarantee of such an end result. In fact, because the landowner is receiving two types of rent, differential and MR2, the changes in the **A** matrix might have different effects on each rental type; for example, differential rent might increase but MR2 decline. In short, because the unpredictable effects of the three strategies that individual landowners pursue are a result of the complexity of the **A** matrix, it is again ignorance of technical relations that is the potential source of counterfinality.

In summary, we have argued in this section that landowners, like capitalists, are for the most part subject to the indirect form of conflict. Specifically, because the strategies that individual landowners pursue to increase their rent levels change the input–output **A** matrix, consequences can ensue that were not intended by anyone, including the possibility of counterfinality and intra-class conflict.

Summary

We have suggested in this chapter that there are both direct and indirect sources of intra-class conflict among, respectively, capitalists, workers and landowners. Direct conflict occurs when an individual or sub-group knowingly undertakes actions that will hurt the interests of other individuals or sub-groups within the same class. Workers are the class most affected by this kind of antagonism. We argued, following Offe and Wiesenthal (1980), that this is because different workers have different interests from one another owing to the very resource that they control. Capitalists and landowners are not similarly divided because the resources that they respectively control are more homogeneous.

Indirect conflict, in contrast, arises when an action is undertaken that is in the immediate material interest of the individual or sub-group involved, but turns out to have deleterious effects for all or some members of the class. We argued that the most common cause of indirect conflict is misjudgement of the technical relations; geographical relationships are so complex that an individual cannot predict the outcome of her/his action. In particular, any change to the inter-regional input–output matrix **A** potentially produces counterfinality. It is for this reason that much of this chapter discussed the unintended consequences of capitalists' actions; after all, capitalists are the class *par excellence* who possess the power to change the **A** matrix. This also explains why counterfinality for workers must be expressed in terms of a response to capitalists. Although achieving higher wages or changed working conditions at the workplace will alter the **A** matrix, the longer-run consequences are borne by the capitalist in terms of lower profit rates. In this sense a changed consumption vector **b** in the *A* matrix for workers is *not* like a technical change causing a changed set of a_{ij}^{mn}'s for capitalists. In the first case there are no indirect effects for workers unless capitalists take action in response, while in the second case, because of endogenous changes in prices, capitalists have to take actions if they are to remain competitive. Finally, the chapter suggested that landowners as a class are much like capitalists in that their resource ownership gives them the power to change the **A** matrix. Therefore, intra-class conflict for landowners is similar to the counterfinality experienced by capitalists: individual landowners acting in their own self-interest are ignorant of the wider effects on the **A** matrix, thus potentially producing an end result that is deleterious for all.

Note

1 There is a large literature on so-called heterogeneous labour. Some central contributions include Rubin (1972), Morishima (1973), Rowthorn (1974), Bowles and Gintis (1977), and Krause (1981).

PART IV

Disequilibrium:
Technical change and organization

13 Strategies for reducing production costs

Introduction

Capitalists continually search for ways to reduce costs. One might think that in the case of full capitalist competition, where the expected rate of profit is equal in all sectors and regions, there is no incentive for capitalists to attempt this, but the opposite is true. The production price of a particular commodity in a region depends on the socially necessary production method. But this is just an average: the dominant method of production is used as the basis for determining the average factory gate price in the sector. It does not follow that all producers use the socially necessary method, and any capitalist who uses a production method that is cheaper than this will gain increased profits by selling at the same price as others while incurring lower costs. This is a strong incentive for capitalists to reduce production costs below the cost of the socially necessary method; they will gain a windfall profit for being more efficient while still selling at the old price. In the long run, as others adopt cheaper production methods, these methods will come to dominate the region, replacing the previous socially necessary method and reducing production prices. The windfall profits will then disappear. This, however, provides a new incentive to find even cheaper production methods in order to make new windfall profits. In this sense, 'competition is not only an equilibrating force but also a force that produces disequilibria, distortions, and misallocation due to the rivalry of capitals ... [M]arket prices fluctuate around their centers of gravity [production prices] and ... profit rates fluctuate around the general rate of profit; they do not converge toward them' (Semmler, 1984, p. 28; see also Farjoun and Machover, 1983).

In this chapter we examine both the means by which capitalists attempt to gain an advantage over their rivals and the impact of these strategies on the welfare of capitalists as a class.

13.1 Cost reduction strategies

Suppose a producer of commodity n in region j currently uses the socially necessary production method in that region. Costs of production are then the sum of capital goods costs, plus the money wage per hour, plus transportation costs for inputs. There are three places where capitalists may intervene to reduce these costs, and thus increase profits. One way is to find a cheaper combination of inputs. If the current combination is given (see section 4.1) by the vector $\mathbf{a_j^n} = [a_j^{1n}, a_j^{2n}, \ldots, a_j^{nn}, \ldots, a_j^{Nn}, l_j^n]$, then individual capitalists seek a combination of inputs that is cheaper than the socially necessary production method at current prices. This may involve a physically superior method, where some of the elements of this vector are smaller and none are larger, which is always cheaper. Alternatively, it may involve the substitution of cheaper inputs for more expensive ones – seeking production techniques using more of the former and less of the latter, in a combination that is cheaper at current prices. It is conventional to define such alterations of the combination of inputs as technical change.

The phrase 'technical change' often carries with it the narrow technical connotation of adopting new machinery, making production more efficient. We argue for a much broader use of the term to refer to any change in the inputs to production adopted by capitalists attempting to increase profit rates. This includes not just the purchase of new machinery or the improvement in the application of current machinery, but also changes in the labour process affecting the direct labour input per unit of production. For example, changes in working conditions that increase the pace of the production line, reduce the length of breaks or vacations, or involve closer supervision of workers on the factory floor also effectively reduce the labour input. Such changes represent political actions, rather than engineering decisions, since the workforce has to be convinced to cooperate (for a more extensive study of shopfloor politics and the labour process, see Friedman, 1977; Burawoy, 1979, 1985; Edwards, 1979; Storper and Walker 1989). Policies that minimize waste of materials or train workers to use machinery more efficiently reduce capital goods requirements. Increasing the capacity utilization rate also effectively reduces inputs of fixed capital.

A second cost reduction strategy is relocation – moving to a region where profits are greater. Ostensibly, one reason for relocation is to reduce costs, by either finding a region where the most expensive inputs are cheaper, or finding a region close to input sources implying lower transportation costs. The profitability of such strategies depends, however, on the mean production methods already in use in the same sector in other regions. Even when the capitalist can lower production costs by relocating, he would still be making a lower than average rate of profit in the new region if the dominant production method there is cheaper than the method already used by this capitalist. More generally, then, relocation will occur only if the production method already in use by the capitalist who is relocating is cheap enough, relative to the socially necessary production method currently employed in the destination region, to imply an increase in the profit rate. Furthermore, the increase in the profit rate must be sufficient to more than offset the cost of relocation.

A third set of cost reduction strategies is to reduce the cost of inputs. Individual capitalists have little direct control over the cost of inputs purchased from other suppliers (i.e. capital goods and transportation), although they can introduce strategies that reduce the capital that must be advanced to pay for this. These include increasing the turnover rate or staggering payment for inputs (section 4.2.1). The costs of labour can, however, be directly influenced by such tactics as union busting or wage reduction – actions that have the effect of reducing the size of the consumption bundle or reducing workers' savings.

13.1.1   Types of technical change[1]

The changes in the inputs used in a particular economic sector in some region may occur as a result of three distinct kinds of behaviour: innovation, imitation and selection. Innovation is the search by individual enterprises for improved production methods that are cheaper than the method currently employed. Imitation is the deliberate adoption of production methods already in use in other firms. Selection is the process whereby firms using cheaper production methods can increase their market share because they can undercut the prices of competitors. Selection, then, involves no change in production methods – just an increase in the relative importance of those firms already using better production methods.

To visualize these different processes, consider Figure 13.1. This figure depicts the range of techniques used by the different firms in a particular sector of a certain region. In order to construct a simple two-dimensional graph, assume that all firms use just two inputs, labour and a single capital good. Each point on the graph represents the combination of labour and the capital good used in a particular firm. In addition, point A represents the socially necessary production method in this region, which we have defined as the mean production method. Assume also that all firms pay the same price for labour and for the capital good. The straight line drawn through A on the figure represents the set of all production methods that costs the same as the socially necessary production

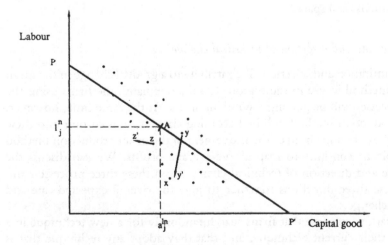

Figure 13.1   Production methods in sector n, region j

method at current prices (line PP'). The slope of this line equals the ratio of the price of the capital good divided by the hourly money wage, and we therefore define it as the price constraint. All firms whose production methods lie above and to the right of this line have production costs exceeding those of the socially necessary production method, and thus face below-average profits. All firms below and to the left of the line have above-average profits.

We conceptualize the process of innovation as a search by firms around the current production method for a new method that is somewhat different and cheaper. In the terminology of David (1975) this is a localized search. This is shown in Figure 13.1 by the movement of firm z to a new, nearby location (z'). The process of imitation is when firms essentially adopt the technique used by some other firm, involving a major movement of the production technology of that firm from its current location to a point very close to the firm being imitated. This is shown by the change by firm y to a production method close to that of firm x (point y'). The process of selection is the growth of the market share of those firms using cheaper production methods, and the decline of those firms using more expensive production methods.

This example is readily generalized to more realistic cases. If firms purchase two capital good inputs plus labour, then Figure 13.1 would be replaced by a three-dimensional diagram, with the mean production located in the midst of a three-dimensional point pattern. The line of production methods costing the same as the mean production method would be replaced by a two-dimensional plane. In the case of N capital goods, the relevant diagram has $N+1$ dimensions, with the set of production methods costing the same as the mean production method represented by an N-dimensional hyperplane. The characterization of technical change in our simple example can be extended to this case without significant change, however, so we will persist with our simpler example. Indeed, this example is best visualized as one two-dimensional slice cut through this $N+1$-dimensional space.

13.1.2 The rate and direction of technical change

Innovation, imitation and selection all contribute to a gradual change in the mean production method in use in the region. On the one hand, the firms using the cheapest processes will be gaining a windfall profit. At the same time, however, the mean production method will be becoming cheaper, thus cutting into these profits. Only by staying ahead of the movement of the mean production method are firms able to continue to reap above-average profits. We can discuss the expected rate and direction of technical change from these three processes, and then add these three directions together to give the overall expected rate and direction of change.

For innovation, suppose that firms search randomly for a new technique in a circle about their current technique, and that they adopt any technique that is cheaper. It is elementary to show that the expected direction of change will be

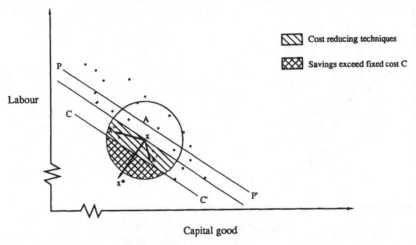

Figure 13.2   Innovation: searching about method x

perpendicular to line PP' on Figure 13.1. To see this, consider Figure 13.2. The line drawn through firm x represents the set of all production methods that cost the same as that currently in use by that firm. This line is also a price constraint parallel to PP', because any line parallel to PP' represents the set of production methods that have the same cost. While firm x may search in all directions about point x, it will adopt a production method only in the semi-circle lying below and to the left of this line because only these techniques are cheaper. While some firms may choose techniques that use more labour and less capital (x''), and others choose techniques using less labour and more capital (x'), the average direction of change will be the average of all possible directions in this semi-circle – which is the line xx^*, perpendicular to the direction of the price constraints.

There is generally a fixed cost, C, associated with innovation – a result, for example, of the need to invest in new machinery and buildings or the need to pay the price of a strike associated with redundancies. With fixed costs, no technique that lies too close to the price constraint for that firm will be adopted, because the savings in production costs do not offset the fixed costs. Such a restriction is represented in Figure 13.2 by the line CC'. This restriction reduces the range of directions in which feasible techniques are likely to be found, and thus makes it more likely that the direction of innovation adopted by any firm lies close to the mean direction of technical change due to innovation.

We conceptualize imitation as the process of adopting the technique used by the most efficient firm at current prices. In this case the most efficient firm is firm s, which is the firm whose price constraint lies below and to the left of the production methods in use by all other firms (Figure 13.3). Imitation by other firms of the most efficient technique implies a shift to a technique lying in the immediate neighbourhood of s as shown by the arrows in Figure 13.3. The mean direction of change can be calculated mathematically. It will depend on the

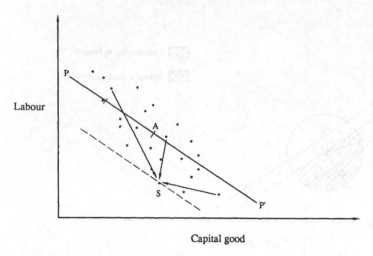

Labour

Capital good

Figure 13.3   Imitation of production method *s*

location of the most efficient firm's production method relative to those of other firms, and on the shape of the distribution of production methods. If this distribution is approximated by an ellipse, as in Figure 13.3, the expected direction of change from imitation lies between the perpendicular to the price constraints and the long axis of this ellipse.

If we assume that the rate of growth of a firm's market share is proportional to the difference between that firm's profit rate and the average profit rate (Metcalfe and Gibbons, 1986), then the growth rate of a firm's market share is proportional to the distance that its production method lies below the line *PP'*, whereas the rate of decline of a firm's market share is proportional to the distance that its production method lies above the line *PP'*. The aggregate effect of this is that the mean production method moves down and to the left, in a direction that again is perpendicular to that of the price constraints.

These three components of change represent the ways used by capitalists to reduce production costs and increase market share. They each lead to a change in the mean production method as firms relocate their production methods downward, and as those firms using the cheapest production methods increase their share of sales. The total expected change in the mean production method will be the sum of these three effects, weighted by their relative importance. The relative importance of each is given by the rate of change from innovation relative to that from imitation and selection. The rate of change through innovation will increase with the financial resources available for, and the effort put into, research and development (R&D), and with the effectiveness of R&D as measured by the expected reduction in production costs achieved per dollar invested. The rate of change from innovation decreases when the fixed costs of innovation are high. The rate of change from imitation increases with the effort put into search, and the information available about production methods used by other firms. It decreases when barriers to imitation, such as patent laws, exist. The rate of change due to selection will increase as competition increases.

Note that the claim that the direction of change for each of the three components depends on the price constraint, whose slope is in turn given by relative prices, does not imply that prices are the proper determinant of technical change. Recall from Chapter 4 (section 4.1.2) that prices depend on the socially necessary production methods and real wages prevailing in all regions, and on the relative location of regions. Prices are thus an intervening variable expressing how production methods, wages and location influence the direction of technical change.

13.1.3 Location and the direction of technical change

The above argument suggests that the direction of change in the mean production method is mostly perpendicular to the price constraints, unless imitation is the most important component of change, because two of the three components contributing to technical change have an expected direction of change determined by the slope of the price constraint. The slope of this frontier, in turn, depends solely on the relative price of capital goods and labour, and not on the quantities used. All the firms in a particular region face the same mean price for an input, regardless of what they produce. This is deduced from equations (4.8) and (4.5) in Chapter 4. Equation (4.8) states that the proportion of an input purchased from region i depends on its delivered price, relative to the delivered price from other regions. Equation (4.5) states that the delivered price does not depend on the quantity demanded. Thus, in any region, every firm using input m purchases the same relative quantity from each other region on average, and pays the same expected production price.

Any sectors in a region that use the same set of capital goods as inputs will therefore have identical price constraints. If the above arguments are true, these sectors will all be expected to adopt the same direction of change due to innovation and selection, regardless of the relative quantities of capital good inputs used. In the terms of Figure 13.3, the mean production method may be found at very different positions initially, but the expected direction of change due to innovation and selection are the same. This suggests the hypothesis that any differences in observed technical change among such sectors is due only to differences in the importance of imitation relative to innovation and selection, and in the direction of change due to imitation. In practice, however, the future direction of change will also depend on the history of past changes, in part due to inertia and in part due to the way in which the fixed capital associated with past changes may constrain future possibilities (Storper and Walker, 1989; see also section 13.2 below).

Sectors using different combinations of inputs will, of course, have price constraints composed of different capital goods, and will have a different expected direction of change from innovation and selection. Yet if this argument is correct there is reason to expect that the direction of technical change will be rather similar for different sectors in the same region. By the same token, the expected direction of technical change varies more between regions because the relative cost of obtaining different inputs differs from one region to the next, implying that the price constraint and the expected direction of innovation and selection also vary. This is shown in Figure 13.4. Points A and B represent the mean production methods for two different sectors in the same region, both of which use capital good 1 and labour as direct inputs to production. The lines running through these points are the price constraints reflecting the relative price of these inputs for firms in this region. Points C and D represent the same two sectors in another region where labour is more expensive relative to capital good 1. The arrows show the expected direction of technical change from innovation and selection. This direction shows inter-regional variation but no inter-sectoral variation. In the region with higher labour costs (points C and D), the expected direction of technical change economizes more on labour, i.e. it is more labour saving.

There is one important exception to the rule that the relative production prices of inputs are the same for all sectors in a region: transportation costs. From equation (4.6), the transportation cost for sector n in region j is the sum of the prices paid for transport from different regions weighted by transportation requirements:

$$\sum_{i=1}^{J} a_{ij}^{tn} \cdot p_i^t,$$

where

$$a_{ij}^{tn} = \sum_{m=1}^{N} a_{ij}^{mn} \cdot \tau_{ij}^m. \tag{4.6}$$

Now, the weights a_{ij}^{tn} differ for different sectors in the same region because the different input requirements imply that some sectors rely more on certain regions for their inputs than do other sectors. The effect of this on the two sectors A and B in the first region of Figure 13.4 is shown in Figure 13.5, where wage and transportation costs are compared. Suppose that sector A faces a higher price for transportation because, in order to obtain its inputs, more transportation is required from regions where transportation is difficult, meaning that its price constraint slopes more steeply. The expected direction of technical change from innovation and selection then implies that sector A is likely to reduce transportation inputs more relative to labour (i.e. adopt more transportation-saving strategies). The way to reduce transportation costs is to relocate closer to the sources of inputs. Thus we would then expect the firms in sector A to be under greater pressure to relocate to other regions. If firms do not relocate, however, they may face increased transportation costs as a result of technical change in

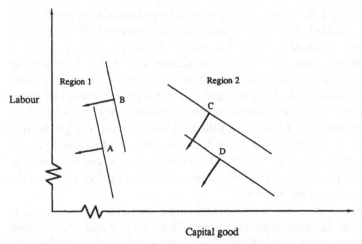

Figure 13.4   Direction of innovation and selection: two regions, two sectors

other dimensions. This is explained more fully below, but the reason is that the demand for transportation is derived from the demand for other inputs and the places from which they are purchased. Thus the demand for transportation is not under the direct control of producers; it may be that one effect of reducing labour costs in Figure 13.5 is to increase the demand for transportation, moving firms to the right on that diagram.

13.1.4   Impact of technical change

When innovation, imitation or selection occur within a particular sector in a certain region, the result is a shift in the socially necessary production method

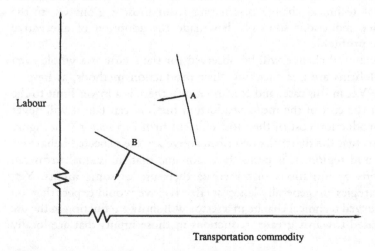

Figure 13.5   Direction of innovation and selection: two sectors, same region

downward and to the left. If the socially necessary production method is the mean production method, this shift is gradual and continual. If the socially necessary production method is defined as the modal method, then change is delayed until so many firms adopt the new technology that it becomes the dominant method of production in the region, at which point the socially necessary production method changes rapidly (see Sheppard, 1989). In either event, the socially necessary production method generally shifts downward, as does the cost of production. There then follows a reduction in the production price for that sector in that region, as a result of the tendency for profit rates to equalize on the average across all regions. This is simply the process whereby the temporary windfall profits accruing to innovators are whittled away as the technology diffuses within the region.

A price reduction in sector n of region j stimulates a diffusion of price decreases unevenly throughout the inter-regional system. Those sectors and regions most intensely using commodity n purchased from region j (i.e. those sectors for which a_{ji}^{nm} is highest) experience the greatest reduction in their production costs and production prices. Those reductions are then passed on to those sectors and regions using these products most extensively, leading in turn to price reductions there. In this way reductions in production prices, stimulated by the cheaper production method adopted by sector n in region j, diffuse through the system along those paths of commodity trade that are most heavily connected, directly and indirectly, with sector n in region j (for a more rigorous discussion, see Sheppard, 1989).

The overall price changes that result from innovation, imitation and selection are the result of such waves of price reduction emanating from each sector and region where technical change is observed. These price reductions in turn affect the profitability of adopting new production methods in other regions, because they reduce the costs of inputs there. Thus a secondary impact is the inducement of new waves of technical change originating from those regions where the diffusion of price reductions suddenly has made the adoption of alternative techniques more profitable.

Notice that technical change will be observed for the sector as a whole even when individual firms are not changing their production methods, as long as selection occurs. Yet in this case (and for imitation), there is a lower limit to the degree to which the cost of the mean production method can fall; it will never fall below the production cost of the most efficient firm in a sector (s in Figure 13.3). Notice also that the dynamics described above are the expected behaviour within a sector and region. It is perfectly reasonable that individual firms can adopt innovations or imitations that increase their use of some inputs. Yet, because such strategies are generally less cost effective, we would expect that, on the average, observed technical change in a sector will show a reduction in the use of all inputs, biased towards greater reductions in those inputs that are locally most expensive.

A second result of technical change is a shift in the equalized rate of profit for capitalists and the possibility of a falling rate of profit. This was discussed in section 12.2.2. We argued there that, particularly in a space economy, the complexity of price changes resulting from technical change makes it unrealistic to expect that individual capitalists are able to predict the long-run effects of their own technical improvements on the collective rate of profit. Their only guide is the immediate effect on profitability – the windfall profits that are the incentive for innovation and imitation – but such cost-reducing changes may actually reduce the mean rate of profit in the long run.

13.1.5 Relocation and wage contracts

It was noted above that firms have recourse to other strategies than technical change – namely, relocation and changing the cost of labour. Several scenarios represent relocation: individual firms leaving a sector in one region to join that sector in another region; individual firms leaving a sector in one region to become the first firm in that sector in another region; or wholesale relocation of an entire sector from one region to another. In the first case, the mean production method is expected to fall in both regions. Such moves are most likely to be profitable for the firm if it moves from a region where it had above-average production costs (meaning that the exit of this firm reduces the cost of the mean production method at current prices), to a region where it will have below-average production costs (thus lowering the cost of the mean production method there). This implies that technical change is observed in both regions, leading to the kind of dynamics of prices and profits referred to in section 13.1.4. In the case of creating a new sector in a region, or a change in the allocation of sectors among regions, a new location pattern results. For reasons similar to those outlined in section 13.1.4, it is again not clear that when such actions are profitable for individual capitalists in the short run they are also profitable for capitalists as a class in the long run (Sheppard and Barnes, 1986).

In sectors and regions where labour inputs or wages are high there is a strong incentive to act to reduce these costs, using the direct influence that capitalists can exert on labour costs by modifying wage contracts made with workers. Yet, particularly where unionization is high or where skilled labour is scarce, there are considerable fixed costs associated with this strategy, owing to labour unrest in response to attempts to lower wages. The expected cost-effectiveness of this strategy will be compared with those of technical change or relocation as capitalists attempt to determine how best to reduce production costs.

13.1.6 The case of resources

The exhaustibility of natural resource exploitation sites discussed in section 9.3.1 implies a dynamic that runs counter to that of technical change and cost reduction described above. As current resources run out, new resources must be identified and brought into production, generally implying a higher cost of

exploitation than at current resource sites. In order to attract investment into
these locations, prices of resources must rise, implying increased differential rent
on those sites currently in production, possible changes in monopoly rent 2, and
price rises diffusing through the space economy in the same way that technical
change induces the diffusion of price decreases (section 13.1.4). The long-run
impact of these changes on the equalized rate of profit representing full capitalist
competition is again uncertain.

13.2   Fixed costs

There is an integral relationship between fixed costs associated with the above
strategies and the rate of reduction of production costs that can be expected. First,
fixed costs are central to all these strategies. Technical change requires new
machinery and buildings, involving an investment in fixed capital. Indeed, there
is evidence that the ratio of fixed capital costs to circulating capital and wages has
increased historically. The costs of searching for innovations also represent a fixed
cost. Second, relocation of production requires a search for alternative locations,
the move itself (including the temporary halt in production) and, in many cases,
the purchase or construction of buildings at the new location, which are all fixed
costs. Third, the reduction of labour costs includes the fixed costs of labour
negotiations and unrest, possible disrupted production, and the costs associated
with pension liabilities and other fringe benefits that the firm is obligated to
continue (see Clark, 1988).

All these fixed costs share two things in common. First, funds must be raised to
pay for fixed costs and production prices must be raised to amortize these costs
within the lifetime of the investment. In the case of fixed capital, this lifetime is
the expected economic life of the machinery and buildings. In the case of
relocation, it is the length of time the firm expects to be at its new location. In the
case of labour costs, it is the length of the new wage contract. Second, the extra
costs associated with amortizing this investment mean that the profit margin
must be that much greater than the current rate of profit. This implies that it is
more difficult to find cost-effective innovations because they represent a more
radical departure from the method of production currently used.

One conclusion from this is that the greater the fixed costs, the slower the rate
at which capitalists reduce production costs. This is because the costs of change
are higher, and the search for viable production methods more difficult. If
capitalists can forecast in advance both the lifetime of a new production method
and the fixed costs of replacing it at the end of its life, then they are able to plan
for the costs of future technical change. This is done by adding the costs of fixed
capital to current production costs in such a way that they accumulate sufficient
funds by the end of its lifetime to replace it. An example is where the costs of
replacing fixed capital equal its original cost; see equation (4.12) in Chapter 4.
Roemer (1981) has shown that, in this case, the effect of technical changes

incorporating fixed capital on the rate of profit in an aspatial economy is the same as for the case of circulating capital. Yet to assume that anybody can foresee the disequilibrium dynamics of capitalism with this degree of accuracy is heroic indeed.

If production methods last at least as long as expected, and if the costs of finding and implementing a replacement are no more than expected, then the inaccuracy of capitalists' expectations is unproblematic. They will accumulate more than enough financial resources to implement the next change in production methods. In practice, however, the opportunities for major technical and locational changes come when least expected. Although the effort invested in technical change, for example, is closely dependent on capitalists' perceived needs for a new production method, the actual timing and nature of the innovation implemented are subject to a high degree of uncertainty. Similarly, the breadth of impact of a new production method is hard to determine.

When new production methods are discovered or are available before previous fixed-capital investments have been paid for, capitalists are faced with three alternatives: to wait, trapped in the current investment until it is paid off; to find a way of rapidly devaluing fixed capital so that it can be profitably abandoned; or to find funds elsewhere to cover the costs of implementing new production methods. An example of the first situation is where capitalists develop a built environment that was highly functional at the time it was developed, but then becomes obsolete before it is paid for. 'As conditions alter, ... what was the crowning glory of a prior stage of accumulation becomes obsolete, literally freezing the past into stone, where it can become a trap for capital and labor' (Walker, 1981, p. 407; see also Harvey, 1982, ch. 13). Accelerated devaluation of fixed capital is achieved with the help of tax regulations introduced by the state, including accelerated depreciation allowances and the ability to write capital losses off against taxes. The third option can be realized with the help of commercial credit from financial markets. Harvey (1982, ch. 9) correctly points to the vital roles of financial credit to finance major fixed-capital investments and of public assistance in devaluing fixed capital as mechanisms for mediating those crises that occur because of pressure to change production methods before previously implemented methods are paid for.[2] In general, given these uncertainties, there is some merit to the hypothesis that large enterprises set prices sufficiently above costs to allow them to accrue the funds necessary for planned growth (Eichner, 1976).

13.3 Reducing circulation time

We saw in Chapter 4 that the transportation system affects the profitability of production in two ways. First, it is a direct monetary cost that is a component of production costs. Second, the time of circulation affects the capital advanced to pay production costs. The joint effect on prices and profit rates of improving transportation is given in the following equation from that chapter:

$$p_j^n = [1 + r] \cdot \left\{ \left[\sum_{i=1}^{J} \sum_{m=1}^{N} a_{ij}^{mn} \cdot p_i^m + \sum_{i=1}^{J} \left(\sum_{m=1}^{N} a_{ij}^{mn} \cdot \tau_{ij}^m \right) \cdot p_i^t + \sum_{i=1}^{J} a_{ij}^{sn} \cdot p_i^s \right] \cdot \left[(T_j^n)^{-1} + \psi_j^t \right] \right.$$

$$\left. + \sum_{i=1}^{J} a_{ij}^{sn} \cdot p_i^s / \Theta_s C_n^s + \sum_{i=1}^{J} \sum_{i=1}^{J} \sum_{k=1}^{K} a_{ij}^{kn} \cdot p_i^k / \Theta_k C_n^k + \left(\sum_{k=1}^{K} a_{ij}^{kn} \cdot \tau_{ij}^k \right) \cdot p_i^t \right\}, \quad (4.14)$$

where

$$\sum_{i=1}^{J} a_{ij}^{sn} = \sum_{m=1}^{N} a_j^{mn} \cdot \sigma_m \cdot \delta_m \text{ for all } m \notin K.$$

This states that the price of production is equal to the sum of capital and wage goods costs

$$\sum_{i=1}^{J} \sum_{m=1}^{N} a_{ij}^{mn} \cdot p_i^m,$$

transportation costs

$$\sum_{i=1}^{J} \left(\sum_{m=1}^{N} a_{ij}^{mn} \cdot \tau_{ij}^m \right) p_i^t,$$

and storage costs

$$\sum_{i=1}^{J} a_{ij}^{sn} \cdot p_i^s,$$

all multiplied by the total time of production and circulation

$$(T_j^n)^{-1} + \psi_j^n;$$

plus the costs of depreciating fixed capital

$$\sum_{i=1}^{J} \sum_{k=1}^{K} a_{ij}^{kn} \cdot p_i^k / \Theta_k c_n^t.$$

Any innovations that reduce the effort and time required for transportation both reduce the size of the transportation inputs, τ_{ij}^m, and reduce the mean circulation time, ψ_j^t. The former is a reduction in the inputs required for production (a form of technical change), whereas the latter leads to a reduction in the capital advanced (which is proportional to production plus circulation time; section 4.2.2). Both changes increase the rate of profit for any firm implementing them. This kind of twofold benefit is much like the benefit that a firm obtains from introducing machinery that simultaneously reduces the quantity of circulating capital and labour required per unit of output and increases the speed of production (i.e. decreasing the production period). In addition, as noted in section 4.2.2, a reduction in circulation time also increases the capacity utilization rate, thus reducing the capital advanced to pay for fixed capital goods. With transportation improvements reducing capital advanced and increasing profit rates in these several ways, it is clear that capitalists stand to gain considerably from transportation innovations.

The transportation commodity also possesses some very special characteristics (see Feldman, 1977). First, it is very difficult for an individual capitalist to appropriate the benefits of transportation improvements. If a capitalist has exclusive right to a new form of transportation that delivers products more quickly than those of any competitors, then the benefits of transportation innovation can be appropriated. This is an unlikely possibility, however. If the transportation industry in a region purchases transportation equipment that improves its efficiency, thus reducing both τ_{ij}^m and ψ_j^n, this is available to all capitalists purchasing goods from that region. Furthermore, if the transportation improvement in a region involves new highways, railways, ports or airports, then these improvements in the inter-regional transportation network not only benefit everyone shipping goods to or from that region, but also reduce transportation effort and time for everyone who ships commodities via that region, whether or not they originate or terminate there. Transportation improvements are thus a public good: all capitalists in a region stand to benefit from transportation improvements in that region, as do many capitalists elsewhere, even if they do not contribute to its development.

We can be a little more precise about the geographical impact of an improvement in transportation in one region. When such an improvement occurs, it is likely that capitalists (and workers) in this region will initially benefit more from the improvement than those in other regions. This is because all inputs and outputs for firms located in that region are reduced in cost, whereas for capitalists and workers located elsewhere this is true only for the fraction of their shipments that passes through the region. Capitalists in the region experiencing improved transportation have an advantage, therefore, over competitors located elsewhere. These savings allow them to sell at a lower price, or to incur greater profits by continuing to sell at the same price as in the past. If a region is relatively isolated, these benefits might be retained for some time, long enough to attract more investment into the region to take advantage of this competitive edge. Yet, in a well-connected transportation and trading system, the benefits of this transportation improvement will rapidly leak to other regions, and the region that implemented the improvement will gain less.

A general improvement in transportation that reduces transport effort, τ_{ij}^m, and circulation time, ψ_j^n, by the same proportion throughout the inter-regional system benefits those sectors and regions for which transportation inputs are a higher proportion of total inputs, and for which transportation is slower. These are the more remote regions and those sectors whose products are expensive or bulky to ship. Such a general improvement increases the average profit rate of full capitalist competition in the long run, because both transportation inputs and circulation time generally fall for capitalists. Recall from Chapter 12, However, that long-run profit rates may fall with technical change, because transportation inputs increase for certain economic sectors in some regions.

The second distinguishing characteristic of transportation is that major transportation improvements historically require very large investments in fixed capital, in order to construct the infrastructure used in providing transportation. This makes innovation expensive, and reduces the likelihood and pace of its occurrence unless cheap credit is available (section 13.2). Third, transportation is both a capital and a wage good. Improvements in transportation thus translate immediately and directly into reduced costs for both capital goods and the real wage. If the institutions of wage bargaining maintain money wage rates, both workers and capitalists gain immediately from transportation improvements.

It is therefore not surprising that the national government has played a major role in the development of transportation innovations. In the United States this ranges from federal credit to the early railroad barons, to the construction of the interstate highway system. The public nature of transportation innovations makes capitalists and even regional alliances reluctant to take the initiative (a free-rider problem). The large fixed capital costs are most easily covered by state-backed credit, and the direct benefit to capitalists and workers makes it relatively easy to assemble support legitimating such state action.

Summary

Capitalists can reduce production costs in three ways; changing the mix of inputs (commonly but misleadingly referred to as technical change), relocating and reducing labour costs. Observed technical change in a sector of a region is itself a result of a combination of three processes: individual capitalists innovating to improve their profits; capitalists imitating one another for the same purpose; and a process of selection whereby more profitable firms increase their market share by having more funds for reinvestment and the capability to undersell less profitable firms.

The likely pace and direction of technical change observed in a sector in a given region suggest the hypothesis that, unless imitation is very important, the expected direction of technical change depends on the relative price of inputs. If this is so, and given that relative production prices for all inputs except transportation are the same for all sectors in the same region, we expect a close correlation between the directions of technical change in the different sectors of a region. The one exception is that sectors with higher transport costs are under greater pressure to relocate in order to reduce these. By contrast, the expected direction of technical change will vary more among regions because the relative cost of inputs varies. For example, in regions where money wages are relatively high, either because the real wage is high or because wage goods are expensive, we expect there to be a greater emphasis on labour-saving technical change in all sectors. Non-produced resources are an exception to the usual stories about attempts to reduce production costs. The exhaustion of individual resource sites often leads to an increase rather than a decrease in the costs of resource exploitation, an increase appropriated by resource owners as differential rent.

Most ways of reducing production costs incur considerable fixed costs, resulting in a deceleration of the rate at which cost reduction is likely to occur. If capitalists could perfectly predict the lifetime of a newly introduced production method and the cost of replacing it, then these fixed costs can be paid for by increasing production costs and production prices. The impossibility of this degree of foresight, however, means that when the opportunities for such change are unexpectedly soon or unexpectedly expensive, then capitalists can become trapped in their fixed capital unless either it can be rapidly devalued or easy and cheap credit is available to finance its replacement.

Transportation improvements represent a special and particularly important case. They can bring about major reductions in production costs since they reduce both transportation requirements and circulation time – thus reducing the capital advanced to pay for fixed and circulating capital and for labour. Yet transportation has some particular characteristics stemming from its geographical structure: individual capitalists and even regional alliances find it difficult to appropriate the benefits of transportation improvement; there are very high fixed capital costs for transportation infrastructure; and transportation improvements benefit workers and capitalists. Given such features it is logical that the national state has played a major rôle in introducing transportation innovations, innovations that are very likely to increase the rate of profit for capitalists.

Notes

1 Much of the argument developed in this and the following section is the joint product of research carried out in cooperation with David Rigby and Michael Webber, and reported more fully in Rigby, Sheppard and Webber (1989).

2 Franke (1988) has shown how finance capital may be integrated into production price models of the kind developed in this book

14 Strategies for organizational restructuring

Introduction

In Chapter 13 we examined the actions that capitalists pursue as they attempt to alter production methods to increase profits and market share – actions that both stimulate instability and respond to it. A second traditional strategy is the centralization and concentration of the ownership of production into fewer, larger corporations (see Chandler, 1962, 1977; Holland, 1976; Hymer, 1972). Extending control in this manner replaces marketplace transactions, over which firms have little control, with internal, planned transactions among corporate affiliates. Production costs are then reduced because corporations can take advantage of scale economies, reduce uncertainties by controlling linkages, increase monopoly power over markets by reducing competition, and increase the degree of influence that individual corporations have in national and local economies. Once again responses aimed at reducing uncertainty and instability engender further disequilibrium behaviour as the large corporations change the terms of inter-capitalist competition.

This process of centralization is the most dramatic example of capitalists altering the organizational structure of production. Yet there are other ways of changing the organizational structure, which may or may not entail centralization and concentration of ownership. These include: the development of multi-plant enterprises (which does not necessarily follow concentration of ownership); the organization of partially integrated linkages such as subcontracting, franchising and production under licence; the development of a division of labour within and among the plants of a corporation; and changes in the rules governing the nature and degree of control that is exercised within a corporation. As with the reduction of production costs, these kinds of changes represent a process of 'strong competition' whereby 'capitalists are driven to revolutionize production in order to gain an edge on their competitors' (Storper and Walker, 1989, p.48).

Doreen Massey (1984a) devotes considerable effort to analysing the organization of production, pointing to the many ways that it differs across sectors and regions (see also Watts, 1980; Clarke 1985; Jenkins, 1987). We will not attempt to replicate the subtlety of her treatment here, but will restrict ourselves to showing how some of these differences may be taken into account within the theoretical framework of this book. We first examine the implications of incorporating inter-sectoral differences in profit rates that reflect differences in organizational structure. We differentiate between the cases of single-plant and multi-plant enterprises, and determine the flow of investment funds within corporations (section 14.1). We then examine some of the alternative ways in which multi-plant firms organize the division of labour within and closely linked to the corporation, such as an internal spatial division of labour, and how they can organize subsidiaries and subcontractors to employ the just-in-time system of production (section 14.2). Finally, we consider the investment and disinvestment strategies available to corporations (section 14.3).

14.1   Oligopoly and the rate of profit

14.1.1   The debate over profit rate differentials

There is an extensive debate over the ability of different sectors of the economy to realize different profit rates as a result of the degree of concentration of ownership. Kalecki (1939–40) argued that the average cost of production per unit produced for large corporations is approximated by the sum of wage and circulating capital costs, a sum that is constant no matter how large the firm. He claimed that these per unit costs are constant because corporations typically operate at less than full capacity, implying that no fixed capital is purchased or scrapped as production levels rise or fall. Kalecki went on to argue that the difference between the production price and per unit costs, otherwise known as the mark-up charged by firms in a sector, is positively related to the degree of oligopoly in that sector. His rationale is that in sectors with fewer numbers of firms collusion is easier; corporations have more influence over prices, and it is easier to erect entry barriers to prevent new competitors from entering the sector.

This mark-up, the difference between costs and production price, is equivalent to the definition of the rate of profit for firms used in section 4.1. Thus one might conclude that Kalecki's hypothesis is that the rate of profit in a sector is proportional to the degree of monopoly. Indeed, this is widely assumed in the economic literature on the post-Keynesian theory of profit rates (for discussion, see Semmler, 1984). As pointed out in section 4.2, however, this mark-up is not necessarily the same as the rate of profit on capital advanced. The two are identical only if there is no fixed capital, all production periods equal one year, all inputs are paid in advance, and no money is reserved for financing technical change.

Eichner (1976) argues that the mark-up charged by corporations includes income to pay for replacing fixed capital, as well as a reserve fund (corporate levy) to provide for planned growth, technical change and cash flow. This implies that corporations in sectors where the rate of growth is high, or where technical change is rapid or expensive, require a higher corporate levy and thus charge a higher mark-up. Furthermore, he argues that the mark-up charged per unit produced depends on production levels. As we saw in section 4.2, when the capacity utilization rate falls the limited lifetime of machinery implies that per unit costs effectively increase in order to pay for prompt replacement of fixed capital (equation 4.12). The costs of financing growth and technical change are also like fixed capital costs in that they are also expenditures that must be made within a fixed period of time. So any reduction in the capacity utilization rate increases costs per unit produced that must be changed in order to pay for these expenditures promptly.

In this view, the mark-up over per unit costs suggested by Kalecki is proportional to the rate of profit only if the sum of the costs of replacing fixed capital, financing growth and realizing technical change is the same, per unit produced, in all sectors. There is no reason for this to be the case. As we argued, even in an individual sector, a change in the capacity utilization rate will change the per unit charges necessary to raise the funds for such activities. Indeed Semmler (1984), in a comprehensive survey of theoretical and empirical research on prices and profit rates, provides evidence that inter-sectoral variation in the mark-up is systematically related to the cost structure of sectors. As he also points out, however, the more relevant question is whether the rate of profit per unit of capital advanced varies systematically with the organizational structure of industry. This is a more accurate measure of corporate profits than Kalecki's calculation, it is perfectly possible for there to be inter-sectoral variations in the mark-up simply as a result of differences in the proportion of fixed capital, the turnover rate and the costs of technical change. As yet, however, there is no evidence about whether observed inter-sectoral differences in the rate of profit are due to these factors or to inter-sectoral differences in the degree of monopoly.

There are two schools of thought on this. Some expect a positive relationship – that rates of profit on capital advanced will increase with an increase in oligopolistic concentration in a sector. This is essentially the Kaleckian argument about collusion and oligopolistic power, as well as that of political economists such as Baran and Sweezy (1966) who stress a growing gap between production and consumption under monopoly capitalism. Webber and Tonkin (1990) provide empirical evidence showing that rates of profit measured in this way do vary significantly among economic sectors in Canada, and that these differences show no statistical tendency to dissipate over time. They do not relate these, however, to inter-sectoral variations in organizational structure.

Others argue that there is no reason for such differences to persist (Semmler, 1984). Entry and exit barriers may allow higher profits in periods of growth, but prevent corporations from disinvesting in periods when growth and profits are low. A reduction in the number of corporations in a sector may lead to collusion, but it may also lead to enhanced rivalry. The latter tendency may be reinforced because firms in one sector compete with those in other sectors; it is all but impossible to define industrial sectors in such a way that the products of one sector are not competing with those of another sector. (see Kreisler, 1987, pp.11–13, Kaldor, 1934). Third, barriers to directly entering a sector are overcome by placing money in the capital market from where it is indirectly invested in such firms. In this way firms in all sectors exploit the tremendous mobility of money capital to share in the high profits earned in particular sectors, implying a tendency for profit rates to equalize. For these reasons, it is argued that concentration of ownership need not reduce the competitiveness of corporations, nor does it lead to secular increases in the rate of profit in oligopolistic sectors. Instead, variations in profit rates are due to other factors, such as differences in the growth rates of industries and their export share. This leads Semmler to conclude, with Harvey (1982, ch.5) and Jenkins (1987, pp.44–6), that corporate concentration is not associated with excess profits and reduced competition, and thus that corporate profit rates are not significantly different between sectors.

14.1.2   Incorporating profit rate differentials

The mixed evidence and contrasting theoretical expositions summarized above imply that the possibility of inter-sectoral variations in the rate of profit on capital advanced cannot be dismissed. The impact of any such variations on the geography of production can be traced in the following way. Suppose, initially, that the profit rate in sector n is equal to r^n. Fixed capital, capacity rates and turnover rates must be included in discussions of corporate pricing, so we use equation (4.12) of section 4.2 to define the rate of profit in this sector and region:

$$p_j^n = [1 + r^n] \cdot \left\{ \left[\sum_{i=1}^{J} \sum_{m=1}^{N} a_{ij}^{mn} \cdot p_i^m + \sum_{i=1}^{J} \left(\sum_{m=1}^{N} a_{ij}^{mn} \cdot \tau_{ij}^m \right) \cdot p_i^t \right] \cdot [(T_j^n)^{-1} + \psi_j^n] \right.$$

$$\left. + \left[\sum_{i=1}^{J} \sum_{j} a_{ij}^{sn} \cdot p_i^s / \theta_s C_n^s + \sum_{i=1}^{J} \sum_{k=1}^{K} a_{ij}^{kn} \cdot p_i^k \middle/ \theta_k C_n^k + \sum_{i=1}^{J} \left(\sum_{k=1}^{K} a_{ij}^{kn} \cdot \tau_{ij}^k \right) \cdot p_i^t \right] \right\}, \quad (14.1)$$

where

$$\sum_{i=1}^{J} a_{ij}^{sn} = \sum_{m=1}^{N} a_{ij}^{mn} \cdot \sigma_m \cdot \delta_m \quad \forall \, m \notin K.$$

Recall that equation (14.1) states that the production price in sector n, region j, equals total capital advanced (the term in curly brackets), incremented by the sectoral profit rate. Total capital advanced equals: the sum of the cost of capital and wage goods and transportation (the first square brackets), multiplied by the length of the production period in years (the second square bracket); plus the costs of inventory and replacing fixed capital (the third square bracket).

We now consider the impact of this on organizational structure. Suppose that the average difference between the mean rate of profit for the economy as a whole and the mean rate of profit in sector n depends on differences between the degree of concentration of capital in the economy as a whole and the degree of concentration of capital in this sector. One measure of the degree of concentration of capital in a sector is the degree of inequality in assets controlled by firms. Many possible measures of inequality could be used, but one has the desirable statistical property that it assumes nothing about the nature of this inequality – the entropy measure (Sheppard, 1976).

If firm k of sector n possesses the fraction S_k of total assets in that sector, then the entropy measure for sector n is defined as:

$$E_n = - \sum_k S_k . ln(S_k), \qquad (14.2)$$

where the summation is over all firms in sector n. A similar measure may be calculated for the entire economy by defining S_k as the share of all assets in the economy, and summation is over all firms in all sectors. Define this measure of national corporate concentration as E. Then a simple hypothesis making profit rates proportional to economic concentration is:

$$r^n - r = \epsilon . (E_n - E). \qquad (14.3)$$

Note that the constant ϵ, measuring the responsiveness of profit rates to corporate concentration, may be positive or negative. For example, it may be negative when exit barriers are preventing disinvestment by large corporations in times of economic crisis, lowering rates of profit. Substituting this equation into (14.1) for r^n:

$$p_j^n = [1 + r + \epsilon . (E_n - E)] . \left\{ \left[\sum_{i=1}^{J} \sum_{m=1}^{N} a_{ij}^{mn} . p_i^m + \sum_{i=1}^{J} \left(\sum_{m=1}^{N} a_{ij}^{mn} . \tau_{ij}^m \right) . p_i^m \right] . $$

$$[(T_j^n)^{-1} + \psi_j^n] + \left[\sum_{i=1}^{J} a_{ij}^{sn} . p_i^s \Big/ \theta_s C_n^s + \sum_{i=1}^{J} \sum_{k=1}^{K} a_{ij}^{kn} . p_i^k \Big/ \theta_k C_n^k + \sum_{i=1}^{J} \left(\sum_{k=1}^{K} a_{ij}^{kn} . \tau_{ij}^k \right) . p_i^k \right] \right\},$$

$$(14.4)$$

14.1.3   Inter-regional profit rate differentials

If the rate of profit differs among sectors, then the mean rate of profit will also vary between regions, if only because regions with a higher concentration of high-profit sectors will show above-average profit rates. Yet there are also variations in the rate of profit among firms within a sector, discussed in section 13.1, which may have an impact on inter-regional differences in the profit rate. The profits of individual firms differ depending on their production costs. Kalecki (1954) suggested in his later work that individual firms' mark-ups depend on both the degree of monopoly and the average production price in the sector. If so, then firms whose costs are low relative to the average production price will have a higher mark-up than those whose costs are high relative to the average production price. Semmler (1984) also reports statistical studies suggesting that the rate of profit made by individual firms increases with firm size and market share within a sector. It is generally assumed that this is because larger firms are more efficient, able to take advantage of economies of scale and thus able to make greater profits. In terms of Figure 13.1, we would expect the firms found to the bottom and left of the distribution of production methods to be larger. Yet the opposite situation also exists; under certain conditions large firms prove to be costly and conservative, and small firms are the innovators (Massey, 1984a).

There is also abundant evidence to suggest that there are significant variations in the proportion of large and small firms within a sector in different regions. For example, the American clothing industry consists of larger firms in the Piedmont area of the south-east, and smaller sweat shops in big cities such as Los Angeles. Massey (1984a) points out similar differences in the British clothing, footwear and electronics industries. Yet even if large firms make different rates of profit than small firms, and some regions have a greater than average proportion of large or small firms, it does not necessarily imply that sectoral profit rates differ among regions. This depends on the extent of inter-regional competition. Suppose there is full competition, and that a particular sector in region i is dominated, say, by small and stagnant firms using less efficient production methods. These production methods dominate that region, and thus represent the socially necessary production method. Under full competition the average rate of profit is the same everywhere, implying that the disadvantage of an inefficient production method in this region will not be realized as a lower profit rate. Instead, the disadvantage will appear in other ways. For example, the profit-maximizing location pattern may be such as to exclude production of this commodity in this particular region, suggesting that there will be pressure for disinvestment. Furthermore, even if production occurs, the output of this sector from this region may be low by comparison with other regions because of low demand (as indicated by the socially necessary division of labour; Chapter 9).

On the other hand, there may not be full competition. The largest corporations may be able to dominate price setting even in those regions where their presence is limited. One example is basing point pricing, whereby the Pittsburgh-based steel corporations could organize prices in such a way that no competitor in any region could undersell the delivered price of Pittsburgh steel. This meant that local producers in any other region would make a profit equal to the difference between capital advanced and the basing point price, irrespective of whether this margin was above or below the going rate of profit in that sector. Corporations also use their market power to set prices at minimal profit rates, selling the product locally at a loss leader price in order to either maintain market share or drive competitors out of business. If full competition does not prevail for any of these reasons, the average rate of profit within a sector may also vary by region, implying that the sectoral rate of profit, r^n, in equation (14.4) should be replaced by a rate of profit varying by region and sector, r_j^n.

14.1.4   The accumulation of investment funds

To understand the basis for investment and growth policies pursued by corporations, it is necessary to account for the profits accruing to corporations in each region and sector. We saw in Chapter 4 (section 4.1.5) how to account for flows of wages, capital costs and profits among regions (for further details, see Sheppard, 1983b, 1987). A region containing the headquarters of large corporations controlling many branch plants in other regions will accumulate more profits in that sector for reinvestment. Similarly, regions whose corporations control plants in regions with high local rates of profit will earn more money. In order to account for the inter-regional flow of profits it is necessary to know the geography of corporate control.

Suppose we know for each region the quantity of output produced in the branch plants of the corporations headquartered there, and where those branch plants are located. This describes both the location of branch plants in the economy, and the locations from which control over those branch plants is exerted.[1] The relative amount of control exerted by corporations from sector n in region i over production in region j equals the output of plants controlled there, divided by the total output from this sector in all regions (X^n). Define this ratio as φ_{ij}^n. The profits accruing to firms of sector n of region i from branch plants owned in region j then equals the value of output controlled from region j ($p_j^n.\varphi_{ij}^n.X^n$) multiplied by the fraction of this value that is profits ($r_j^n/(1 + r_j^n)$). The total profits accruing to corporations of sector n headquartered in region i, π_i^n, are then the sum of profits made from branch plants owned in all regions:

$$\pi_i^n = X^n . \sum_{j=1}^{J} [r_j^n/(1 + r_j^n)].p_j^n.\varphi_{ij}^n. \tag{14.5}$$

In this way we can account for how the organizational structure of production influences the profits available to corporations by sector and region, profits that then form the foundation for new direct or portfolio investment.

14.2   The organization of production

By the organization of production, we mean the hierarchical organization of production that characterizes a corporation. Massey (1984a) points to two salient characteristics of this. The first is the rules of control within a corporation that define the power possessed by different plants over the process of production (referred to by Massey as relations of production) and over the reinvestment of funds and capital accumulation (referred to by Massey as relations of economic ownership). These rules vary significantly among corporations. The second is the organization of production activities among plants. This ranges from an internal spatial division of labour, whereby different parts of the production process are allocated to different locations according to comparative advantage (see Hymer, 1972; Westaway, 1974; Watts, 1981; Stanback and Noyelle, 1982), to a corporation whose branch plants are clones of one another with each carrying out essentially the same activities (Massey, 1984a, p.76).

These aspects of the organization of production are purely internal to the corporation. An alternative way of organizing production, however, is to employ other firms that are so dependent on the corporation that they find it difficult to survive as autonomous economic enterprises. Storper and Walker (1989) refer to these as industry governance systems. This encompasses all firms developing such a close contractual arrangement with a corporation that the resulting interactions cannot be approximated by market relationships. Certain forms of subcontracting, licensing and franchising fall into this category, but not all. For example, Holmes (1986), in his useful survey of varieties of subcontracting, refers to what is termed 'Supplier contracting . . . where the subcontractor is in many respects an independent supplier with full control over the development, design and fabrication of its product' (Holmes, 1986, p.86). This is a case in which the interactions between subcontractor and corporation are approximated by market relationships, and thus falls outside our definition of a production system organized by a corporation. Similarly, franchises where the franchise owner takes full responsibility for setting prices and marketing the product, and where the parent corporation provides no special support services, would also fall outside our definition.

The rules governing the relationship between a corporation and subcontractors, franchisees and licensees are highly variable, ranging from virtual central control to virtual market relationships, and many will fall in a grey area where it is not clear whether they are best represented by market interactions or by an organized production system. One case, however, that is clearly a corporate strategy to organize production is the kind of 'just-in-time' system used by Toyota and some other Japanese companies (Sayer, 1986; Kaplinsky, 1988). Consider the following description: 'in Toyota's case, it produces cars in Toyota City with suppliers located in close proximity, and delivery on an hourly basis on purpose-built roads' (Kaplinsky, 1988, p.458). We are not suggesting that all firms adopting a just-in-time system are part of the organizational strategy of a large corporation, but this is clearly the case for the recognized innovator in this field, Toyota.

For further analysis we will select just three paradigmatic cases from a large range of ways of organizing production: the cases of cloning, an internal division of labour, and a just-in-time system. Take, first, the case of cloning. This is the easiest to incorporate within the theory developed in this book, because all affiliates are producing the same commodity, often using similar production methods. An example would be the franchises of a supermarket chain. It is sufficient in this case simply to identify the geography of corporate control as described in section 14.1.4 and to include this in analyses of the location of production, pricing and the accumulation of capital as described in that section. If all branches have full control over production methods and pricing decisions, and take full responsibility for making a profit, then the behaviour in individual plants is equivalent to that of autonomous capitalists, and there is nothing further to add to the model. If, however, the parent corporation influences these decisions, then the prices charged could be administered prices, reflecting the goals of the corporation, rather than production prices. It then becomes necessary to discuss whether such administered prices will vary significantly from production prices, and if so to determine the processes through which they are set.

In the case of an internal division of labour, the various plants making up a corporation are engaged in producing different products. In the case of a horizontally integrated corporation they are different products from different sectors of the economy. In a vertically integrated corporation these different products represent various stages in the production of a certain group of commodities, such as the stages involved in making finished steel products or computer hardware. We will examine each possibility in turn.

The case of horizontal integration may be captured by defining organizational control linkages that cross both sectoral and regional boundaries. Define φ_{ij}^{mn} as the proportion of all production from branch plants producing commodity n in region j that are owned by a corporation headquartered in region i and classified as being in sector m. Clearly in the case of horizontally integrated corporations it can be difficult to decide which sector a corporation belongs to. This is the case for such highly diverse holding companies as Litton Industries. Yet for the purposes of determining where profits accumulate this is not critical. Coefficients φ_{ij}^{mn} can be introduced in place of ϕ_{ij} in equation (14.4) in order to calculate the accumulation of profits for reinvestment. As in the case of cloning, strategies for setting administered prices must again be taken into account.

In the case of a vertically organized production system, those plants making an intermediate product for shipment to another branch plant for the next stage of manufacture are actually not engaged directly in commodity production. It is only at the final stage of production that a commodity is manufactured. Despite this, each plant producing intermediate products generally charges a mark-up on capital advanced, which counts as the profit accumulated at that stage of the production process. These profit levels are, however, frequently manipulated via corporate accounting systems to ensure that profits accumulate at those locations where the corporation can retain them most effectively. Jenkins (1984, p.9) points to two ways in which corporations can disguise or geographically redistribute profits: by manipulating transfer prices between subsidiaries, and by over- or undervaluing capital equipment and technology provided to subsidiaries.

Given these considerations, the most appropriate strategy for including such vertical integration into our paradigm is to assign all the costs and profits associated with the various stages of production in the calculation of profits made at the location where the final product is finished. For example, if partial assembly of personal computers occurs in one region where unskilled labour is cheap, and final assembly then occurs in a second region, all of the costs and profits associated with both operations should be assigned to the latter location. It is useful to separate out a branch plant as a separate production facility only if its manager has autonomous control over reinvestment of part of the profits made at that plant. In that case, however, once again it is necessary to determine the adminstrative rules under which prices and profits are determined at that location. The inclusion of organizational control links connecting branch plants to the parent corporation would then account for the accumulation of profits for reinvestment as in section 14.1.4.

The just-in-time system of interaction between a corporation and its sub-contractors is a system for minimizing, indeed virtually eliminating, storage costs that must be paid to hold large inventories of inputs. If it is also associated with flexible production systems that allow immediate response to new orders, then storage costs for keeping inventories of the produced commodity on hand can also be minimized, as it will no longer be necessary to keep large numbers of different models of a product in stock in order to cope with unexpected demands. Consider equation (14.4). The right-hand side of that equation equals capital advanced incremented by the rate of profit. One component of capital advanced is inventory storage costs. If one sector or region implements a just-in-time system, it minimizes storage costs. In the absence of full competition this leads to a greater rate of profit for that sector or region. These savings can be significant. In 1984 Toyota of Japan had storage requirements that were just one-sixth of those for American automobile manufacturers (Kaplinsky, 1988). By taking account of storage costs in the way done in equations (14.1) and (14.4), and their impact on profit rates (r_j^n), it becomes possible to determine the profits accumulating in a region as a result of its corporations adopting a just-in-time system (π_i^n) using equation (14.5)

14.3 Investment strategies

In sections 14.1 and 14.2, we detailed a method for determining the impact of organizational structure on profit rates, and calculating the total profits accumulating in each sector of each region as a result of this organizational structure. These profits provide the internal resources available for new investment, part of which may involve changing the organizational structure. In this section we will point out some of the investment strategies available, and their impact on organizational structure. Depending on the strategy chosen by a corporation, the impact on organizational structure and on the accumulation of profits for further investment can then be worked out. We identify three different strategies: direct investment, direct disinvestment and investment in the capital market (or portfolio investment). A fourth strategy involves mergers and take-overs, but these are intimately related to financial markets and finance capital –

aspects that we do not address in this book, although they can be included (Franke, 1988).

Two strategies are identified in theories of direct investment: investment in locations where rates of profit are expected to be high, and investment in locations where the corporation invested in the past. The former is ostensibly economically rational, whereas the latter recognizes that previous investments provide a knowledge and history of past success and inertia that attract further investment (Hymer, 1972; Sheppard, 1983b). The relative weight attached to these two alternatives varies between sectors and regions. More conservative and less profit-seeking capitalists will follow the latter strategy, perhaps even when profit rates are not very high. Massey (1984a) suggests that this is the case for smaller firms of the British clothing and footwear industries. It is a strategy where corporations continue to invest in those sectors and regions where they have previously invested. More knowledgeable and profit-seeking firms are expected to give higher priority to the former strategy. If corporations are prepared to consider the rate of profit in any sector and region, this leads to radically different patterns of organizational control.

Direct investment involves expanded levels of production in branch plants, changes in patterns of organizational control (such as opening and closing plants), and changes in the rate of accumulation of profits. Any changes in organizational control also affect the relative profitability of different sectors and regions. The new patterns of organizational control and profit rates then form the basis for new direct investment patterns. This dynamic is sketched out by Sheppard (1983a).

Alternatively, direct disinvestment may occur if corporations determine that directly owned branch plants are less profitable than other forms of investment. This disintegration of ownership is extensively analysed by Scott (1988b, c), using the analogy of markets versus hierarchies (Williamson, 1975). In this view, branch plant production is profitable when the economies of scope within a corporation (defined as the economies that accrue from joint organization and operation of potentially divisible production processes) are high. Production is then most efficiently organized hierarchically within a corporation, rather than being left to autonomous producers interacting through the marketplace. Scott further suggests that the factors reducing economies of scope include: the increased possibility of separating stages of the production process from one another, which many argue has become cheaper owing to the availability of flexible production processes that reduce economies of scale (Scott, 1988b,c); high uncertainty in the marketplace, making it difficult to take advantage of scale economies that result from producing large numbers of the identical model of some commodity; economies from separate producer services, more cheaply provided by an autonomous firm to a number of clients than by each corporation for itself; higher labour costs in corporate labour markets (Friedman, 1977); and spatial clustering, reducing the interaction costs among autonomous or quasi-autonomous firms (as in Toyota's just-in-time system).

When such conditions prevail, then they lead to lower rates of profit in those areas dominated by larger firms. When these profits are low by comparison with the rate of interest available in capital markets, direct disinvestment (such as the sale, closure or buy-out of branch plants) is expected to follow. Disinvestment will also change ownership patterns and profit rates, affecting the relative profitability of further direct disinvestment. Note that direct disinvestment in some places may well be accompanied by direct investment in other regions or sectors, depending on the profitability of that strategy by comparison with portfolio investment.

The relative profitability of direct investment and disinvestment also depends on the fixed costs associated with these strategies. When large quantities of fixed capital are purchased as a part of direct investment, or depreciated and written off in the case of direct disinvestment, then this will slow the rate of direct investment and disinvestment. Indeed, empirical research by Sheppard, Tödtling and Maier (1990) shows that there is more inertia to direct investment and disinvestment behaviour in sectors where high fixed costs imply high entry and exit barriers. This parallels the impact of fixed costs on changes in production methods outlined in section 13.2.

The third strategy available is portfolio investment. The profitability of this strategy depends on how the mean rate of interest in the capital market compares with the mean rate of profit obtained by a corporation in a given region. It is to be expected that, when the ratio of the rate of interest to the rate of profit is much greater than one, then firms are more likely to engage in portfolio investment. Once again, however, there will be important differences among firms, sectors and regions. Massey (1984a) observes that small firms in the British clothing and textile industries reinvest relatively little in growth and changes in production methods, suggesting that they have a relatively high propensity to make portfolio investments. She makes a similar observation for British industry as a whole by comparison with other industrialized nations. The impact of portfolio investment on organizational structure depends on the industries to which this money is lent. Firms with higher profit rates are more likely to have the collateral and prospects to borrow money successfully in the capital market, suggesting that the effect of portfolio investment will increase the importance of those regions and sectors with higher rates of profit, and of the organizational strategies they pursue.

It is clear that firms can follow a variety of investment strategies, depending on the relative importance they place on the three alternatives sketched above, and on the detailed courses of action possible within these alternatives. At this point we know too little about these processes even to guess the degree to which systematic generalizations can be made about the kinds of investment behaviours adopted under various circumstances. We have attempted to show, however, that it is possible to work out the consequences of investment strategies adopted in one period on the organizational structure and the rates of profit prevailing in the next time period. These changes in turn affect the relative attractiveness of the different alternatives. One conclusion can be drawn relatively safely, given the range and complexity of the investment strategies available, it is unlikely that rates of profit are equal in different sectors and regions even in a profit-maximizing space economy in dynamic equilibrium. Webber (1987c) shows this theoretically, even under very simple investment rules where just two alternatives (direct and portfolio investment) are available.

Summary

This chapter has attempted to examine the influence of organizational structure on profits, and the influence of investment strategies using those profits on organizational structure. The debate about the impact of the degree of oligopoly – one aspect of organizational structure – on profit rates is as yet unresolved. If there is a systematic relation between the two, however, it is possible to include this in our model of production prices to determine the prices charged and the rates of profit prevailing in different sectors and regions. An additional level of complexity is introduced if we consider the possibility that rates of profit in the same sector vary between regions, because the actions of major corporations affect the profit rates that are obtained by their competitors – even in regions where the corporations themselves have no production facilities. If it is possible, however, to determine production prices and prevailing rates of profit in different sectors and regions, then, by taking into account the current geographical structure of ownership of branch plants, it is possible to determine how profits accumulate in different sectors and regions. These ownership patterns are a second aspect of the organizational structure.

A third aspect of organizational structure is the way in which corporations organize the production process itself. This includes the nature of the hierarchy of decision making; the division of labour extant within a corporation; and the way in which that corporation organizes its interaction with those other firms largely dependent on the corporation. We identified three of many possible cases – cloning, an internal division of labour and just-in-time systems – and discussed how the effects of this organization on production prices and profit rates, and thus on the accumulation of profits, can be worked out.

Having determined the accumulated profits available for reinvestment, we examined different investment strategies available to firms, and the effect of pursuing these on organizational structure. Three strategies were identified. Direct investment extends the scope of organizational control to those places where the corporation already directly invests as well as to those places where profit rates are higher. Direct disinvestment occurs when the profitability in branches of large integrated corporations is low. The capital thus released is directly invested in other branches or put into portfolio investment, changing the organizational structure and the relative profitability of the different alternatives. Portfolio investment occurs when rates of interest are high relative to rates of profit. Money thus placed in the capital market tends to be lent to more profitable sectors and/or regions of production, reinforcing the organizational structures being developed in those branches.

We have deliberately presented strategies with respect to organizational structure separately from those associated with changing methods of production for pedagogic purposes. In reality, of course, the strategies adopted by capitalists in their attempts to increase profit rates or avoid impending crises are a mixture of production method changes and alterations in organizational structure. The distinctions between these are blurred, and in many cases all but impossible to find. Thus Kaplinsky (1988) argues that the just-in-time system adopted in Japan entails methods of production that require flexible production, rigorous quality control, and factory line work involving increased responsibility and multiple tasks and skills. Yet even this is not a universal combination. Just-in-time inventory control systems can increase profits even in sectors where production runs are very large and inflexible. In fact, the multiple options available to firms, and the variety of histories underlying the development of different industries in different regions, are so complex as to be still resistant to the development of general theories predicting the strategies adopted by capitalists.

Despite this, the accounting carried out in a model of this kind can still be useful in determining the profits available for reinvestment and the relative economic attractiveness of alternative investment strategies. This model also helps us understand why capitalists face such a complex combination of situations in a market economy, because it suggests why the capitalist space economy is generally out of equilibrium and riven with conflict and uncertainty.

Note

1 For an empirical analysis of such a pattern of corporate control, see Sheppard *et al.* (1990).

15 *Conclusions*

In this book we have attempted a synthetic overview of the economic geography of industrialized capitalist society. We have chosen an analytical framework to do this because it allows us to develop the links between the various elements that make up the capitalist space economy. In this conclusion we will draw some of these threads together, discuss some elements still missing from the framework, and speculate on some implications of the analysis for research in economic geography.

15.1 Summary of the argument

The location of economic activities in capitalist society reflects both the production and circulation of commodities, and the social distribution of the surplus produced. Production processes are represented by the production requirements in each place, the labour requirements and the workers' standard of living. These may seem to be simply technological data, but they are far more than this. They are the socially necessary production relations, reflecting not only the state of technical knowledge but also the working conditions negotiated in the workplace and the standard of living that workers have achieved for themselves and their families. Circulation represents the movement of produced commodities from producer to consumer through the operation of markets. Production and circulation must be coordinated in order for capitalists to realize and reinvest profits, a *sine qua non* for the successful reproduction of the entire system. Furthermore, it must be possible to produce a surplus in the economy over and above the commodities available prior to production. Capitalists, workers and landlords then struggle over this surplus, each group trying to maximize its share.

Production and circulation can be looked at in three ways: exchange value, labour value and the physical quantities or use values. We began with exchange value – the price circuit (Chapters 4–7). We showed that, given the geographical distribution of socially necessary production relations, there is a determinate set of production prices charged at each place and patterns of commodity flow that will allow all capitalists to make, on the average, the same rate of profit. Certain location patterns will maximize capitalists' profits. At the same time, however, the struggle for higher wages by workers can cut into profits, and changes in this social distribution of income in turn influence prices, commodity flows and profit-maximizing location patterns. Profits, prices and locational configurations also depend on the time taken for the production and circulation of commodities, because this affects the capital that must be advanced prior to production.

We demonstrated also how non-produced goods, notably land and physical resources, must be taken into account. These are largely ignored in economic models, but geographers cannot afford to do this because they are among the fundamental features that differentiate places from one another. Although scarcity cannot be read from the physical environment, the existence of non-produced goods does mean that there are limits on the levels of production that can be achieved in a place, because it is not just produced commodities that are required in order to produce commodities (*pace* Sraffa). For non-produced goods, social necessity is defined by the most costly location that is required in order to provide sufficient quantities of the good for economic reproduction. The 'price' of these goods can be measured by rent, but rents are not an index of naturally occurring scarcity since they also reflect the ability of resource owners to obtain a piece of the surplus.

Production and circulation can also be examined as a circuit of labour values (Chapters 3 and 8). We showed that labour values can be precisely calculated, both theoretically and empirically. We also showed that, although the relationship between labour values and exchange values remains in dispute, empirical evidence points to an astonishingly close correlation between the two. The location of economic activities has a significant impact on questions of exploitation, since it is possible for workers in different places to be exploited at different rates (they may even be negative in some places) as long as the average rate of exploitation is positive – itself a necessary condition for there to be positive money profits. Class alliances formed within a place then become an understandable outcome of geographical differentiation, rather than some perverse result to be attributed to false consciousness.

Third, we examined the circulation of physical commodities and accumulation of capital as a measure of the success of economic reproduction on an expanded scale (Chapter 9). We showed that a geography of commodity production and trade does exist that will allow the smooth accumulation of capital, but that it seems unlikely that capitalist entrepreneurs can achieve or maintain this desirable configuration (for them) simply by reinvesting their profits in growth, because of the formidable coordination problems that face an economy of interdependent producers.

The ability to specify theoretically both a profit-maximizing space economy, and the possibility of economic growth belies the fact that the contradictions of capitalist society make it very difficult to believe that any such equilibrium is readily attained, or that it is socially desirable. Conflicts between social classes over the distribution of the surplus are at the heart of the structure of capitalist society, and cannot be resolved by maximizing either social welfare or efficiency. Resolution of these conflicts is in the realm of politics, and thus, even if capitalists were to achieve a desirable space economy, other classes would be continually challenging and attempting to redefine it. For this reason equilibrium will always be resisted from within the system by those classes whose interests are not served by the current state of affairs.

The introduction of a spatial dimension to this discussion has two contrasting effects. On the one hand, the reality of the communities in which we live provides a plausible rationale for the occurrence of collective class action (Chapter 10). This is missed in rational choice analyses of behaviour, where the attendant position of methodological individualism makes it difficult to understand the very existence of collective class action. On the other hand, the additional complexity of a spatially extensive economy makes it very difficult for classes to be sure that actions that appear immediately beneficial will benefit the class in the long run (Chapter 11). It also makes it very difficult for individuals pursuing their self-interest to be sure that what they do will be good for their class as a whole (Chapter 12). Issues that seemed resolved in an aspatial approach, such as comparative advantage and trade and the falling rate of profit, are called into question again when the reality of location is brought into the discussion. We also examined how the complexities of a spatial economy influence the strategies available to capitalists as they attempt to deal with the uncertainties of capitalism – the development, adoption and imitation of new labour processes and production methods, and alternative ways of institutionally organizing production (Chapters 13 and 14).

Throughout, we have attempted to differentiate our interpretation from the neoclassical approach that dominates most writing in the analytical tradition of economic geography and economics (see particularly Chapters 2, 5 and 6). We have shown that the conditions for neoclassical macroeconomic theory to be correct are identical to those necessary for a consistent solution to Marx's transformation problem – a fact generally overlooked by those quick to criticize Marx for his logical errors. In order to obtain a number of its important results, neoclassical macroeconomic theory assumes a homogeneous entity called capital. Once this assumption is replaced by a more realistic alternative, much conventional wisdom about the social desirability of markets, of the free movement of labour and 'capital' and of free trade becomes questionable.

15.2 Extensions

There are still some important aspects of the economy that have not been integrated into the above account, and we pause briefly to indicate some of the more important of these and how they may be incorporated. We consider in turn complexities of the labour market, finance capital and credit, and the state.

While labour has played a central role in our account, we have paid little attention to its heterogeneity, other than to point out how heterogeneity can compromise collective action. Given our insistence on the inappropriateness of assuming that capital is homogeneous, it behoves us to examine the effect on our argument of heterogeneous labour forces. This has been a subject of some interest, despite the argument about abstract versus concrete labour summarized in Chapter 3 (see Bowles and Gintis, 1977; Krause 1981; Zalai, 1981). Zalai (1981) argues that skilled labour may be treated in a way that is consistent with unskilled labour by calculating the wages of skilled workers as their cost of reproduction including their higher costs of training. In addition, Krause '(1981) shows that assuming heterogeneous labour does not invalidate the fundamental Marxian theorem. On these grounds, therefore, there is good reason to believe that the introduction of heterogeneous labour will not be as devastating for the general logic of political economic theory as heterogeneous capital has been for neoclassical macroeconomic theory.

Bowles and Gintis (1977) show that heterogeneous labour can lead to segmented labour markets, making collective action by workers more problematic because capitalists can negotiate separately with members of different labour sub-markets. One kind of labour market segmentation is geographic, so it is not surprising that Bowles and Gintis' arguments parallel our own (Chapter 12). There is tremendous debate about the degree of mobility between labour sub-markets. Higher mobility would tend to equalize wages in the economic view, reducing segmentation. An important dimension here, which should be included in a geographical analysis, is the geographical mobility of labour. While we have not done so, it would be straightforward to link the framework of this book to models of population growth and migration in order to examine the interaction of labour supply and demand with commodity production, taking into account the different possibilities available to different groups of workers. It would then be possible to investigate the contrasting effects of mobility. Another key issue in this area that is virtually ignored in economic geography is the geography of union formation and negotiations by workers for higher wages. The evidence we have about the influence of unions and other forms of workers' collective action on geographical wage differentials is largely anecdotal (but see Clark, 1986, and Peet, 1987).

A second area of considerable importance but largely neglected in economic geography is the role of finance capital and credit markets. Harvey (1982) has laid out the broad dimensions of the influence of finance capital and credit but supplies few details. A more precise examination of this issue is possible by linking financial markets to the models developed here. To do that involves: developing a theory of the relation between interest rates and returns on stocks and bonds and the profit rate; an examination of the outside financing of the production of commodities (for both producers and consumers); and understanding the process by which profits from commodity production enter the money market. In short, the operation of money markets must be linked to the flow of money between these markets and the circuits of commodity capital. Some steps have been taken in this direction within the paradigm of economics used in this book (Liossatos, 1983, 1988; Lianos, 1987; Schutz, 1987; Franke, 1988). There remains the task of integrating these advances with a geographical model, but given the very high mobility of money and credit this should not be too difficult.

The third area is one where there has been considerable recent work in geography – the role of the state apparatus in capitalism. The geographical issues here are the powers and autonomy of different parts of the state apparatus at different geographical scales, and also the competition both among local states, and between them and other hierarchical levels of the state. Such concerns overlap with those in economic geography because state institutions raise revenue, redistribute money and provide certain goods and services. The more significant role of the state, though, is in underwriting and directing private commodity production. Given the inherent instabilities of capitalism and the conflicting class interests, it is clear, given the argument of this book, why the state must be concerned with the location of private commodity production and investment, and balancing the concerns of the different classes. We have seen that the market cannot solve all problems, and the political institutions of any territory depend on private investment to provide economic security for its residents (and voters) under capitalism. State insitutions must therefore be concerned that capitalists make sufficient profits to continue to invest in the territory. Subsidies for private investment take resources away from workers, however, so the state must balance the requirements of accumulation against those of continued political legitimacy with voters and interest groups. We do not insist that state actions are determined by economic considerations, yet it is important to identify the economic forces influencing state action and the economic options available to the state.

When the space economy is in disequilibrium, state intervention may well be necessary, but it is impossible to generalize about the kind of intervention to expect. In a region with above-average capital accumulation the state clearly faces different options than in a stagnating or declining region. One way to proceed, however, is to take certain prototypical situations that urban regions may face, and analyse the options available to the local state under varying scenarios about the actions and policies pursued at higher tiers of the state apparatus. It is also possible to conceive of an empirical analysis. Given the state of the inter-regional economy, profit levels and rates of growth in a particular region can be calculated and compared with the average. The possibilities for (and consequences of) state action in this context can then be analysed.

15.3 Reflections

From this summary of the book, and discussion of the topics that could have been included, we hope it is clear that despite its formal language our analytical framework allows treatment of many of the facets of a capitalist space economy. Indeed, its strength is the ability to place these various issues in a precise relationship with one another, and to identify logical consequences and inconsistencies. Many of our arguments have developed from demonstrating the inconsistencies that stem from unnecessary and unrealistic assumptions, and pursuing their logical implication. While this results in an argument that is often quite formalistic, it does place that argument on a solid theoretical foundation.

One of the principal theoretical points that we have tried to establish is that the inclusion in our analysis of the relative location of economic activities, as represented in the economic interactions between places, does make a significant difference to general theoretical conclusions about the operation of a capitalist economy. In this sense, space does indeed make a difference. It is now widely suggested that space makes a difference in human geography because the particular characteristics of places entail unique consequences. In other words, the particularities of a place represent contingent factors that alter the outcome of general causal (necessary) relationships (Sayer, 1984, 1985). As such, predictions are rendered difficult in the social sciences, which suggests that case studies should be pursued rather than extensive statistical studies. We do not wish to deny the importance of the particularities of place, but we do argue that location is important in a deeper way, in influencing the very relationships that should be taken as necessary. For example, the conflict between capitalists and workers, embodied in theories of exploitation and wage–profit frontiers, can be significantly compromised by geographical differences in wage and exploitation rates.

In Marxist analysis, and in economic theory in general, the effects of location are frequently assumed away in order to identify essential economic relations. We argue that this exercise may be fundamentally misleading. While many other complexities of real societies, such as heterogeneous labour forces, can be usefully ignored as a first approximation (as long as such simplifications do not materially affect an explanation), this is harder to justify for location. While it is logically possible for a society to be composed of labour that is of identical skill, it is not logically possible for a society to exist on the head of a pin. Furthermore, the inclusion of relative location, as our book demonstrates, makes a significant difference to the theory.

This does not mean that we can draw neat geometrical diagrams of capitalist society, geographically organized in an economically rational world (see Lösch, 1954). Indeed, the reader has no doubt noticed the lack of any diagrams of spatial configurations in the book. Such diagrams suggest a neatly ordered spatial structure reflecting the immutable effects of distance, *ceteris paribus*. In fact, distances are not some physical constraint to which all realizations of a process are subject in identical ways, as in the laws of physics, because spatial structures are socially constructed and far more complex than the isotropic and stationary spaces that are generally relied on in geographical analysis. To avoid such idealizations, we represent location as the interaction among places – for example, as inter-regional input–output coefficients. When location is so represented, we can examine its rôle without having to analyse particular spatial configurations. The changeable nature of such coefficients is also a good example of the continual reconstruction of spatial relations, since those coefficients change as a result of the actions of capitalists, workers and landlords trying to improve their situation (see Chapters 11 and 12). We insist, however, on the importance of including these relations in this general way.

We also do not suggest that the ability to apply analytical tools means that we can or wish to make precise predictions of the kind sought in positivist and critical rationalist paradigms. First, while the quantitative nature of economic relationships makes it possible to represent many aspects of an economy in an analytical model, many other crucial aspects are very difficult to discuss in detail. While there may be a logic to class relations, the historical actions of individuals and classes are susceptible to many non-economic relationships, and are arguably better examined from a more flexible viewpoint. Such actions are none the less central to the working out of economic relationships. It is also difficult to discuss in more than a very broad manner certain economic phenomena, such as fixed capital and the built environment, because their discontinuities are not easily treated in an insightful way with the analytical languages currently available.

Second, despite our ability to lay out what is possible and logical under certain conditions, we are still a very long way indeed from predicting what will actually occur. There are several reasons for this. In order to use the models developed here for empirical predictions of some kind, we face a formidable task because the data necessary to specify these models accurately are horrendous. The models are also based on certain conditions that, while they may plausibly be general tendencies that hold on average, they will in all likelihood never be true of any particular case being studied. Finally, as we have argued throughout the book, the added complexity of space actually makes it harder to determine the eventual outcome of even carefully conceived economic decisions in a well-specified situation.

Our particular analytical strategy reflects these uncertainties and ambiguities in that we have tried to emphasize that we are discussing not deterministic relationships, but the averages of stochastic processes. For example, in stating that under full competition we expect rates of profit to be equal everywhere, we do not mean that no capital flows from one place to another because every capitalist makes the same profit. Rather, we imply that outward capital flows, from enterprises with low profit rates, are balanced by an equally large inward flow, because in every place there are going to be some who are doing better than average and others doing worse. Our conclusions must be similarly qualified, and there is no reason to believe at present that precise predictions about events in individual places will be possible. Yet, with these qualifications and limits, our dicsussion does represent an analysis of the general relationships expected in a capitalist space economy, arguments that are susceptible both to logical criticism and extension, and to examination on the grounds of their empirical plausibility.

Glossary

absolute rent Rental payments received by landlords because of their ability to appropriate the difference between the *labour value* of a commodity and its (lower) *production price*.

animal spirits Term coined by Keynes to capture the idea that investment decisions by capitalists are subject to a wide range of non-economic influences.

autarky A region in autarky is self-sufficient in economic production, exchanging no commodities with any other region.

base Also known as the infrastructure, the base is the part of the *mode of production* made up of the forces and relations of production. In fundamental Marxism the base plays a pivotal role in both shaping society and creating historical transformation (it is 'the engine of history').

basic/non-basic goods A distinction first suggested by Piero Sraffa. Basic goods are those commodities that are directly or indirectly required to produce all other goods. In contrast, non-basic commodities, often conceived as luxury goods, do not enter as inputs into all other goods.

capacity constraint Fixed limits on the total output possible from a particular location.

capacity utilization rate The quantity of commodities actually produced from a machine expressed as a proportion of the total possible output from the machine.

capital goods Commodities purchased as inputs for the production of other commodities.

capital intensity Expressed in dollars, it is the ratio of capital to labour used in the production process.

capital logic approach Describes those Marxist approaches that prioritize the needs of capital over other interests. It is often associated with *functionalism*.

capital reversing The case where an increase in the rate of profit results in a more capital-intensive method of production being more profitable than a less capital-intensive method.

circulating capital *Capital goods* that are entirely consumed during a *production period*.

class exploitation correspondence principle (CECP) Associated with John Roemer, this asserts that there is a one-to-one relationship between class and *exploitation*: capitalists are exploiters, workers are exploited, and petty bourgeois are neither.

class-for-itself The recognition by a class, and the acting upon it, of its objective *class interests*.

class-in-itself A set of *class interests* that objectively defines a group of people within society. Individuals themselves may not consciously know their own objective interests.

class interests A set of material interests defined by exploitation, domination and asset ownership. Class interests can be immediate or fundamental.

class wealth correspondence principle (CWCP) Associated with John Roemer, this asserts that there is a one-to-one relationship between class status and the magnitude of wealth holdings: capitalists have the highest status and most assets, workers have the lowest status and least assets.

cloning A division of labour within a corporation whereby each plant of the corporation carries out the same activities as each other plant (example: supermarket chains).

commodity Something that is produced with the intention of exchanging it for money or some other commodity.

comparative advantage Usually used when discussing trade relationships. The principle of comparative advantage states that profits are maximized when regions specialize and trade in those goods in which they have the greatest ratio of advantage compared with other regions. Advantage is usually expressed in terms of cost per unit.

constant capital The *labour value* of all capital good inputs.

consumption bundle A set of physically defined goods that represents the *real wage* received by a worker.

consumption–growth frontier The relationship between the rate of growth in an economy and the quantity of goods consumed by workers.

corn model Originating with Ricardo, the corn model represents a one-commodity world. Corn is used as seed for capital investment, and is also the metric in which wages, profits and land rents are expressed.

counter-factual A hypothetical reconstruction of what might have happened if the actual historical record had been different (example: a situation where private wealth holdings under capitalism were distributed equally among all members of society).

counterfinality Coined by Sartre, it refers to the deleterious unintended consequences when individuals act collectively.

dead labour Past living labour embodied in capital equipment (equivalent to constant capital).

differential rent I The difference in rents paid on two plots of land when the same production method is used on both, as a result of differences in fertility or location.

differential rent II The difference in rents paid on two plots of land of the same fertility and locational advantage, as a result of differences in the production methods used.

diffusion (of innovation) The spread of a method of production among plants.

direct investment Investment of funds in the activities of plants owned by the investor.

dynamic equilibrium (of production) An economy growing indefinitely at a constant rate, where all products are sold and the same rate of growth is achieved in all sectors and regions.

economic lifetime The age at which a piece of fixed capital is no longer profitable (see also *physical lifetime* and *truncation period*).

economic rationality The view that humans make those choices that maximize their satisfaction (utility) given the limited set of resources at their disposal.

embodied labour values The total quantity of labour actually used in the production of a commodity, including the labour actually used in the production of *capital goods* incorporated in that commodity.

equilibrium A state of balance in which there is no incentive to change.

exchange value See *production price*.

exploitation The diffeence between the labour contributed by a worker and the *labour value* of that worker's *real wage*.

extensive rent Sraffian version of *differential rent I*.

externalities Economic costs and benefits that are not paid for through the market (example: pollution).

fallacy of composition A philosophical concept stating that what is true for the parts of a whole need not be true of the whole itself.

fictitious capital Associated with David Harvey, it represents the property right associated with some future revenue. That property right can vary from landownership to holding stock.

fixed capital *Capital goods* that are employed during a *production period* without being used up (examples: durable machinery; buildings).

flexible labour or production techniques Labour practices or *techniques of production* that allow firms easily to change the amount, quality and kinds of commodities produced.

f.o.b. pricing The delivered price of a good, which equals the sum of the price at the factory gate plus transportation costs.

forces of production One of the two components that make up the *base*/infrastructure within the mode of production. They are usually presented as all those elements that assist in the actual physical production of commodities (technology, natural resources, intermediate inputs, and so on).

Fordism A type of production predominant between the 1950s and 1970s. Associated with large, capital-intensive firms employing Taylorist work principles that produce a standard, mass-produced output. It is also associated with mass consumption.

franchising An organizational arrangement whereby production facilities operate under the name of a larger corporation (example: McDonalds).

free-riders Individuals who do not engage in collective action (thereby avoiding potential costs), but who none the less receive the benefits from such action.

functionalism A form of explanation that suggests that the cause of an event is explained by its effect. For example, the actions of the capitalist state are not explained by the beliefs of the politicians carrying out such actions (cause explains effect); rather, the beneficial consequences for capitalism of such actions explain why politicians hold such beliefs (effect explains cause).

game theory   Associated with the work of Oscar Morgenstern and John von Neumann, this is a theory of interdependent decision making in which there are two or more 'players', where each player receives a different pay-off depending upon both the strategy they choose and the strategies chosen by others.

general economic equilibrium   An economic state such that all producers and all consumers have no incentive to change from their present position.

homo economicus   See *economic rationality*.

ideology   A set of beliefs that represent a distorted (biased) view of society, polity and economy.

imitation (of production method)   The copying of a production method already in use.

innovation   The discovery of a new way of combining labour and *capital goods* to produce commodities.

input–output coefficient   Specifies the amount of one commodity required to produce one unit of another commodity.

input–output equation   An equation specifying the commodity inputs required to produce a given level of output.

input–output matrix   A table describing how much of each commodity is used in the production of each other commodity.

intensive rent   Sraffian version of *differential rent II*.

intermediate goods   Commodities that are required to produce other goods (example: iron is an intermediate input in making steel).

invisible hand   Associated with Adam Smith, this is the idea that economic resources are guided to their best (most efficient) uses through price changes in the market.

joint production   The case where a single production process results in the simultaneous production of two or more commodities (example: oil refining).

just-in-time   An organizational arrangement whereby suppliers of *capital goods* deliver them immediately before they are required in the production process, minimizing storage time and inventory costs.

labour aristocracy   Generally used to denote stratification in the working class between one set of workers who are better paid and/or work in better conditions than another set.

labour contributed   The number of hours of labour worked by workers.

labour power   The capacity of workers to work (the commodity sold by workers in the labour market).

labour theory of value   The theory that holds that the value of commodities is determined by the *socially necessary labour* time required to produce them.

labour value   The hours of *socially necessary labour* invested in the production of a *commodity*, including the hours of socially necessary labour invested in producing the *capital goods* incorporated in that commodity.

law of value The theory that holds that the *exchange value* of commodities is governed by their *labour value*.

licensing An organizational arrangement whereby production facilities are licensed by a corporation to produce the corporation's products.

marginal producer The producer whose costs of production are greatest.

mark-up The difference between the *per unit cost* of a commodity and the price it is sold for.

marginal productivity (of a *production factor*) The increase in output that results from employing one extra unit of a production factor.

methodological individualism The view that society is best explained by reducing it to the (usually rational) beliefs and actions of the individuals who compose it.

mode of production A complete system of economic and social relationships - for example, feudalism or capitalism. Within fundamental Marxism the mode is broken down into two main components: *base* and *superstructure*, where the former determines the nature of the latter.

monopoly rent I Rental payments that occur because the capitalist occupying that land is able to sell her/his product at a monopoly price - a price greater than the production price.

monopoly rent II Rental payments due to the ability of landlords as a class to demand a minimum rent on all plots of land, no matter how marginal.

neoclassical An economic paradigm that reduces economic activities to market interactions.

neo-Fordism A term associated with the French regulationist school of economics that suggests that both production and consumption relations have been transformed over the last two decades. Production relations have changed with the emergence of smaller production units employing computerized flexible technologies and labour practices. Consumption has moved from mass consumption to more specialized products.

neo-Ricardian A school of economics associated with Cambridge University. It is heavily influenced by Piero Sraffa's work and his interpretation of the nineteenth-century English economist, David Ricardo.

non-produced commodities *Commodities* such as land that were never produced by *intermediate goods* or labour inputs.

numeraire The standard adopted in which to express prices.

organic composition The ratio of *constant capital* to *variable capital* used in a production process.

output matrix A table showing how much of each commodity is produced in each *socially necessary production method*.

Pareto optimum The economic state where no one can be made better off without making someone else worse off.

perfect competition A market with many producers and consumers, each rational and possessing perfect information.

per unit costs The cost of *circulating capital* and wages, per unit of a commodity.

physical lifetime The age at which a piece of *fixed capital* is scrapped because of wear and tear.

portfolio investment Investment of money in the capital market.

preference function A mathematical function specifying in ordinal terms the preferences of an individual.

price gradient The direction and rate of change of *production prices* per unit of distance on a map.

prisoner's dilemma A special case of *game theory*, which shows for a certain set of conditions the gains to cooperative behaviour and losses from selfish behaviour.

producer services Services that are intermediate inputs for producers (example: accountants).

production coefficient See *input–output coefficient*.

production factor An input to production (example: labour, capital, technology).

production function A mathematical function describing the relationship between the quantity of *production factors* used and the quantity of *commodities* produced.

production period The length of time between advancing capital to pay for production inputs and the receipt of revenues from sale of the product.

production price The expected long-run factory gate price for a commodity that reflects full capitalist competition and an equal rate of profit on all *socially necessary production methods*.

profit-maximizing location pattern The location pattern of production (an allocation of economic activities to regions) that maximizes the equalized rate of profit that would prevail under full capitalist competition.

rate of exploitation The ratio of uncompensated to compensated labour time measured in *labour values*.

rate of surplus value See *rate of exploitation*.

real wage The bundle of *wage goods* consumed by a worker and her/his family per week.

relative prices The price of one good measured relative to the price of another good.

reswitching The situation that occurs when one production method is most profitable for high and low rates of profit, whereas another production method is most profitable for intermediate profit rates.

selection (of production methods) The process whereby more profitable firms increase their market share over time compared with less profitable firms, leading to a shift in the *socially necessary production method* towards the methods favoured by the more profitable firms.

socially necessary division of labour The allocation of labour to sectors and regions that guarantees that each place produces exactly the amount of each commodity that can be sold to other producers, implying no over- or under-supply of any product.

socially necessary labour The amount of labour incorporated in producing a product using the *socially necessary production method*

socially necessary production method The production method that dominates production and price setting in a sector or region.

socially necessary real wage The average *real wage* that workers expect to be able to purchase in a region at a certain point in time.

social relations of production Form part of the *base* along with the *forces of production*. The social relations of production are the class relations within the *mode of production*; for example, the relationships between landowners and serfs, or between capitalists and workers.

spatial divisions of labour The geographical location of different segments of the labour force.

Sraffian See *neo-Ricardian*.

stable equilibrium The point to which a system will return if circumstances force it slightly away from *equilibrium*. Stable equilibrium is brought about through market forces.

subcontracting An organizational arrangement whereby 'the firm offering the subcontract requests another independent enterprise to undertake the production or carry out the processing of a material, component, part or subassembly for it according to specifications or plans provided by the firm offering the subcontract' (Holmes, 1986, p.84).

superstructure The non-economic part of society that constitutes, for example, the religious, cultural, political and legal institutions. In fundamental Marxism the form of the superstructure is determined by the economic *base*.

surplus The physical quantity of commodities produced in a year over and above the quantity needed to reproduce the same amount the following year.

surplus value The *surplus* measured, in *labour values*.

switching The change from one production method, or set of production methods, to another.

technical change Any observed change in the relative proportion of *capital goods* and labour used up in a production method.

technique of production The precise combination of capital and labour inputs that is necessary to produce a given level of output. A technique of production is represented by the *input–output equation*.

trip distribution model A mathematical model representing the geographical travel patterns of given individuals for given tasks.

transformation problem Associated with Marx, this represents his attempt to find the correct procedure for moving from exchange ratios expressed in *labour values* to exchange ratios expressed in prices.

truncation period   Used in connection with the analysis of *fixed capital*.   It represents the profit-maximizing age for scrapping an item of fixed capital. That age may well be less than the *physical lifetime* of the fixed capital.

turnover rate   The number of *production periods* per year.

unstable equilibrium   Once out of *equilibrium*, the system cannot return to it because the initial deviation is followed by ever-increasing deviations.

unequal exchange   Associated with the work of Emmanuel, this is defined as a situation in which the exchange ratio for two commodities that are traded for one another measured in prices is different from the exchange ratio measured in labour values.

use value   The usefulness of a commodity to a potential purchaser.

utility   The mental satisfaction derived from undertaking a particular act.

value rate of profit   The ratio of surplus to capital advanced, measured in *labour values*.

variable capital   The *labour value* of the wages consumed by labour in the production of a commodity; the *labour value* of *labour power*.

vertical integration/disintegration   Vertically integrated production units are those where several stages in a production process are carried out by the same firm. Single vertically integrated production units tend to be large and capital intensive. Vertically disintegrated production units are those where only one or a few tasks of the production process are carried out. Such units tend to be small and labour intensive.

wage goods   Commodities consumed by workers as a part of their *real wage*.

wage–profit frontier   The trade-off between the rate of profit that capitalists earn and the level of wages that are paid to workers.

Weberian class categories   Classes defined with respect to market rather than production relations.

withdrawal rules   Associated with the work of John Roemer, these represent the *counter-factual* situation where individuals take their share of society's resources and decide whether it is to their advantage either to remain in or opt out of the economy.

zero sum game   A game in which the pay-off is fixed and does not change. As a result, whatever one player wins, another must lose.

Bibliography

Abraham-Frois, G. and Berrebi, E. (1979) *Theory of values, prices and accumulation: A mathematical integration of Marx, von Neumann and Sraffa.* Cambridge: Cambridge University Press.

Aglietta, M. (1979) *A theory of capitalist regulation – the US experience.* London: New Left Books.

Agnew, J. (1981) 'Homeownership and the capitalist social order', in M. Dear and A.J. Scott (eds), *Urbanization and urban planning in capitalist society.* London: Methuen, pp. 457–80.

Alonso, W. (1964) *Location and land use.* Cambridge, MA: Harvard University Press.

Amin, S. (1974) *Accumulation on a world scale.* New York: Monthly Review Press.

Amin, A. and Smith, I. (1986) 'The internationalization of production and its implications for the UK', in A. Amin and J. Goddard (eds), *Technological change, industrial restructuring and regional development.* London: Allen & Unwin, pp. 41–76.

Anderson, W.H.L. and Thompson, F.W. (1988) 'Neoclassical Marxism', *Science and Society*, 52, 215–28.

Arnott, R. (1980) 'A simple urban growth model with durable housing', *Regional Science and Urban Economics*, 10, 53–76.

Badcock, B. (1989) 'An Australian view of the rent gap hypothesis', *Annals, Association of American Geographers*, 79, 125–45.

Baldone, S. (1980) 'Fixed capital in Sraffa's theoretical scheme', in L. L. Pasinetti (ed.), *Essays on the theory of joint production.* London: Macmillan, pp. 88–137.

Ball, M. (1977) 'Differential rent and the role of landed property', *International Journal of Urban and Regional Research*, 1, 380–403.

Ball, M. (1980) 'On Marx's theory of agricultural rent: A reply to Ben Fine', *Economy and Society*, 9, 304–26.

Ball, M. (1985) 'The urban rent question', *Environment and Planning A*, 17, 503–25.

Bandyopadhyay, P. (1981) 'Critique of Wright: 2. in defence of a post-Sraffian approach', in I. Steedman *et al.* (eds), *The value controversy.* London: Verso Editions, pp. 100–29.

Baran, P. and Sweezy, P. (1966) *Monopoly Capital.* New York: Monthly Review Press.

Barbalet, J. M. (1987) 'The "labor aristocracy" in context', *Science and Society*, 51, 133–53.

Barnes, T.J. (1983) 'The geography of value, production, and distribution: Theoretical economic geography after Sraffa', unpublished PhD dissertation, Department of Geography, University of Minnesota.

Barnes, T.J. (1984) 'Theories of agricultural rent within the surplus approach', *International Review of Regional Science*, 9, 125–40.

Barnes, T.J. (1985) 'Theories of interregional trade and theories of value', *Environment and Planning A*, 17, 729–46.

Barnes, T.J. (1988) 'Scarcity and agricultural land rent in light of the capital controversy: Three views', *Antipode*, 20, 207–38.

Barnes, T.J. (1989) 'Place, space and theories of economic value: Contextualism and essentialism in economic geography', *Transactions, Institute of P ish Geographers*, 14, 299–316.

Barnes, T. and Sheppard, E. (1984) 'Technical choice and reswitching in space economies', *Regional Science and Urban Economics*, 14, 345–62.

Bhaduri, A. (1969) 'On the significance of recent controversies on capital theory: A Marxian view', *Economic Journal*, 79, 532–9.

Blaug, M. (1974) *The Cambridge revolution: Success or failure?* London: Institute of Economic Affairs.

Bleaney, M. (1976) *Underconsumption theories*. London: Lawrence & Wishart.

Booth, D. (1978) 'Collective action, Marx's class theory and the union movement', *Journal of Economic Issues*, 12, 163–85.

Borts, G. H. and Stein, J. L. (1964) *Economic growth in a free market*. New York: Columbia University Press.

Bowles, S. and Gintis, H. (1977) 'The Marxian theory of value and heterogeneous labour: A critique and reformulation', *Cambridge Journal of Economics*, 1, 173–92.

Bowman, J. (1982) 'The logic of collective action', *Social Science Information*, 21, 571–604.

Braudel, F. (1982) *The wheels of commerce*. New York: Harper & Row.

Braverman, H. (1974) *Labor and monopoly capitalism*. New York: Monthly Review Press.

Brueckner, J. K. (1980a) 'A vintage model of urban growth', *Journal of Urban Economics*, 8, 389–402.

Brueckner, J. K. (1980b) 'Residential succession and land use dynamics in a vintage model of urban housing', *Regional Science and Urban Economics*, 10, 225–40.

Brueckner, J. K. (1982) 'Building ages and urban growth', *Regional Science and Urban Economics*, 12, 197–210.

Burawoy, M. (1979) *Manufacturing consent*. Chicago: University of Chicago Press.

Burawoy, M. (1985) *The politics of production*. London: Verso.

Burgess, R. (1976) 'Marxism in geography'. Occasional Papers, Department of Geography, University College London.

Burgess, R. (1985) 'The concept of nature in geography and Marxism', *Antipode*, 17, 68–78.

Burmeister, E. (1980) *Capital theory and dynamics*. Cambridge: Cambridge University Press.

Calhoun, C. (1987) 'Class, space and industrial revolution', in N. J. Thrift and P. Williams (eds), *Class and space: The making of urban society*. London: Routledge & Kegan Paul, pp. 51–72.

Carling, A. (1986) 'Rational choice marxism', *New Left Review*, 160, 24–62.

Carling, A. (1987) 'Exploitation, extortion and oppression', *Political Studies*, 35, 173–88.

Chandler, A. (1962) *Strategy and structure*. Cambridge, MA: MIT Press.

Chandler, A. (1977) *The visible hand*. Cambridge, MA: Harvard University Press.

Clark, G. L. (1986) 'The crisis of the mid-west auto industry', in A. J. Scott and M. Storper (eds), *Production, work, territory: The geographical anatomy of industrial capitalism*. Boston: Allen & Unwin, pp. 127–48.

Clark, G. L. (1988) 'Location, management strategy and workers' pensions', Pittsburgh: Carnegie Mellon University, School of Urban and Public Affairs, WP 88–46.

Clark, G., Gertler, M. and Whiteman, J. (1986) *Regional dynamics: Studies in adjustment theory*. Boston: Allen & Unwin.

Clark, J. B. (1891) 'Distribution as determined by a law of rent', *Quarterly Journal of Economics*, 5, 289–318.

Clarke, I. (1985) *The spatial organization of multinational corporations*. New York: St Martin's Press.

Cohen, G. A. (1978) *Karl Marx's theory of history: A defence*. Oxford: Oxford University Press.

Cohen, G. A. (1981) 'The labour theory of value and the concept of exploitation', in I. Steedman *et al.* (eds), *The value controversy*. London: Verso Editions, pp. 202–23.

Cox, K. R. and Mair, A. (1988) 'Locality and community in the politics of local economic development', *Annals, Association of American Geographers*, 78, 307–25.

D'Agata, A. (1983) 'The existence and unicity of cost-minimizing systems in intensive rent theory', *Metroeconomica*, 35, 147–58.

D'Agata, A. (1986a) 'Non-produced means of production in Sraffa's system: Basics, non-basics and quasi-basics. A comment', *Cambridge Journal of Economics*, 10, 379–86.

D'Agata, A. (1986b) 'Non-produced means of production: Neo-Ricardians vs. Fundamentalists, a comment', *Review of Radical Political Economics*, 18, 93–9.

David, P. A. (1975) *Technical change, innovation and economic growth*. Cambridge: Cambridge University Press.

Desai, M. (1979) *Marxian economics*. Totowa, NJ: Littlefield & Adams.

de Vroey, M. (1981) 'Value, production and exchange', in I. Steedman *et al.* (eds), *The value controversy*. London: Verso Editions, pp. 130–62.

Dobb, M. (1970) 'The Sraffa system and critique of the neo-classical theory of distribution', *De Economist*, 4, 347–62.

Dobb, M. (1973) *Theories of value and distribution since Adam Smith*. Cambridge: Cambridge University Press.

Dobb, M. (1975–6) 'A note on the Ricardo–Marx–Sraffa discussion', *Science and Society*, 39, 468–70.

Duménil, G. and Lévy, D. (1987) 'The dynamics of competition: a restoration of the classical hypothesis', *Cambridge Journal of Economics*, 11, 133–64.

Dutt, A. (1987) 'Competition, monopoly power and the uniform rate of profit', *Review of Radical Political Economics*, 19(4), 55–72.

Eatwell, J. (1975) 'Mr. Sraffa's standard commodity and the rate of exploitation', *Quarterly Journal of Economics*, 89, 543–55.

Edel, M. (1976) 'Marx's theory of rent: Urban applications', *Kapitalstate*, 4(5), 100–24.

Edel, M. (1979) 'A note on collective action, Marxism, and the prisoner's dilemma', *Journal of Economic Issues*, 13, 751–61.

Edwards, R. (1979) *Contested terrain*. New York: Basic Books.

Eichner, A. S. (1976) *The megacorp and oligopoly*. Cambridge: Cambridge University Press.

Elson, D. (1979) 'The value theory of labour', in D. Elson (ed.), *The representation of labour in capitalism*. London: CSE Books, pp. 115–80.

Elster, J. (1985) *Making sense of Marx*. Cambridge: Cambridge University Press.

Emmanuel, A. (1972) *Unequal Exchange*. New York: Monthly Review Press.

Evans, A. (1975) 'Rent and housing in the theory of urban growth', *Journal of Regional Science*, 15, 113–25.

Farjoun, E. (1984) 'The production of commodities by means of what?' in E. Mandel and A. Freeman (eds), *Ricardo, Marx, Sraffa*. London: Verso Editions, pp. 11–41.

Farjoun, E. and Machover, M. (1983) *Laws of chaos: A probabilistic approach to political economy*. London: Verso Editions.

Feldman, M. (1977) 'A contribution to the critique of urban political economy: The journey to work', *Antipode*, 9(1), 30–49.

Ferrão, J. (1985) 'Regional variations in the rate of profit in Portuguese industry', in R. Hudson and J. Lewis (eds), *Uneven development in southern Europe: Studies of accumulation, class, migration and the state*. London: Methuen, pp. 211–45.

Fine, B. (1979) 'On Marx's theory of agricultural rent', *Economy and Society*, 8, 241–78.

Fine, B. (1986) *The value dimension*. London: Routledge & Kegan Paul.

Fine, B. and Harris L. (1979) *Rereading Capital*. New York: Columbia University Press.

Flaschel, P. and Semmler, W. (1986) 'The dynamic equalization of profit rates for input–output models with fixed capital', in W. Semmler (ed.), *Competition, instability and nonlinear cycles*. New York: Springer-Verlag, pp. 1–34.

Foot, D. (1981) *Operational urban models*. London: Methuen.

Foot, S. P. H. and Webber, M. J. (1983) 'Unequal exchange and uneven development', *Environment and Planning D: Society and Space*, 1, 281–304.

Foster, J. (1974) *Class struggle and the industrial revolution*. London: Methuen.

Franke, R. (1988) 'Integrating the financing of production and a rate of interest into production price models', *Cambridge Journal of Economics*, 12, 257–71.

Freeman, A. (1984) 'The logic of the transformation problem', in E. Mandel and A. Freeman (eds), *Ricardo, Marx, Sraffa*. London: Verso Editions, 221–64.

Friedman, A. (1977) *Industry and labour*. London: Macmillan.

Fröbel, F. Heinrichs, J. and Kreye, O. (1980) *The new international division of labour: Structural unemployment in industrialised countries and industrialisation in developing countries*. Cambridge: Cambridge University Press.

Garegnani, P. (1966) 'Switching of techniques', *Quarterly Journal of Economics*, 80, 554–67.

Garegnani, P. (1970) 'Heterogeneous capital, the production function and the theory of distribution', *Review of Economic Studies*, 37, 407–36.

Garin, R. A. (1966) 'A matrix formulation of the Lowry model for intrametropolitan activity allocation', *Journal of the American Institute of Planners*, 32, 361–4.

Gertler, M. (1984) 'Regional capital theory', *Progress in Human Geography*, 8, 50–81.

Gibson, B. (1980) 'Unequal exchange: Theoretical issues and empirical findings', *Review of Radical Political Economics*, 12(3), 15–35.

Gibson, B. and Esfahani, H. (1983) 'Non-produced means of production: Neo-Ricardians vs. Fundamentalists', *Review of Radical Political Economics*, 15, 83–105.

Gibson, B. and MacLeod, D. (1983) 'Non-produced means of production in Sraffa's systems: Basics, non-basics and quasi-basics', *Cambridge Journal of Economics*, 7, 141–50.

Gibson, K., Graham, J., Horvath, R. and Shakow, D. (1986) 'Toward a Marxist empirics', manuscript.

Goodwin, R. (1976) 'Use of normalized general coordinates in linear value and distribution theory', in K. Polenski and J. Skolka (eds), *Advances in input–output analysis*. Cambridge, MA: Ballinger, pp. 581–602.

Goodwin, R. (1987) 'Macrodynamics', in R. Goodwin and L. Puzo, *The dynamics of a capitalist economy*. Boulder, CO: Westview Press, pp. 3–162.

Gudeman, S. (1986) *Economics as culture*. London: Routledge & Kegan Paul.

Hadjimichalis, C. (1987) *Uneven development and regionalism: State, territory and class in southern Europe*. London: Croom Helm.

Hahn, F. (1982) 'The neoRicardians', *Cambridge Journal of Economics*, 6, 353–74.

Harcourt, G. C. (1972) *Some Cambridge controversies in the theory of capital*. Cambridge: Cambridge University Press.

Harcourt, G. C. (1975) 'The Cambridge controversies: The afterglow', in M. Parkin and A. Norbay (eds), *Contemporary issues in economics*. Manchester: Manchester University Press, pp. 305–36.

Harcourt, G. C. (1976) 'The Cambridge controversies: Old ways and new horizons – or dead end?' *Oxford Economic Papers*, 28, 25–65.

Harcourt, G. C. (1982) 'The Sraffian contribution: An evaluation', in I. Bradley and M. Howard (eds), *Classical and Marxian political economy*. London: Macmillan, pp. 255–75.

Harcourt, G. C. (1986) 'Some Cambridge controversies in the theory of capital', in O. Hamounda (ed.), *Controversies in political economy: Selected essays by G. C. Harcourt*. New York: New York University Press, pp. 145–206.

Hard, G. (1988) *Die Störche und die Kinder, die Orchideen und die Sonne*. Berlin: Walter van Gruyter.

Harris, D. J. (1977) *Capital accumulation and income distribution*. Stanford, CA: Stanford University Press.

Harris, D. J. (1980) 'A postmortem on the neoclassical "parable"', in E. J. Nell (ed.), *Growth, profits and property.* Cambridge: Cambridge University Press, pp. 43–63.

Harris, R. (1983) 'Space and class: A critique of Urry', *International Journal of Urban and Regional Research,* 7, 115–21.

Harris, R. (1984) 'Residential segregation in the capitalist city: A review and directions for research', *Progress in Human Geography,* 8, 26–49.

Harris, R. and Pratt, G. (1987) 'Housing tenure and social class: Introduction', in R. Harris and G. Pratt (eds), *Housing tenure and social class.* Research report SB:10, The National Swedish Institute for Building Research, pp. 9–26.

Hartwick, J. (1976) 'Intermediate goods and the spatial differentiation of land uses', *Regional Science and Urban Economics,* 6, 127–45.

Harvey, D. (1972) *Society, the city and the space economy of urbanism.* AAG Resource Paper. Washington DC: Association of American Geographers.

Harvey, D. (1973) *Social justice and the city.* London: Edward Arnold.

Harvey, D. (1974) 'Class monopoly rent, finance capital, and the urban revolution', *Regional Studies,* 8, 239–55.

Harvey, D. (1982) *The limits to capital.* Chicago: University of Chicago Press.

Harvey, D. (1985a) *Consciousness and the urban experience.* Oxford: Basil Blackwell.

Harvey, D. (1985b) *The urbanization of capital.* Oxford: Basil Blackwell.

Harvey, D. (1985c) 'The geopolitics of capitalism', in D. Gregory and J. Urry (eds), *Social relations and spatial structures.* London: Macmillan, pp. 128–63.

Harvey, D. (1987) 'Flexible accumulation through urbanization: Reflections on "post-modernism" in the American city', *Antipode,* 19, 260–86.

Hausman, D. (1981) *Capital, profits and prices.* New York: Columbia University Press.

Himmelweit, S. (1984) 'Value relations and divisions with the working class', *Science and Society,* 48, 323–43.

Himmelweit, S. and Mohun, S. (1981) 'Real abstractions and anomalous assumptions', in I. Steedman *et al.* (eds), *The value controversy.* London: Verso Editions, pp. 224–65.

Hodgson, G. (1982) *Capitalism, value and exploitation.* Oxford: Oxford University Press.

Holland, S. (1976) *Capital versus the regions.* London: Macmillan.

Holmes, J. (1986) 'The organization and locational structure of production subcontracting', in A. J. Scott and M. Storper (eds), *Production, work, territory: The geographical anatomy of industrial capitalism.* Boston: Allen & Unwin, pp. 80–105.

Hudson, R. and D. Sadler (1986) 'Contesting work closures in Western Europe's old industrial regions: Defending place or betraying class', in A. J. Scott and M. Storper (eds), *Production, work, territory: The geographical anatomy of industrial capitalism.* Boston: Allen & Unwin, pp. 172–93.

Huriot, J. M. (1983) 'Rentes différentielles et rente absolue: Un réexamen', Working paper number 62, Institut de Mathématiques Économiques, University of Dijon.

Huriot, J. M. (1984) 'Size of plants and distance to city center: A new theoretical approach', Paper presented to the Polish–French meeting, Poznan, Poland.

Huriot, J. M. (1985) 'Land rent, production and location', Paper presented at the annual meeting of the Canadian Association of Geographers, Trois Rivières, Canada.

Hymer, S. (1972) 'The multinational corporation and the law of uneven development', in J. Bhagwati (ed.), *Economics and world order: From the 1970's to the 1990's.* New York: The Free Press, pp. 113–40.

Jenkins, R. (1984) *Transnational corporations and industrial transformation in Latin America.* London: St Martin's Press.

Jenkins, R. (1987) *Transnational corporations and uneven development.* London: Methuen.

Jevons, W. (1970) *The theory of political economy.* Harmondsworth: Penguin.

Johnson, H. (1974) 'The current and prospective state of economics', *Australian Economic Papers,* 13, 1–27.

Kaldor, N. (1934) 'Mrs Robinson's Economics of imperfect competition', *Economica*, N.S. 1, 335–41.

Kalecki, M. (1938) 'The determinants of distribution of national income', *Econometrica*, 6, 97–112.

Kalecki, M. (1939–40) 'The supply curve of an industry under imperfect competition', *Review of Economic Studies*, 7, 91–112.

Kalecki, M. (1943) *Studies in economic dynamics*. London: Allen & Unwin.

Kalecki, M. (1954) *Theory of economic dynamics*. London: Allen & Unwin.

Kaplinsky, R. (1988) 'Restructuring the capitalist labour process: some lessons from the car industry', *Cambridge Journal of Economics*, 12, 451–70.

Katz, S. (1986) 'Towards a sociological definition of rent: Notes on David Harvey's *Limits to Capital*', *Antipode*, 18, 64–78.

Keat, R. and Urry, J. (1982) *Social theory as science*, 2nd edn. London: Routledge & Kegan Paul.

Krause, U. (1981) 'Heterogeneous labour and the fundamental Marxian theorem', *Review of Economic Studies*, 48, 173–8.

Kreisler, P. (1987) *Kalecki's microanalysis*. Cambridge: Cambridge University Press.

Kurz, H. D. (1978) 'Rent theory in a multisectoral model', *Oxford Economic Papers*, 32, 16–37.

Kurz, H. D. (1979) 'Sraffa after Marx', *Australian Economic Papers*, 18, 52–70.

Lash, S. and Urry J. (1984) 'The new Marxism of collective action: A critical analysis', *Sociology*, 18, 33–50.

Lauria, M. (1982) 'Selective urban redevelopment: A political economic perspective', *Urban Geography*, 3, 224–39.

Lauria, M. (1985) 'Implications of Marxian rent theory for community controlled redevelopment strategies', *Journal of Planning, Education and Research*, 4, 16–24.

Lebowitz, M. A. (1988) 'Is "analytical Marxism" Marxism?' *Science and Society*, 52, 191–214.

Leitner, H. (1990) 'Cities in pursuit of economic growth: The local state as entrepreneur', *Political Geography Quarterly*, 9, 146–70.

Leitner, H. and Sheppard, E. (1989) 'The city as locus of production: The changing geography of commodity production within the capitalist metropolis', in R. Peet and N. J. Thrift (eds), *The new models in geography*, Vol. 2. Boston: Unwin Hyman, pp. 55–83.

Lenin, V. (1947[1916]) *Imperialism: The highest stage of capitalism*. Moscow: Foreign Languages Publishing House.

Lenin, V. (1947) *The essentials of Lenin*. London: Lawrence & Wishart.

Leonardi, G. (1982) 'The structure of random utility models in the light of the asymptotic theory of extremes', Laxenburg, Austria: International Institute of Applied Systems Analysis, WP–82–91.

Ley, D. F. (1987) 'Reply: The rent gap revisited', *Annals, Association of American Geographers*, 77, 465–8.

Lianos, T. (1987) 'Marx on the rate of interest', *Review of Radical Political Economics*, 19(3), 34–55.

Liossatos, P. (1980) 'Unequal exchange and regional disparities', *Papers of the Regional Science Association*, 45, 87–103.

Liossatos, P. (1983) 'Commodity production and interregional transfers of value', in F. Moulaert and P. Salinas (eds), *Regional analysis and the new international division of labor*. Hingham, MA: Kluwer-Nijhoff, pp. 57–76.

Liossatos, P. (1988) 'Value and competition in a spatial context: A Marxian model', *Papers of the Regional Science Association*, 64, 1–10.

Lipietz, A. (1982) 'The so-called "transformation problem" revisited', *Journal of Economic Theory*, 2, 59–88.

Lipietz, A. (1986) 'New tendencies in the international division of labour: regimes of accumulation and modes of regulation', in A. J. Scott and M. Storper (eds), *Production, work, territory: The geographical anatomy of industrial capitalism*. London: Allen & Unwin, pp. 16–40.

Lipietz, A. (1987) *Mirages and miracles: The crises of global fordism*. London: Verso.

Lösch, A. (1954) *The economics of location*. New Haven, CT: Yale University Press.

Lovering, J. (1978) 'The theory of the "internal colony" and the political economy of Wales', *Review of Radical Political Economy*, 10(3), 55–67

Mainwaring, L. (1974) 'A neo-Ricardian analysis of international trade', *Kyklos*, 27, 537–53.

Mainwaring, L. (1984) *Value and distribution in capitalist economies*. Cambridge: Cambridge University Press.

Manara, C. F. (1980) 'Sraffa's model for the joint production of commodities by means of commodities', in L. L. Pasinetti (ed.), *Essays on the theory of joint production*. London: Macmillan, pp. 1–15.

Mandel, E. (1984) 'Introduction', in E. Mandel and A. Freeman (eds), *Ricardo, Marx, Sraffa*. London: Verso Editions, pp. ix–xvi.

Mandel, E. and Freeman, A. (1984) *Ricardo, Marx, Sraffa*. London: Verso Editions.

Marelli, E. (1983) 'Empirical estimation of intersectoral and interregional transfers of surplus value: The case of Italy', *Journal of Regional Science*, 23, 49–70.

Markusen, A. R. (1978) 'Class, rent and sectoral conflict: Uneven development in Western U.S. boomtowns', *Review of Radical Political Economics*, 10, 117–29.

Marx, K. (1867[1972]) *Capital*, vol. 1. Harmondsworth: Penguin.

Marx, K. (1885[1972]) *Capital*, vol. 2. Harmondsworth: Penguin.

Marx, K. (1896[1972]) *Capital*, vol. 3. Harmondsworth: Penguin.

Marx, K. (1959) *Theories of surplus value, part II*. London: Lawrence & Wishart.

Marzi, G. and Varri, P. (1977) *Variazioni de proddutivita nell' economia Italiana: 1959–1967*. Bologna: Mulino.

Massey, D. B. (1973) 'A critique of industrial location theory', *Antipode*, 5(3), 33–9.

Massey, D. B. (1984a) *Spatial divisions of labour: Social structures and the geography of production*. London: Macmillan.

Massey, D. B. (ed.) (1984b) *Geography matters!* Cambridge: Cambridge University Press.

Massey, D. B. and Meegan, R. A. (1982) *The anatomy of job loss. The how, where and why of employment decline*. London: Methuen.

Matthaei, J. (1984) 'Rethinking scarcity: Neoclassicism, neoMalthusianism, and neo-Marxism', *Review of Radical Political Economics*, 16, 81–94.

Metcalfe, J. S. and Gibbons, M. (1986) 'Technological variety and the process of competition', *Économie Appliquée*, 39, 493–520.

Metcalfe, J. S. and Steedman, I. (1979a) 'Heterogeneous capital and the Hecksher–Ohlin–Samuelson theory of trade', in I. Steedman (ed.), *Fundamental issues in trade theory*. New York: St Martin's Press, pp. 64–76.

Metcalfe, J. S. and Steedman, I. (1979b) 'Reswitching and primary input use', in I. Steedman (ed.), *Fundamental issues in trade theory*. New York: St Martin's Press, pp. 15–37.

Montani, G. (1975) 'Scarce natural resources and income distribution', *Metroeconomica*, 27, 68–101.

Morishima, M. (1973) *Marx's economics: A dual theory of value and growth*. Cambridge: Cambridge University Press.

Morishima, M. and Catephores, G. (1978) *Value, exploitation and growth*. London: McGraw Hill.

Morishima, M. and Seton, F. (1961) 'Aggregation in Leontief matrices and the labour theory of value', *Econometrica*, 29, 203–20.

Morris, J., Davies, A. and Thompson, A. (1988) *Labour market responses to industrial restructuring and technical change*. Brighton: Wheatsheaf.
Moss, S. (1980) 'The end of orthodox capital theory', in E. J. Nell (ed.), Growth, profits and property. Cambridge: Cambridge University Press, pp. 63–79.
Muth, R. (1973) 'A vintage model of the housing stock', *Papers of the Regional Science Association*, 30, 141–56.

Napoleoni, C. (1978) 'Sraffa's tabula rasa', *New Left Review*, 112, 75–7.
Nell, E. (1972) 'The revival of political economy', *Australian Economic Papers*, 11, 19–31.
Nikaido, H. (1968) *Convex structures and economic theory*. New York: Academic Press.
Nikaido, H. (1983) 'Marx on competition', *Zeitschrift für Nationalökonomie*, 43, 83–102.

Ochoa, E. (1984) 'Labor values and prices of production: An interindustry study of the U.S. economy, 1947–1972', PhD dissertation, New School for Social Research, New York.
O'Connor, J. (1984) *Accumulation crisis*. Oxford: Basil Blackwell.
Offe, C. and Wiesenthal, H. (1980) 'Two logics of collective action: Theoretical notes on social class and organizational form', in M. Zeitlin (ed.), *Political power and social theory*. Greenwich, CT: JAI Press, pp. 67–115.
Ohlin, B. (1933) *Interregional and international trade*. Cambridge, MA: Harvard University Press.
Okishio, N. (1961) 'Technical changes and the rate of profit', *Kobe University Economic Review*, 7, 85–99.
Okishio, N. (1963) 'A mathematical note on Marxian theorems', *Weltwirtschaftliches Archiv*, 91, 287–99.
Olson, M. (1965) *The logic of collective action: Public goods and the theory of groups*. Cambridge, MA: Harvard University Press.

Parys, W. (1982) 'The deviation of prices from labor values', *American Economic Review*, 72, 1208–12.
Pasinetti, L. L. (1962) 'Rate of profit and income distribution in relation to the rate of economic growth', *Review of Economic Studies*, 29, 267–79.
Pasinetti, L. L. (1966) 'Changes in the rate of profit and switches of techniques', *Quarterly Journal of Economics*, 80, 503–17.
Pasinetti, L. L. (1977) *Lectures in the theory of production*. London: Macmillan.
Pasinetti, L. L. (ed.) (1980) *Essays on the theory of joint production*. London: Macmillan.
Pasinetti, L. L. (1981) *Structural change and economic growth*. Cambridge: Cambridge University Press.
Pasinetti, L. L. (1988) 'Growing sub-systems, vertically hyper-integrated sectors and the labour theory of value', *Cambridge Journal of Economics*, 12, 125–34.
Pavlik, C. (1990) 'Technical reswitching: A spatial case', *Environment and Planning A* (forthcoming).
Peet, R. (1981) 'Spatial dialectics and Marxist geography', *Progress in Human Geography*, 5, 105–10.
Peet, R. (1987) 'The geography of class struggle in the relocation of United States manufacturing industry', in R. Peet (ed.), *International capitalism and international restructuring*. Winchester, MA: Allen & Unwin, pp. 40–71.
Perelman, M. (1979) 'Marx, Malthus and the concept of natural resource scarcity', *Antipode*, 11 (2), 80–90.
Polanyi, K., Arensberg, C. and Pearson, H. (1957) *Trade and market in the early empires*. Illinois: University of Illinois Press.
Pratt, G. (1986) 'Housing tenure and social cleavages in urban Canada', *Annals, Association of American Geographers*, 76, 366–80.

Przeworski, A. (1985a) *Capitalism and social democracy*. Cambridge: Cambridge University Press.

Przeworski, A. (1985b) 'Marxism and rational choice', *Politics and Society*, 14, 379–409.

Puchinger, K. (1979) 'Der Einfluss regionaler Warenumlaufzeitdifferenzen auf die Standortwahl von Industrie und Gewerbe', *Wiener Beiträge zur Regionalwissenschaft*, 2. Vienna: Institut für Stadt und Regionalforschung der technischen Universität Wien.

Quadrio-Curzio, A. (1980) 'Rent, income distribution and orders of efficiency and rentability', in L. L. Pasinetti (ed.), *Essays on the theory of joint production*. London: Macmillan, pp. 218–39.

Resnick, S. A. and Wolff, R. D. (1987) *Knowledge and class: A Marxian critique of political economy*. Chicago: University of Chicago Press.

Review of Radical Political Economics (1986) Special issue on 'Empirical work in Marxian crisis theory', vol. 18, nos 1&2.

Richardson, H. W. (1973) *Regional growth theory*. London: Macmillan.

Richardson, H. W. (1977) *The new urban economics*. London: Pion.

Rigby, D. L. (1990) 'Technical change and the rate of profit. An obituary for Okishio's theorem', *Environment and Planning A* (forthcoming).

Rigby, D. L., Sheppard, E. and Webber, M. J. (1989) 'Technical change', manuscript.

Robinson, J. V. (1953–4) 'The production function and the theory of capital', *Review of Economic Studies*, 21, 81–106.

Robinson, J. V. (1964) *Economic philosophy*. Harmondsworth: Penguin.

Robinson, J. V. (1965) *Collected economic papers, volume III*. Oxford: Basil Blackwell.

Roemer, J. (1978) 'Neoclassicism, Marxism, and collective action', *Journal of Economic Issues*, 12, 147–61.

Roemer, J. (1979) 'Continuing controversy on the falling rate of profit: fixed capital and other issues', *Cambridge Journal of Economics*, 3, 379–98.

Roemer, J. (1981) *Analytical foundations of Marxian economic theory*. Cambridge: Cambridge University Press.

Roemer, J. (1982a) *A general theory of exploitation and class*. Cambridge, MA: Harvard University Press.

Roemer, J. (1982b) 'New directions in the Marxian theory of exploitation and class', *Politics and Society*, 11, 253–87.

Roemer, J. (ed.) (1986) *Analytical Marxism*. Cambridge, MA: Harvard University Press.

Roemer, J. (1988) *Free to lose*. Cambridge, MA: Harvard University Press.

Roncaglia, A. (1978) *Sraffa and the theory of prices*. London: John Wiley.

Rowthorn, R. (1974) 'Neoclassicism, neo-Ricardianism and Marxism', *New Left Review*, 86, 63–87.

Rubin, I. I. (1972) *Essays on Marx's theory of value*, translated by M. Samardzija and F. Perlman. Detroit: Black and Red.

Ruccio, D. (1988) 'The merchant of Venice or Marxism in the mathematical mode', *Rethinking Marxism*, 1, 36–68.

Salvadori, N. (1983) 'On a new variety of rent', *Metroeconomica*, 35, 73–85.

Samuelson, P. (1962) 'Parable and realism in capital theory: The surrogate production function', *Review of Economic Studies*, 19, 193–206.

Samuelson, P. A. (1966) 'A summing up', *Quarterly Journal of Economics*, 80, 568–83.

Samuelson, P. A. (1971) 'Understanding the Marxian notion of exploitation: A summary of the so-called transformation problem between Marxian values and competitive prices', *Journal of Economic Literature*, 9, 399–431.

Saunders, P. (1978) 'Domestic property and social class', *International Journal of Urban and Regional Research*, 2, 233–51.

Saunders, P. (1979) *Urban politics: A sociological approach*. London: Hutchinson.

Sayer, A. (1976) 'A critique of urban modelling', *Progress in Planning*, 6, 167–254.

Sayer, A. (1984) *Explanation in social science: A realist approach*. London: Hutchinson.

Sayer, A. (1985) 'The difference that space makes', in D. Gregory and J. Urry (eds), *Social relations and spatial structures*. London: Methuen, pp. 49–66.

Sayer, A. (1986) 'New developments in manufacturing: The just-in-time system', *Capital and class*, 30, 43–72.

Schefold, B. (1980) 'Fixed capital as a joint product and the analysis of accumulation with different forms of technical progress', in L. L. Pasinetti (ed.), *Essays on the theory of joint production*. London: Macmillan, pp. 138–217.

Schutz, E. (1987) 'Non-produced inputs, differential profit rates and the Okishio theorem', *Review of Radical Political Economics*, 19(2), 43–60.

Schweizer, U. (1978) 'A spatial version of the non-substitution problem', *Journal of Economic Theory*, 19, 307–20.

Schweizer, U. and Varaiya, P. (1976) 'The spatial structure of production with a Leontief technology', *Regional Science and Urban Economics*, 6, 231–52.

Schweizer, U. and Varaiya, P. (1977) 'The spatial structure of production with a Leontief technology II: substitute techniques', *Regional Science and Urban Economics*, 7, 293–320.

Scott, A. J. (1976) 'Land and land rent: An interpretive review of the French literature', *Progress in Human Geography*, 9, 101–46.

Scott, A. J. (1979) 'Commodity production and the dynamics of land-use differentiation', *Urban Studies*, 16, 95–104.

Scott, A. J. (1980) *The urban land nexus and the state*. London: Pion.

Scott, A. J. (1982) 'Production system dynamics and metropolitan development', *Annals, Association of American Geographers*, 72, 185–200.

Scott, A. J. (1988a) *Metropolis: from the division of labor to urban form*. Berkeley, CA: University of California Press.

Scott, A. J. (1988b) *New industrial spaces*. London: Pion.

Scott, A. J. (1988c) 'From Fordism to flexible accumulation', *International Journal of Urban and Regional Research*, 12, 171–86.

Scott, A. J. and Storper, M. (eds) (1986) *Production, work, territory: The geographical anatomy of industrial capitalism*. Boston: Allen & Unwin.

Semmler, W. (1984) *Competition, monopoly and differential profit rates*. New York: Columbia University Press.

Seton, F. (1957) 'The "transformation problem"', *Review of Economic Studies*, 24, 149–60.

Shaikh, A. (1977) 'Marx's theory of value and the "transformation problem"', in J. Schwartz (ed.), *The subtle anatomy of capitalism*. Santa Monica, CA: Goodyear Publishing, pp. 106–39.

Shaikh, A. (1980) 'Marxian competition versus perfect competition: Further comment on the so-called choice of technique', *Cambridge Journal of Economics*, 4, 75–83.

Shaikh, A. (1981) 'The poverty of algebra', in I. Steedman *et al.* (eds), *The value controversy*. London: Verso Editions, pp. 266–300.

Shaikh, A. (1984) 'The transformation from Marx to Sraffa', in E. Mandel and A. Freeman (eds), *Ricardo, Marx, Sraffa*. London: Verso Editions, pp. 43–84.

Sheppard, E. (1976) 'Entropy, theory construction and spatial analysis', *Environment and Planning A*, 8, 741–52.

Sheppard, E. (1978) 'Spatial interaction and geographical theory', in G. Olsson and S. Gale (eds), *Philosophy in Geography*. Dordrecht: D. Reidel, pp. 361–78.

Sheppard, E. (1979) 'Geographical potentials', *Annals, Association of American Geographers*, 69, 438–47.

Sheppard, E. (1983a) 'Commodity trade, corporate ownership and urban growth', *Papers of the Regional Science Association*, 52, 175–86.

Sheppard, E. (1983b) 'Pasinetti, Marx and urban accumulation dynamics', in D. Griffith and A. Lea (eds), *Evolving geographical structures*. The Hague: Martinus Nijhoff, pp. 293–322.

Sheppard, E. (1984) 'Value and exploitation in a capitalist space economy', *International Regional Science Review*, 9, 97–108.

Sheppard, E. (1987) 'A Marxian model of the geography of production and transportation in urban and regional systems', in C. Bertuglia, G. Leonardi, S. Occelli, G. Rabino, R. Tadei and A. Wilson (eds), *Urban systems: contemporary approaches to modelling*. London: Croom Helm, pp. 189–250.

Sheppard, E. (1989) '"Technical change" in a Marxian space economy', manuscript.

Sheppard, E. (1990) 'Transportation in a capitalist space economy: Transportation demand, circulation time and transportation innovations', *Environment and Planning A* (forthcoming).

Sheppard, E. and Barnes, T.J. (1986) 'Instabilities in the geography of capitalist production: Collective vs. individual profit maximization', *Annals, Association of American Geographers*, 76, 493–507.

Sheppard, E., Tödtling, F. and Maier, G. (1990) 'The geography of organizational control: Austria 1973–1981', *Economic Geography* (forthcoming).

Siebert, H. (1969) *Regional economic growth: Theory and policy*. Scranton, PA: International Textbook Company.

Smith, N. (1979) 'Toward a theory of gentrification. A back to the city movement of capital, not people', *Journal of the American Planning Association*, 45, 538–48.

Smith, N. (1981) 'Degeneracy in theory and practice: Spatial interactionism and radical eclecticism', *Progress in Human Geography*, 5, 111–18.

Smith, N. (1984) *Uneven development: Nature, capital and the production of space*. Oxford: Basil Blackwell.

Smith, N. (1987) 'Gentrification and the rent gap', *Annals, Association of American Geographers*, 77, 462–78.

Smith, N. and O'Keefe, P. (1985) 'Geography, Marx and the concept of nature', *Antipode*, 17, 79–88.

Smolinski, L. (1973) 'Karl Marx and mathematical economics', *Journal of Political Economy*, 81, 1189–204.

Soja, E.W. (1980) 'The socio-spatial dialectic', *Annals, Association of American Geographers*, 70, 207–25.

Soja, E.W. (1989) *Post-modern geographies: The reassertion of space in social theory*. London: Verso.

Sraffa, P. (1960) *The production of commodities by means of commodities*. Cambridge: Cambridge University Press.

Sraffa, P. (ed.) (1951–73) *Works and correspondence of David Ricardo*, 11 vols. Cambridge: Cambridge University Press.

Stanback, T.M. and Noyelle, T.J. (1982) *Cities in transition*. Totowa, NJ: Allenheld, Osmun & Co.

Steedman, I. (1975) 'Positive profits with negative surplus values', *Economic Journal*, 85, 114–23.

Steedman, I. (1977) *Marx after Sraffa*. London: New Left Books.

Steedman, I. (1979) *Trade amongst growing economies*. Cambridge: Cambridge University Press.

Steedman, I. (1980) 'Basics, non-basics and joint production', in L.L. Pasinetti (ed.), *Essays on the theory of joint production*. London: Macmillan, pp. 44–50.

Steedman, I. (1981) 'Ricardo, Marx, Sraffa', in I. Steedman et al. (eds), *The value controversy*. London: Verso Editions, pp. 11–19.

Steedman, I. and Metcalfe, J.S. (1979) 'Reswitching, primary inputs and the Hecksher–Ohlin–Samuelson theory of trade', in I. Steedman (ed.), *Fundamental issues in trade theory*. New York: St Martin's Press, pp. 38–46.

Stoecker, R. (1988) 'From concrete to grass roots: A case study of successful urban insurgency in Cedar-Riverside', unpublished PhD dissertation, Department of Sociology, University of Minnesota.

Storper, M. (1985) 'The spatial and temporal constitution of social action: a critical reading of Giddens', *Society and Space*, 3, 407–24.

Storper, M. and Walker, R. (1989) *The capitalist imperative: Territory, technology and industrial growth*. Oxford: Basil Blackwell.

Sweezy, P. (1942) *The theory of capitalist development*. New York: Monthly Review Press.

Sweezy, P. (1981) 'Marxian value theory and crisis', in I. Steedman *et al.* (eds), *The value controversy*. London: Verso Editions, pp. 20–35.

Thrift, N. J. (1983) 'On determination of social action in space and time', *Environment and Planning D: Society and Space*, 1, 23–57.

Thrift, N. J. (1985) 'Flies and germs: A geography of knowledge', in D. Gregory and J. Urry (eds), *Social relations and spatial structures*. London: Macmillan, pp. 366–403.

Thrift, N. J. (1987) 'Introduction: The geography of late twentieth-century class formation', in N. J. Thrift and P. Williams (eds), *Class and space: The making of urban society*. London: Routledge & Kegan Paul, pp. 207–53.

Thrift, N. J. and Williams, P. (eds) (1987) *Class and space: The making of urban society*. London: Routledge & Kegan Paul.

Urry, J. (1981) 'Localities, regions and class', *International Journal of Urban and Regional Research*, 5, 455–74.

Urry, J. (1986) 'Capitalist production, scientific management and the service class', in A. J. Scott and M. Storper (eds), *Production, work, territory: The geographical anatomy of industrial capitalism*. Boston: Allen & Unwin, pp. 43–66.

Van Parijs, P. (1986–7) 'A revolution in class theory', *Politics and Society*, 15, 453–82.

Von Bortkiewicz, L. (1906[1952]) 'Value and price in the Marxian system', *International economic papers*, 2, 5–60.

Von Weizsäcker, C. C. (1971) *Steady state capital theory*. New York: Springer Verlag.

Walker, R. (1975) 'Contentious issues in Marxian value and rent theory: A second and longer look', *Antipode*, 6(1), 31–54.

Walker, R. (1981) 'A theory of suburbanization: Capitalism and the construction of urban space in the United States', in A. J. Scott and M. Dear (eds), *Urbanization and urban planning in capitalist society*. London: Methuen, pp. 383–430.

Walker, R. (1985) 'Class, division of labour and employment in space', in D. Gregory and J. Urry (eds), *Social relations and spatial structures*. London: Macmillan, pp. 164–89.

Watts, H. (1980) *The large industrial enterprise: Some spatial perspectives*. London: Croom Helm.

Watts, H. (1981) *The branch plant economy*. London: Longman.

Webber, M. J. (1983) 'Location of manufacturing and operational urban models', in F. E. I. Hamilton and G. J. R. Linge (eds), *Spatial analysis, industry and the industrial environment*. London: John Wiley, pp. 141–202.

Webber, M. J. (1984) *Explanation, prediction and planning: The Lowry model*. London: Pion.

Webber, M. J. (1986) 'Survey 9. The theory of prices, profits and values', *Society and Space*, 4, 109–16.

Webber, M. J. (1987a) 'Profits, crises and industrial change 1: Theoretical considerations', *Antipode*, 19, 307–28.

Webber, M. J. (1987b) 'Quantitative measurement of some Marxist categories', *Environment and Planning A*, 19, 1303–22.

Webber, M. J. (1987c) 'Rates of profit and interregional flows of capital', *Annals, Association of American Geographers*, 77, 63–75.

Webber, M. J. (1988) 'Profits, crises and industrial change 2: The experience of Canada 1950–81', *Antipode*, 20, 1–18.

Webber, M. J. (1990) *Value crisis and regional development* (forthcoming)

Webber, M. J. and Foot, S. P. H. 'The measurement of unequal exchange', *Environment and Planning A*, 16, 927–47.

Webber, M. J. and Rigby, D. L. (1986) 'The rate of profit in Canadian manufacturing, 1950–1981', *Review of Radical Political Economics*, 18, 33–55.

Webber, M. J. and Tonkin, S. (1990) 'Profitability and capital accumulation in Canadian manufacturing industries', *Environment and Planning A* (forthcoming).

Westaway, J. (1974) 'The spatial hierarchy of business organizations and its implications for the British urban system', *Regional Studies*, 8, 145–55.

Wheaton, W. C. (1982) 'Urban spatial development with durable but replaceable capital', *Journal of Urban Economics*, 12, 53–67.

Williams, H. C. W. L. (1977) 'On the formation of travel demand models and economic evaluation measures of user benefit', *Environment and Planning A*, 9, 285–344.

Williamson, O. (1975) *Markets and hierarchies*. New York: The Free Press.

Willis, P. (1978) *Learning to labour: How working class kids get working class jobs*. Farnborough: Saxon House.

Wilson, A. G. (1974) *Urban and regional models in geography and planning*. London: John Wiley.

Wolff, R. (1982) 'Piero Sraffa and the rehabilitation of classical political economy', *Social Research*, 49, 209–38.

Wolff, R. D. and Resnick, S. A. (1987) *Economics: Marxian versus neoclassical*. Baltimore, MD: Johns Hopkins University Press.

Wolfstetter, E. (1976) 'Positive profits with a negative surplus value. A comment', *Economic Journal*, 86, 864–72.

Wright, E. O. (1978) *Class, crisis and the state*. London: New Left Books.

Wright, E. O. (1980) 'Varieties of Marxist conceptions of class structure', *Politics and Society*, 9, 323–70.

Wright, E. O. (1981) 'Reconsiderations', in I. Steedman *et al.* (eds), *The value controversy*. London: Verso Editions, pp. 130–62.

Wright, E. O. (1984) 'A general framework for the analysis of class structure', *Politics and Society*, 13, 383–423.

Wright, E. O. (1985) *Classes*. London: Verso Editions and New Left Books.

Zalai, E. (1981) 'Heterogeneous labour and the determination of value', *Acta Oeconomica*, 25, 259–75.

Name index

Abraham-Frois, G. 65
Aglietta, M. 172
Alonso, W. 137, 156

Baldone, S. 147
Ball, M. 111–14, 121, 237, 239
Baran, P. 284
Barbalet, J. M. 254–5
Barnes, T. J. 88, 94, 105
Berrebi, E. 65
Booth, D. 215
Bowles, S. 298–9
Bowman, J. 246
Braverman, H. 209
Burawoy, M. 226–7

Calhoun, C. 217
Carling, A. 12, 208
Castells, M. 138
Clark, G. L. 15, 227, 257
Clark, J. B. 19
Cohen, G. A. 8, 11, 55
Cox, K. R. 220

David, P. A. 268
Dobb, M. H. 9, 108
Duménil, G. 59

Edel, M. 218
Eichner, A. S. 283
Elson, D. 55
Elster, J. 8, 12, 203–5, 207, 209, 213, 216, 221–3, 244–5, 248, 255, 257
Emmanuel, A. 168–74, 176, 256, 310
Engels, F. 254
Esfahani, H. 108, 132

Farjoun, E. 10–11, 49, 53, 56
Ferrão, J. 173–4
Fine, B. 111–12
Foot, S. P. H. 174–5
Franke, R. 59

Gertler, M. 15, 28, 227
Gibson, B. 108, 132, 171–3, 175
Gibson, S. 50
Gintis, H. 298–9
Goodwin, R. 199

Hadjimichalis, C. 174–5
Harcourt, G. C. 23–4
Hard, G. 17
Harris, R. 217

Harvey, D. 1–2, 15, 55, 62–3, 82, 122, 133, 135, 138, 141, 155, 163, 182–3, 220, 222, 226–7, 231, 233, 235–8, 240–2, 246, 248, 254, 256–7, 277, 285, 299, 305
Himmelweit, S. 43
Holmes, J. 289, 309
Huriot, J. M. 141

Jenkins, R. 285, 290
Jevons, W. S. 7
Johnson, H. 8

Kalecki, M. 15, 283–4, 286
Kaplinsky, R. 289, 295
Katz, S. 109, 129, 134, 231–2, 234
Kautsky, K. 202
Keat, R. 210
Keynes, J. M. 227, 303
Krause, U. 299

Lafargue, K. 5
Lauria, M. 234
Leitner, H. 106–7
Lenin, V. I. 202, 254
Lévy, D. 59
Liossatos, P. 172–3

Machover, M. 10–11, 49, 56
Maier, G. 293
Mainwaring, L. 145, 171
Mair, A. 220
Mandel, E. 10
Marelli, E. 39, 174–5
Marshall, A. 7
Marx, K. 3–5, 7, 27, 31, 33–7, 39, 41, 44–6, 48, 54–6, 73, 102, 105–6, 109, 113–14, 122, 128–9, 131–3, 170, 174, 181, 187, 201–3, 213, 217, 219, 221–2, 245, 248, 254–5, 298
Massey, D. B. 15, 216, 282, 287–8, 292–3
Meegan, R. 15
Metcalfe, J. S. 97
Mills, E. 137
Mohun. S. 43
Morgenstern, O. 306
Morishima, M. 49, 164
Muth, R. 137, 142

Nell, E. J. 7
Nikaido, H. 84

Offe, C. 253–4, 256, 260
Ohlin, B. 100
O'Keefe, P. 105

Okishio, N. 248
Olson, M. 213–14, 218

Parys, W. 172–3
Pasinetti, L. L. 4, 6–7, 138, 187
Pavlik, C. 88, **5.1**
Peet, R. 217, 222
Perelman, M. 107
Pratt, G. 236
Przeworski, A. 219, 223, 226–7

Resnick, S. A. 203, 208
Ricardo, D. 3–5, 7, 31, 34–5, 45, 307
Rigby, D. 80, 281
Robinson, J. V. 5, 9, 23, 27
Roemer, J. 8, 11–12, 50, 55, 162, 165, 184, 191, 205–10, 215–16, 230, 242, 248–9, 276, 303, 310
Roncaglia, A. 117
Rubin, I. I. 37
Ruccio, D. 13

Samuelson, P. A. 26–7, 47
Sartre, J.-P. 221, 304
Saunders, P. 235
Schefold, B. 146
Scott, A. J. 15, 28, 98–9, 131, 135, 138, 150, 156, 241–2, 292
Semmler, W. 39, 50, 265, 284–6
Shaikh, A. 38, 49–51, 162, 183
Sheppard, E. S. 88, 94, 106–7, 238–9, 281, 292
Smith, A. 27–8, 31, 103, 105, 221, 306

Smith, N. 15, 105–6, 138, 222, 259
Soja, E. W. 222
Sraffa, P. 7–10, 19, 23–4, 108, 142–5, 159, 185, 239, 297, 303, 306–7, 309
Steedman, I. 5, 7–10, 47–8, 51–4, 97, 183, 250–1
Stigler, G. 3
Storper, M. 282, 289
Sweezy, P. M. 284

Thrift, N. J. 215
Tödtling, F. 293
Tonkin, S. 284

Urry, J. 210, 256

von Neumann, J. 306
van Parijs, P. 205

Walker, R. 15, 138, 277, 282, 289
Webber, M. J. 64, 77–8, 80, 148, 174–5, 227, 281, 284, 294
Whiteman, J. 15, 227
Wiesenthal, H. 253–4, 256, 260
Williams, P. 215
Willis, P. 218
Wilson, A. G. 153
Wolff, R. 203, 208
Wolff, R. D. 7–8
Wright, E. O. 203–7, 210, 212–13, 255

Zalai, E. 298

Subject index

accumulation **9.1.1**
  and wages 182–3, 226
analytical Marxism 8, **1.2.3**, 201, 219
  class and exploitation 205–8, 215
  and rationality 11–13, 205–6, 208, 218, 223, 305–6
analytical methods 1–2, 5, 7–8, 10, 12–13, 201
  and space 298, 300–2

basic goods 114, 116–17, 119, 303

capacity utilization 77, 80, 82, 279, 283–4, 303
capital advanced **4.2**, **4.2.1**, 81–2, 279
capital controversy 7, chapter 2, 112, 142
  and economic geography **2.2**, 99–101
capital logic approach 72, 231, 303
capital reswitching 23–5, 29, 308
  in a space economy 27, chapter 5
  spatial numerical example 90–2
capital reversing 23–7, 303
capitalists
  and luxury consumption 184–8
  and savings 184–8
circulation 33, 61–2, 83, 160, 296
  labour values 160, **8.1.4**, 297
  prices **4.1**, **4.1.4**, 83, 160. 296–7
  time **4.2.2**, 82
  use values 177, 181, 297
class chapter 10
  alliance 165–6, 191, 219–20
  class-for-itself **10.1**, **10.3**, 223, 303
  class-in-itself **10.1**, **10.2.1**, 212–13, 215, 223, 304
  consciousness 213–18
  and contradictory locations 203–4, 207, 255
  defence of place 220
  defined by analytical Marxists 205–8
  definitions 202–9
  factionalization 220, 235–6, 254–6
  formation 213–16, 218
  and space 209–10
  structure **10.2.2**
class interests **10.2.3**, 213, 216, 218, 304
  fundamental 202, 212–13, 218, 304
  immediate 202, 212, 218, 304
  objective 202, 204, 212
collective action
  difficulties for workers 253–4
  free-riders 213–18, 223, 244, 257, 280, 305
  and individual interests 213–18
  and the prisoner's dilemma 214, 218, 308
  in space 215–18
  unintended consequences 101, 220–2, **12.1**

commodity exchange 32–3, 37, 178–9
commodity production 32–3, 61, 296
constant capital 5, 41, 167, 304
consumption-growth frontier **9.2.1**, 225, 304
corporations
  and branch plants 288–9, 291–3
  and cloning 289–90, 304
  and direct investment 292–3, 304
  and disinvestment 292–3
  geography of control **14.1.4**
  internal division of labour 288, 290
  and just-in-time 78, 289, 291, 295, 306
  organization of production **14.2**
  and portfolio investment 293, 308
  and subcontracting 289, 309
cost reduction strategies 192, chapter 13
  relocation **13.1.5**, 266
  technical change 266
counterfinality 221–2, 224, 244–6, 248–9, 251–2, 257, 260, 304
  and spatial relations 245–6, 253, 298

dynamic instability **9.3**, **9.3.1**, 197, 310

equilibrium 13, 15, 62–3, 70, 74, 84–6, 160, 253, 297, 305–6, 309
  dynamic 63, **9.1**, 194–7, 305
  general spatial 13–14
exchange value 34, 305
exploitation **3.3.3**, 55, 230, 305
  and class definitions 202–3, 205
  fundamental Marxian theorem 40, 163–4
  and geographical variation **8.1.2**, **8.1.3**, 209–10
  rate of exploitation 5, 40, 308

fallacy of composition 221, 244, 305
fictitious capital 231, 240–2, 305
finance capital 277, 299
fixed capital 77–8, 80, 276–7, 280, 284, 305, 308
  and the built environment 142–7
  and truncations 145–7, 309
Fordism 183, 256, 305, 307
functionalism 11, 32, 236, 255, 303, 305
fundamental Marxism **1.2.2**, 255, 303, 307, 309

game theory 206–7, 210, 218, 306, 308, 310
Garin-Lowry model **7.4.1**, 157–8
  and housing **7.4.4**
  and retailing 150–2

inter-class conflict 4–5, 107, 155, 213, **10.3.2**, chapter 11

among workers, capitalists and landlords 102, 122, 134–5
capitalists vs. landlords 219, **11.3**
capitalists vs. workers **11.1**, 219
and community resistance strategies **11.2.2**
over growth or consumption **9.2.1**, **11.1.1**
over labour process **11.1.2**
over land ownership **11.3.3**
over land tenure **11.3.2**
over property relations **11.1.3**
over rent levels **11.2.1**, **11.3.1**
workers vs. landlords **11.2**
inter-regional trade 28, 68–70, 100
accounting relationships **4.1.5**
empirical measurement of unequal exchange **8.2.3**
and intra-class conflict **12.2.3**
unequal exchange **8.2**, 310
intra-class conflict chapter 12
capitalists **12.2**
and housing classes **11.2.3**
landowners **12.4**
over differential rent **12.4.1**
over monopoly rent 2, **12.4.2**
over specialization and trade **12.2.3**
and relocation **12.2.1**
and technical change **12.2.2**
workers 154–6, **12.3**

joint production 51–4, 119, 136, 142, 144, 146, 306

labour aristocracy 165–6, 176, 210, 235, 254, 306
labour power 39–40, 306
labour theory of value 3–4, 10–11, 26, **3.3**, 306
critiques 8, **3.5.3**
embodied labour values 34–5, 305
empirical measurement **3.3.2**, 49–51
and geography **3.8**, **8.1.1**
negative values 51–3
qualitative arguments **3.7**
responses to criticisms 47–51
law of value 38, 48, 50, 162, 184, 307

Marxian theory *see* Marx, K. in Name Index
rent theory 108–9, **6.2.1**, 123
scarcity 104–9
methodological individualism 13, 208, 215, 223, 298, 307
mode of production 31–2, 206–7, 212, 303, 305, 307, 309

neoclassical economics 1, 5–7, 19, **5.3**, 298, 307
capital intensity 21–2, 25, 27, 29, 303
capital theory 19–22
choice of technique 21–2
general equilibrium model 5
and nature 104–5

and the new urban economics 137, 156
problems of measuring capital 23
production function 20–1, 26–7, 308
rationality 5–6, 305–6
rent 28, 101, 105, 107–8
scarcity 20–2, 100, 104–5, 107–8
theory of distribution 6–7
theory of marginal productivity 20–1, 29, 100–1, 307
theory of profits 19–21, 23, 27, 29
theory of wages 19–21, 29
neo-Fordism 226, 256–7, 307
neo-Ricardianism **1.2.1**, 10–11, 19, 48–9, 101–2, 138, 201, 307, 309
rent theory 108, **6.2.2**, 122–3
and scarcity 108, 114, 117, 119
non-basic goods 114, 116–17, 119, 185, 303

oligopoly
and differential profit rates **14.1.2**
and price mark-up 283–4, 286, 290, 307
and the rate of profit **14.1**
organic composition of capital 41–2, 45, 109, 129–32, 169, 221, 307

political economy 1–5
compared with neoclassical economics 6–7
and nature 104–6
rent 106–8
theory of distribution 4–5, 27, 29
potential theory 86
production prices 10–11, 34, **3.5.1**, 47–8, 58, 65, 82, 85–6, 305, 308
effect on trade patterns 68–70
and natural resources 106, 110–11
profit
debate on differential rates **14.1.1**
differential rates and oligopoly **14.1.2**
falling rate 221, 245, 248, 275
interregional differentials **14.1.3**
profit-maximizing location pattern 72–3, 103, 225, 247, 287, 308
intra-urban **7.2**

rent 93–4, **6.1**, **12.4**
absolute **6.5.1**, 303
on buildings *7.3*
differential rent I 109–11, 232, 234, **12.4.1**, 276, 304–5
differential rent II 111–14, 232, 234, **12.4.1**, 276, 304, 306
extensive 115–17, **6.3.1**, 237–40, 305
intensive 117–22, **6.3.2**, 136, 237–40, 306
intra-urban **7.2.2**
and Marx 108–9, **6.2.1**, 119–23
monopoly rent 1 **6.5.2**, 307
monopoly rent 2 **6.5.3**, 138, 141, 155, 232–4, 237–8, 258, **12.4.2**, 276, 307
and neo-Ricardians 108, **6.2.2**, 119–23

on non-land resource **6.4**
political economic review 106–7
and transportation cost differentials 95–7,
  6.3
Ricardian theory, *see* Ricardo, D. in Name
  Index
  corn model 4–5, 9, 304
  land rent theory 4
  value theory 4

socially necessary
  division of labour 37–8, 46, 49–51, 62, 163,
    178–9, 182–4, 309
  labour values 4–5, 34, **3.3.1**, 51, 53, 55, 163,
    309
  production technique 36, 38, 53–4, 56–7,
    111, 265, 267–8, 271, 273–4, 287, 309
  profits 227
  rents 237–8
  wages 225–7, 309
spatial division of labour 180, 227, 256, 309
Sraffian theory, *see* Sraffa, P. in Name Index
  rent **6.2.2**
  critique of neoclassical capital theory 23–7
  theory of production 9–10
state apparatus 193, 280, 299–300
surplus value 5, 41–2, 167, 183, 202–3, 208,
  219, 240, 308

technical change 228–9, **12.2.2**, 266, **13.1.4**,
  309
  and circulation time **13.3**
  and fixed costs **13.2**
  and just-in-time 291, 295, 306

imitation 267–71, 273–4, 306
innovation 267–74, 306
and location **13.1.3**, **13.1.5**
rate and direction **13.1.2**
selection 267–8, 270–4, 308
and transportation 192, 277–80
types **13.1.1**
transportation 66–7, 76, 80–3, 97–8, 161,
  248–9, 251–2, 272–3, 279–80
  and technical change 277–80
turnover time 77–81, 310

use value 33–4, 226, 310

value controversy 8, chapter 3
  and joint production 51–4
  transformation problem 5, **3.5.2**, 132, 162,
    169–71, 175, 309
  values compared to prices **3.5**, 46, 49–51, **3.7**
value rate of profit 41–2, 45, 310
variable capital 5, 41, 167, 310

wages 40, 44, 162–3, 166, 304, 308, 310
  intra-urban variation 154–6
wage-growth frontiers 182, 184, 187
wage-profit frontiers 20, 22, 24–6, **4.1.2**, 109,
  111–15, 120, 136, 142, 145–7, 224–5, 247,
  259, 310
  and location 71–2, 88, 91
  wage-profit-rent frontiers 97–8, 133–4
workers 39–40
  heterogeneous labour 298–9
  intra-urban location **7.4.2**, 232
  and savings 184–79